Books and the Sciences in History

The history of the sciences and the history of the book are complementary, and there has been much recent innovative research in the intersection of these lively fields. This accessibly-written, well-illustrated volume is the first systematic general work to do justice to the fruits of scholarship in this area.

The twenty specially commissioned chapters, by an international cast of distinguished scholars, cover the period from the Carolingian renaissance of learning to the mid-nineteenth-century consolidation of science. They examine all aspects of the authorship, production, distribution, and reception of manuscripts, books and journals in the various sciences. An editorial introduction surveys the many profitable interactions of the history of the sciences with the history of books. Two afterwords highlight the relevances of this wide-ranging survey to the study of the development of scientific disciplines and to the current predicaments of scientific communication in the electronic age.

MARINA FRASCA-SPADA is an Affiliated Lecturer in the Department of History and Philosophy of Science, University of Cambridge, and a Fellow of St Catharine's College

NICK JARDINE is Professor of the History and Philosophy of the Sciences, University of Cambridge, and a Fellow of Darwin College

Books and the Sciences in History

EDITED BY

MARINA FRASCA-SPADA
and NICK JARDINE

ADVISORY PANEL   Silvia De Renzi, Anthony Grafton, Lisa Jardine,
Adrian Johns, Sachiko Kusukawa, Elisabeth Leedham-Green,
David McKitterick, James Secord, E. C. Spary

CABRINI COLLEGE LIBRARY
610 King of Prussia Road
Radnor, PA 19087

PUBLISHED BY THE PRESS SYNDICATE OF THE UNIVERSITY OF CAMBRIDGE
The Pitt Building, Trumpington Street, Cambridge, United Kingdom

CAMBRIDGE UNIVERSITY PRESS
The Edinburgh Building, Cambridge CB2 2RU, UK   http://www.cup.cam.ac.uk
40 West 20th Street, New York, NY 10011-4211, USA   http://www.cup.org
10 Stamford Road, Oakleigh, Melbourne 3166, Australia
Ruiz de Alarcón 13, 28014 Madrid, Spain

© Cambridge University Press 2000

This book is in copyright. Subject to statutory exception
and to the provisions of relevant collective licensing agreements,
no reproduction of any part may take place without
the written permission of Cambridge University Press.

First published 2000

Printed in the United Kingdom at the University Press, Cambridge

Typeface Plantin Light MT 10.5/12.5pt   *System* QuarkXPress™   [SE]

A catalogue record for this book is available from the British Library

Library of Congress Cataloguing in Publication data

Books and the sciences in history / edited by Marina Frasca-Spada and Nick Jardine.
      p.   cm.
ISBN 0 521 65063 1 (cloth) – ISBN 0 521 65939 6 (pbk)
1. Science – History.   2. Books – History.   I. Frasca-Spada, Marina.   II. Jardine, Nicholas.
Q125.H664   2000
509–dc21   99-087281

ISBN 0 521 65063 1 hardback
ISBN 0 521 65939 6 paperback

Contents

Acknowledgements	*page* vii
List of illustrations	ix
Introduction: books and the sciences Marina Frasca-Spada and Nick Jardine	1

I TRIUMPHS OF THE BOOK

1	Books and sciences before print Rosamond McKitterick	13
2	Printing the world Jerry Brotton	35
3	Geniture collections, origins and uses of a genre Anthony Grafton	49
4	Annotating and indexing natural philosophy Ann Blair	69
5	Illustrating nature Sachiko Kusukawa	90
6	Astronomical books and courtly communication Adam Mosley	114
7	Reading for the philosophers' stone Lauren Kassell	132
8	Writing and talking of exotic animals Silvia De Renzi	151

II LEARNED AND CONVERSABLE READING

9	Compendious footnotes Marina Frasca-Spada	171
10	On the bureaucratic plots of the research library William Clark	190
11	Encyclopaedic knowledge Richard Yeo	207

12	Periodical literature *Thomas Broman*	225
13	Natural philosophy for fashionable readers *Mary Terrall*	239
14	Rococo readings of the book of nature *E. C. Spary*	255
15	Young readers and the sciences *Aileen Fyfe*	276
16	The physiology of reading *Adrian Johns*	291

III PUBLICATION IN THE AGE OF SCIENCE

17	A textbook revolution *Jonathan Topham*	317
18	Useful knowledge for export *Eugenia Roldán Vera*	338
19	Editing a hero of modern science *Lisa Jardine and Alan Stewart*	354
20	Progress in print *James Secord*	369

Afterwords

| | Books, texts, and the making of knowledge
Nick Jardine | 393 |
| | The past, present, and future of the scientific book
Adrian Johns | 408 |

| | *Notes on contributors* | 427 |
| | *Index* | 432 |

Acknowledgements

The Department of History and Philosophy of Science at Cambridge, in which both the editors work, is a lively centre of research on the history of books in relation to the history of the sciences. The idea for an edited volume originated in a series of discussions on 'History of the Sciences / History of the Book' organised by the editors and colleagues in the Cambridge Historiography Group. We are very grateful to all members of the Group as well as to the participants in Jim Secord's informal Book History Reading Group.

For expert assistance in planning this work we warmly thank our advisory panel: Silvia De Renzi, Anthony Grafton, Lisa Jardine, Adrian Johns, Sachiko Kusukawa, Elisabeth Leedham-Green, David McKitterick, Jim Secord and E. C. Spary. Special thanks for moral as well as intellectual and practical support to Joanna Ball, Kate Fletcher, Nick Hopwood, Lauren Kassell, Joad Raymond, Jon Topham and Paul White.

For unfailing helpfulness and efficiency we are indebted to Bill Davies, Jo North, Caroline Murray and all those at Cambridge University Press involved in the production of this book.

Finally, as editors we offer heartfelt thanks to our contributors for the promptness with which they delivered their splendid essays and responded to editorial and copy-editorial suggestions.

Illustrations

Title-page to Part 1	Book-wheel, from Agostino Ramelli, *Le diverse et artificiose machine*... (Paris, 1588).	*page* 11
1.1	Titus writing a letter to St Paul, illustration redrawn from a ninth-century Frankish manuscript (Düsseldorf, Universitätsbibliothek, MS A 14, fol. 119v).	14
1.2	Pages from a Carolingian gromatic and geometrical compilation (Cambridge, Trinity College Library, MS R.15.14, fols. 13v–14r).	16
1.3	A section of the *Nomina lapidum finalium et archarum positiones* with figures (Cambridge, Trinity College Library, MS R.15.14, fols. 12v–13r).	21
1.4	A page from a Carolingian astronomical collection containing Hyginus and Cicero's version of the *Aratea* and Abbo of Fleury's works on astronomy, from Abbo's *De cursu planetarum per Zodiacum circulum* (Cambridge, Trinity College Library, MS R.15.32, fol. 6v).	22
1.5	*Cursus lunae* (Cambridge, Trinity College Library, MS R.15.32, fol. 7v).	23
1.6	Page from a thirteenth-century manuscript containing Alhacen's *Perspectiva* (Cambridge, Trinity College Library, MS O.5.30, fol. 121v).	29
2.1	Cordiform world map, attr. Hajji Ahmed, Venice, c. 1560.	36
2.2	World map, Francesco Berlinghieri, Florence, 1482.	40
3.1	Pope Julius II's geniture annotated by Gabriel Harvey, from his copy of Luca Gaurico, *Tractatus astrologicus* (Venice, 1552).	63
4.1 and 4.2	Index fingers, from Hippolytus de Marsiliis, *Brassea* (Milan, 1522).	76, 77
4.3	A page from the index to Gregor Reisch, *Margarita philosophica* (Strasburg, 1508, first publ. 1503).	78
4.4	Title-page of Ptolemy's *Geography* (Basel, 1552).	79
4.5 and 4.6	Indexes to the German and Latin editions of Sebastian Münster's *Cosmographia*, both dated 1550, in Frank Hieronymus, *1488 Petri—Schwabe 1988* (Basel: Schwabe, 1997), I, pp. 624–25.	80, 81

4.7	A page from the indexes in Conrad Gesner, *Historia Animalium* (Zurich, 1551).	83
4.8	A page from the index to Johann Heinrich Alsted, *Encyclopedia* (Herborn, 1630).	84
5.1	Diagrams produced with bent metal on plaster, Euclid, *Elements* (Venice: E. Ratdolt, 1482), fols. 114r–115v.	91
5.2	Jan van Calcar, skeleton, illustration to A. Vesalius, *Seven Books on the Fabric of the Human Body* (Basle: J. Oporinus, 1543), p. 203.	93
5.3	Producing illustrations for printed books, L. Fuchs, *History of Plants* (Basle: M. Isengrin, 1542), fol. 897r.	94
5.4	'Prunus Sylvestris', L. Fuchs, *History of Plants* (Basle: M. Isengrin, 1542), fol. 404.	95
5.5	A copy of Dürer's image of the rhinoceros, C. Gesner, *Histories of Animals* (Zurich: C. Froschauer, 1551), p. 953.	96
5.6	'Gauchblum', O. Brunfels, *Live Images of Plants* (Strasburg: J. Schott, 1530), p. 218.	98
5.7	C. Scheiner, *Rosa ursina* (Bracciani, 1630), title-page.	100
5.8 and 5.9	The medicinal plants aloe and anacardus represented by jars, T. Dorsten, *Botanicon* (Frankfurt: C. Egenolff, 1540), fol. 25r (aloe) and fol. 23v (anacardus).	102, 103
5.10	'Aloe', L. Fuchs, *History of Plants* (Basle: M. Isengrin, 1542), fol. 138.	104
5.11	Alphabets as reference index, A. Vesalius, *Seven Books on the Fabric of the Human Body* (Basle: J. Oporinus, 1543), pp. 224–25.	105
5.12	Co-ordinates as a reference system, showing a variety of kidneys, B. Eustachio, *Opuscula anatomica* (Venice, 1564), fol. 1v.	106
6.1	Diagram of the Tychonic world-system from T. Brahe, *De recentioribus phaenomenis* (Uraniborg, 1588).	116
6.2	Woodcut of one of Tycho's sextants, T. Brahe, *Astronomiae instauratae mechanica* (Wandsbeck, 1598), sig. D2 v.	120
6.3	Back and front of the binding of a presentation copy of the *Epistolae astronomicae* (Uraniborg, 1596) in the Herzog August Bibliothek, Wolfenbüttel (8 Astron.).	126
7.1	An emblem from Heinrich Khunrath's *Amphitheatrum sapientiae aeternae* (Hanover, 1609).	134
7.2	Title-page of Ashmole's copy of 'The Epitome of the Treasure of Health' (Ashm. 1419, fol. 57).	137
7.3	A page from one of Dee's notebooks recording his angelic conversations, 'Liber Mysteriorum' (Sloane 3188), fol. 103v.	138
7.4	The entry for 'lapis' in Newton's 'Index Chemicus' (Keynes 30/2, fols. 2v–3r).	145
7.5	The title-page of Dee's *Monas Hieroglyphica* (Antwerp, 1564).	146

8.1	Title-page of *Rerum medicarum Novae Hispaniae thesaurus . . . ex Francisci Hernandez* (Romae, Ex Typographeio Vitalis Mascardi, 1651).	152
8.2	Diagram of the dissemination of Hernandez's work in the 17th century.	155
8.3	Woodcut of the Mexican civet, *Thesaurus* (Romae, 1651), p. 538.	157
8.4	Woodcut of the *amphisbena*, *Thesaurus* (Romae, 1651), p. 797.	159
8.5	Woodcuts of the European civet and its genital organs, *Thesaurus* (Romae, 1651), p. 580.	160
8.6	A Franciscan preaching, D. Valades, *Rhetorica christiana* . . . (Perusiae, 1579), p. 111.	163
Title-page to Part 2	Book-wheel, from Gaspard Grollier de Servière, *Recueil d'ouvrages curieux de mathématique et de mécanique* . . . (Lyon, 1719).	169
9.1	Title-page of the first edition of E. Law's translation of W. King, *The Origin of Evil* (Cambridge, 1731).	172
9.2	Edmund Law in 1777, mezzotint by William Dickinson after a painting by George Romney.	178
9.3 and 9.4	Pages from *The Origin of Evil* (Cambridge, 1731).	182, 184
10.1	Visiting the Library at the University of Altdorf, Johann G. Puschner, *Amoenitates Altdorfinae* (Nuremberg, ca. 1715), plate 16.	191
10.2	The Göttingen University Library, from Georg D. Heumann, *Wahre Abbildung der Köngl. Groß-Britan. u. Churfürstl. Braunschweigisch-Lüneburgische Stadt Göttingen* (Göttingen, 1747) (E:GöttUB, gr. 2° H.Hann. V, 29 *rara*.).	197
10.3 and 10.4	The layout of the Göttingen Library, from Johann Stephan Pütter, F. Saalfeld and G. H. Oesterley, *Versuch einer academischen Gelehrten-Geschichte der Georg-Augustus Universität zu Göttingen*, 4 vols. (Göttingen/Hanover, 1765–1838), vol. 1.	198, 199
11.1	Title-page of John Harris' *Lexicon Technicum*, 2nd edn (London, 1708).	209
11.2	Title page of E. Chambers' *Cyclopaedia*, 4th edn, 2 vols. (London, 1741).	211
11.3	Portrait of John Harris in the frontispiece of his *Lexicon Technicum*, 2nd edn (London, 1708).	214
11.4	The 'View of Knowledge' in the Preface of Chambers' *Cyclopaedia*, 4th edn, 2 vols. (London, 1741), vol. I, p. iii.	217
11.5	Note to the 'View of Knowledge' in the Preface of Chambers' *Cyclopaedia*, 4th edn, 2 vols. (London, 1741), vol. I, p. iv.	218

11.6	Illustration of John Locke's Index in the entry for 'Common-Place-Book' in Chambers' *Cyclopaedia*, 4th edn, 2 vols. (London, 1741), vol. I.	219
13.1	The cosmos, Bernard de Fontenelle, *Entretiens sur la pluralité des mondes* (Paris, 1686), frontispiece.	242
13.2	Didactic conversation between brother and sister, Benjamin Martin, *The Young Gentleman and Lady's Philosophy* (London, 1759), frontispiece.	246
13.3	Working planetarium marketed by Benjamin Martin, from B. Martin, *The Young Gentleman and Lady's Philosophy* (London, 1759).	247
13.4	The stars as centres of gravitational force and light, Pierre-Louis de Maupertuis, *Discours sur les différentes figures des astres*, 2nd edn (Paris, 1742), frontispiece.	251
14.1	Grid-like arrangement of shells, from Henry Augustus Pilsbry, 'A study of the variation and zoogeography of Liguus in Florida', *Journal of the Academy of Natural Sciences of Philadelphia*, 15 (1912), plate 39.	256
14.2	Symmetrical and decorative arrangement of shells, Jacques Mesnil after Jacques de Favanne, in Dezallier d'Argenville, *La Conchyliologie* . . ., 3rd edn (Paris, 1780), plate 18.	257
14.3	Variously orientated shells, Martin Lister, *Historiae sive Synopsis methodicae conchyliorum* (Oxford, 1685), plates 118–120.	259
14.4	An early symmetrical shell plate by F. Ertinger, in Claude du Molinet, *Le Cabinet de la Bibliothèque de Sainte Genevieve* (Paris, 1692), plate 44.	260
14.5	Quintin-Pierre Chedel after Francois Boucher, frontispiece for Dezallier d'Argenville, *L'Histoire naturelle éclaircie* (1742).	263
14.6	Shell plate, sponsored by the collector Bonnier de La Mosson, artist unknown, in Dezallier d'Argenville, *L'Histoire naturelle éclaircie* (1742), plate 20.	265
14.7	Broken symmetry in a plate by Marie-Therese Reboul, in Michel Adanson, *Histoire naturelle du Sénégal* (Paris, 1757), plate 13.	269
14.8	The natural history cabinet of Bonnier de La Mosson, from K. Scott, *The Rococo Interior* (New Haven and London, 1995).	270
14.9	A 'page' from Michel Adanson's shell collection, from E. Fischer-Piette, 'Les mollusques d'Adanson', *Journal de Conchyologie*, 85 (1942): 103–377.	271
15.1	The first four volumes of *Evenings at Home*, 2nd edn, 6 vols. (London, 1794–8).	277
15.2	Mother and Charlotte examining something Henry has found, S. Trimmer, *An Easy Introduction to the Knowledge of Nature and the Holy Scripture* (London, 1780), frontispiece.	279

15.3	Three late-nineteenth-century editions of *Evenings at Home*.	281
15.4	The gathered family, frontispiece by George Cruikshank to James Jennings' *Family Cyclopaedia* (London, 1821).	285
16.1	A representation of alchemy, J. C. Barchusen, *Elementa chemiae* (Leyden, 1718), p. 503.	298
16.2	An alchemist's representation of experience, from M. Maier, *Atalanta Fugiens* (Oppenheim, 1618).	299
Title-page to Part 3	The new magazine machine, from George Cruikshank, *The Comic Almanac*, 1846.	315
17.1	The 'reading' man, aquatint by Francis Jukes, engraving by J. K. Baldrey, from Richard Corbould Chilton, '*Helluones librorum*'.	318
17.2	John Nicholson, engraving by James Caldwell from a portrait by Phillip Reinagle (1790).	321
17.3	Deighton's shop in Trinity Street in an 1870s photograph.	323
17.4	The West Room and the Dome Room of Cambridge University Library, watercolour by Thomas Rowlandson, 1809.	324
17.5	Lecture bill for the Jacksonian Professor's course of chemical lectures, 1796 (Cambridge University Archives, University Papers, UP1 fol. 160).	325
17.6	Title-page of the *Memoirs of the Analytical Society* (Cambridge, 1813).	327
17.7	Typographical complexity, page from *Memoirs of the Analytical Society* (Cambridge, 1813), p. 54.	328
17.8	The notice of the translation of Lacroix, *Elementary Treatise on the Differential and Integral Calculus* (1816), in the *Cambridge University Calendar* (1818), p. [330].	330
18.1	Two pages of Ackermann's *Catecismo de astronomia* (Londres, 1825).	341
18.2	Title-page of Pinnock's *Catechism of Agriculture* (London, 1823).	346
18.3	Title-page of Ackermann's *Catecismo de agricultura* (Londres, 1824).	347
20.1	'The man wots got the whip hand of 'em all', hand-coloured engraving by William Heath, 1829.	371
20.2	An Applegarth and Cowper rotary 'Printing-Machine', from 'The commercial history of a penny magazine', *Penny Magazine* (31 Dec. 1833), p. 509.	372–3
20.3	'A Lady of Scientific Habits', hand-coloured lithograph of the early nineteenth century (author's collection).	376

20.4	'A Box of Useful Knowledge', hand-coloured wood engraving, c. 1832.	378
20.5	The geological record as a series of books, hand-coloured plate from [James Rennie], *Conversations on Geology* (London, 1828).	381
20.6	Nebulae supported by books, wood engraving from T. Milner, *The Gallery of Nature: A Pictorial and Descriptive Tour through Creation* (London, 1846), p. 192.	382
20.7	A large book caricatures itself, wood engraving from a drawing by George Cruikshank in J. Bateman, *The Orchidaceae of Mexico and Guatemala* (London: 1837–43), p. 8.	385
Title-page to Afterwords	The Owl of Minerva (device of Les Belles Lettres, Paris).	391

MARINA FRASCA-SPADA AND NICK JARDINE

Introduction: books and the sciences

> Now, happier lot! enlighten'd realms possess
> The learned labours of the immortal Press;
> Nursed on whose lap the births of science thrive,
> And rising Arts the wrecks of Time survive.
> (Erasmus Darwin)[1]

From classical times great books have stood as landmarks and book lists served as charts of the growth of the arts and sciences. In the earlier twentieth century they played major roles in consolidating the history of science as a discipline. Paul Tannery, George Sarton, Lynn Thorndike and other positivist historians, many of them passionate book-collectors, presented synthetic bibliography ('selective, critical and constructive', as opposed to merely descriptive and analytical, according to Sarton)[2] as the foundation for science history. After the Second World War Alexandre Koyré and his emulators concentrated more on the ideas at work in great books than on accumulating bibliographical detail. University teaching of the history of science should, they insisted, be centred on the critical reading of canonical texts, especially those emanating from the 'Scientific Revolution', itself a twentieth-century historiographical construction. Herbert Butterfield's account of the importance of 1543 nicely conveys the bookish flavour of this didactic history of science:

The year 1543, which saw the publication of Copernicus's great work and of the important translation of Archimedes, is a date of considerable significance in the scientific revolution, because it saw also the publication of the *magnum opus* of Vesalius, namely the *De Fabrica*, the work which stands as the foundation of modern anatomy.[3]

In recent years the disciplines of history of science and history of the book have been greatly expanded and transformed. Where a couple of decades ago these fields were still relatively specialised and isolated, they are now more generously conceived, and more closely integrated with general, social and cultural history. Where once historians of science were at pains to distinguish the activities and products of science itself from its social contexts and uses, nowadays they emphasise rather that scientific activities are (by and large) social activities, and they foreground the interplay of the sciences with other disciplines. Likewise, where history of the book once focused primarily on publication and bibliography, paying relatively little attention to the contents or uses of books, today it is widely accepted that we should study texts and their interpretations hand in hand with books and their uses.

There are further notable parallels between history of science and history of the book. Both fields have moved away from models of diffusion or dissemination of information from active producers to passive consumers. In the history of the book the history of reading has become a lively sub-field. Here books no longer figure as mere vehicles or packaging of texts; rather their material constitution – *mise-en-livre* – and the layout and typography – *mise-en-page* – are recognised as crucial in recruiting readers and conditioning the ways in which they read. Readers emerge as active recipients, variously constituting meanings as they appropriate works. Historians of science, similarly, emphasise readers' active roles in communications among experts, between teachers and pupils, and between elites and popular constituencies. Both fields have shifted their focus from canonical authors and their elite reception to the full range of writings and readerships. Moreover, in both fields the exemplary status of authors and works has come to be recognised not as a given, but as the historical product of often protracted canonisation through the efforts of followers, reviewers and commentators, not to mention the would-be classic authors themselves. Finally, both historians of science and historians of the book have become increasingly alert to the dangers of anachronism – of unreflectively imposing our categories on to past activities, of focusing on precursors to currently valued practices and doctrines. A measure of such imposition is inevitable for purposes of analysis, explanation and communication with present-day readers; but few nowadays can write with an altogether clear conscience about Renaissance 'scientific research' or licensing as a 'precursor' of copyright.

The present volume deals with the interactions between these flourishing fields. To set the stage let us glance at some of the ways in which the history of the sciences and the history of the book can complement and reinforce one another.

Writing in 1606, Johannes Kepler attributed the proliferation of printed books to the effects of planetary conjunctions on that human faculty which makes men social by nature, so that 'the minds of many men may come together in an undertaking'; and he memorably celebrated the effects of the flood of books:

Through them there has today been created a new theology and a new jurisprudence; the Paracelsians have created medicine anew and the

Copernicans have created astronomy anew. I really believe that at last the world is alive, indeed seething, and the stimuli of these remarkable conjunctions did not act in vain.[4]

Two centuries later, in the verses quoted above, Erasmus Darwin hailed the press as nursemaid of the arts and sciences. Similarly, in her seminal, if widely criticised, *The Printing Press as an Agent of Change* (1979) Elizabeth Eisenstein argued that Butterfield's paradigmatic achievements of the Scientific Revolution, the Copernican system and Vesalian anatomy, were made possible by the multiplication, standardisation and fixity of printed books.[5] And books figure as 'immutable mobiles' in Bruno Latour's account of the ways in which 'centres of calculation' – museums, academies, observatories, laboratories – recruit, delegate and control peripheral agents.[6]

The history of such grand pronouncements about the impact of print on the sciences would make an interesting study in its own right; but today few are happy with such claims. In their chapters in this volume, McKitterick notes that the replication of medieval texts was by no means as unreliable and erratic as Eisenstein implies, and Kassell emphasises how manuscripts continued to play major roles in the sciences far into the early modern period; by contrast, the stabilisation of printed texts was a long and painful business, far from completed even by the end of the sixteenth century. In any case, the impacts of the press on learning have not always been unambiguously beneficial – as Erasmus Darwin observed in a footnote to the verses heading this introduction, it has fostered 'new impositions' along with 'the arts of detecting them'. And print has been no less apt to aid and abet what later generations have deemed to be fads and crazes deleterious to the sciences – Lavaterian physiognomy and quinarian taxonomy, for example – than to encourage the sound sobrieties of Newtonian mechanics or analytical geometry. Nor, as Brotton here urges in the case of sixteenth-century Ottomans, should we uncritically take their failure to adopt the press as a sign of backwardness or decadence.

But there are more fundamental problems. The quest for generalisations about books and science is surely doomed by the fact that there was no such discipline as science, in our sense, in the early modern period. Natural history, astrology, alchemy, natural philosophy, physiology and mixed mathematics, to name just a few well represented in this volume, did not, even approximately, form a natural kind. Furthermore, the whole image of manuscript, print and electronic communication as media which variously facilitate or hinder the growth of knowledge is misleading. More profitable, as the chapters of this volume amply reveal, are approaches which attend minutely to the roles of authorship, production, distribution and reception of works within particular sciences in particular periods. What is remarkable is the sheer specificity of many of these roles: take, for example, the operations of reading ambivalent and esoteric texts (on occasion in relation to visionary dreams) in an alchemical adept's initiation, as tellingly described by Kassell and Johns; or the ways in which eighteenth-century encyclopaedias used cross-referencing to convey the systematic linkages between sciences, as spelt out by Yeo. Equally noteworthy are the transformations in the roles of books over time. Thus there have been major shifts in the location of authority and credibility in the sciences: from stationers and booksellers to authors and publishers (Johns' afterword), from commentaries to encyclopaedias (Yeo), from books to articles in journals (Broman), and so on. Another striking case is the metamorphosis of the library catalogue from the Baroque to the Romantic era wittily exhibited by Clark.

For all this diversity the chapters of this book do bring to light some general functions of books across wide ranges of sciences, places and periods. Let us briefly consider some of them. Genre links book history to central concerns of the history of the sciences. Particular sciences are in a given period often associated with, indeed partly constituted by, particular genres of writing – as demonstrated, for example, in Grafton's account of geniture collections and in Mosley's discussion of letter-books. Such genres, often embodied in characteristic material types of books (from formal commentaries in massive folios, to students' textbooks in pocketable duodecimos) constitute the 'systems of expectations' or 'implicit covenants' that link authors with readers. For genre provides at once norms of composition and guidelines for reading; and it is above all through genre that writers and readers take their places in traditions of writing and reading.[7] A grasp of the relevant genres is thus essential for the historian out to understand the works of past sciences, and to appreciate the ways in which those works were addressed to and appropriated by readers.

As brought out by many contributors (McKitterick, De Renzi, Yeo, Fyfe, Roldán Vera), studying genres all too often dismissed as derivative or secondary, such as commentaries, anthologies, editions, textbooks, popularisations and translations, is crucial to understanding the ways in which knowledge and the sciences have been handed on from generation to generation and from place to place. Frasca-Spada and L. A. Jardine and Stewart indicate how the humble work of editors may be of paramount importance in normalising doctrines and creating paradigms, and in establishing authors as canonical. And, as Topham shows, the attempt to found a whole new disciplinary school may be centred on writing and publishing a single textbook. Moreover, as N. Jardine argues, the maintenance of discipline in many of the sciences is heavily dependent on the handbooks, manuals and protocols through which their practices are standardised, calibrated and replicated.

Another active area of book history is the study of the organisation and layout of books, and in particular of 'paratexts' – tables of contents, postillae, footnotes, indices, illustrations, etc. – and of the varied ways in which they guide readers through texts and condition their responses to them.[8] Paratexts have been centrally involved in many of the practices of the sciences. Thus, as Blair and Frasca-Spada show, indices, glosses and footnotes have been crucial in the teaching of natural philosophy, rendering material accessible to students and intimating to them connections between the sciences. And, as Kusukawa, De Renzi and Spary demonstrate, study of the illustration of natural historical books yields substantial insights into the varied conventions linking natural objects with their visual and textual representations, and these, in turn, with the readers to whose tastes they appeal.

It is not only through its useful findings, but also by its historiographical example that book history is inspiring to historians of the sciences. The works of Henri-Jean Martin, Don McKenzie, Robert Darnton, Roger Chartier, Martha Woodmansee, Carla Hesse and Mark Rose, to mention but a handful of distinguished practitioners, offer models that historians of the sciences may profitably emulate. Thence we may draw on writings exemplary in their integration of history of texts with history of books, and of history of ideas with history of material culture (McKenzie, Chartier), in their appreciation of the inextricability of legal and political realities from their representations and ideologies (Hesse, Woodmansee, Rose), in their balancing of respect for actors' categories with use of robust analytic and explanatory categories (McKenzie), and in their narrative ingenuity in conveying past lived experience (Martin, Darnton).

What does the history of science offer in return? To start with, there are many topics central to the history of the book for which the history of the sciences may provide important methodological insights and materials. Two of these, the credibility of books and literary genre, we have already touched on. Practitioners of the sciences have in all periods been much committed to establishing the credentials of their own works and, on occasion, discrediting those of others; and they have long engaged in debates about the proper grounds for acceptance or rejection of testimony. Thus the processes by which credibility was assessed and secured have a high degree of historical visibility. De Renzi, in exploring the relations between types of testimony in natural history – eye-witness accounts *vs.* hearsay, described *vs.* depicted, oral *vs.* written *vs.* printed testimony – contributes to the vigorous current debate about the ways in which the credibility of reports of natural phenomena has been established and adjudicated. The flourishing state of the history of genres and persuasive strategies in the sciences is attested by the contributions of Grafton, Yeo, Broman, Topham and Secord. Here again the advantage to the book historian is visibility. Sometimes exponents of the sciences have been content to adopt and adapt their genres from other fields; but on many occasions they have run neck and neck with the most avant-garde of literary authors in creating new genres: works composed *more geometrico*, new types of encyclopaedias, textbooks, journals, etc. Where with established genres and their adaptations the conventions and modifications are rarely explicitly articulated (and then often in highly simplified and misleading didactic forms) the conventions of new and controversial genres of the sciences are often explicitly articulated, contested and defended.

In a widely cited article of 1982, Darnton introduced the notion of the 'communications circuit', which:

> runs from the author to the publisher (if the bookseller does not assume that role), the printer, the shipper, the bookseller, and the reader. The reader completes the circuit because he influences the author both before and after the act of composition. Authors are readers themselves. By reading and associating with other readers and writers, they form notions of genre and style and a general sense of the literary enterprise, which affects their texts, whether they are composing Shakespearean sonnets or directions for assembling radio kits. A writer may respond in his writing to criticisms of his previous work or anticipate reactions that his text will elicit. He addresses implicit readers and hears from explicit reviewers. So the circuit runs full cycle.[9]

The publications of the sciences provide an ideal field for exploring the most problematic parts of the communications circuit, those relating to the reading and appropriation of books and, most difficult of all, the feedbacks from readers to authors. As Secord has noted elsewhere, for certain types of scientific books the responses of readers are peculiarly traceable:

> for the handful of scientific books that became sensations have left more identifiable traces than comparable works of fiction, history, and poetry; references to fossil footprints and nebular fire-mists have a specificity that makes their source relatively obvious. Because of this, a widely-read scientific work is a good 'cultural tracer': it can be followed in a greater variety of circumstances than almost any other kind of book.[10]

Moreover, as Grafton, Blair, Kassell, Frasca-Spada, Terrall, Fyfe and Roldán Vera show, there is a series of practices relating to the uses and receptions of books – citation, footnoting, 'mnemotechnics', conversation about books – for which the sciences provide a wealth of wonderfully apposite material. As for the obscure processes by which readers' responses and authors' expectations of such responses interact, they are often relatively explicit in the case of scientific works: for in the sciences there have arisen elaborately formalised, and hence researchable, conventions for scrutinising, refereeing and reviewing.

The history of the sciences can also make a major contribution to the history of the book by exploring the ways in which the 'bookish practices' (to use Blair's phrase) that make up the communications circuit are linked to more general cycles of production and consumption. In the sciences there is indeed an extraordinary variety and richness in the relations of books to other objects. Often, as Mosley indicates, the production, use and privileging of astronomical instruments, models and manuals proceeded hand in hand. Books may act as substitutes for objects, and vice versa: for example, the illustrated natural history book being a virtual collection, the collection being organised in imitation of a book. Or the object and the book may be one and the same, as with Linnaeus' 'Plantae Lapponicae', a herbarium bound as a book and donated with Benjamin Delesserts' library to the Institut de France.[11] As theorists of interpretation have long insisted, readers are trapped in the hermeneutic circle: needing to grasp the genre and purpose of a work as a whole in order to get to grips with its parts, and vice versa; conditioned as critics in their reading habits by the very traditions of interpretation they are out to criticise. The interactions of books with objects are sometimes of little help in this predicament, piling mystery on mystery, as with the hieroglyphic alchemical books and invisible substances explored by Kassell. But typically they are godsends. Thus instruments are hermeneutic keys to their manuals, the manuals in turn keys to the interpretation of the instruments and their uses; and as in the cases of Linnaeus' published *Flora Lapponica* and its accompanying 'Plantae Lapponicae' and Adanson's collection of shells and his book on shells (described by Spary) collections may be keys to the reading of books and books to the understanding of collections.

Some areas of the history of the sciences are integral parts of the history of the book. Thus, as Johns shows, the physical and psychological effects of reading and its place in a healthy regimen formed a substantial chapter in the history of physiology. The history of property rights over instruments and discoveries in the sciences is another such field, for, as indicated by Mosley and Johns, the rights of makers over their instruments and of natural philosophers over their discoveries have served as touchstones in the long battle to secure literary property rights for authors. Finally, there is the major branch of the history of technology concerned with the production of books that forms the very foundation of book history. The present work does not venture into the history of technology; however, the contributions of McKitterick, Secord and others are variously indicative of the appropriations of technologies of book production in the transmission of the sciences.

The present volume is intended as introductory, as a work of first resort for all those interested in the history of the sciences in relation to the history of the book. To make their work accessible, contributors have kept footnotes to a minimum and provided 'further reading' lists on their topics. In planning the volume, the editors have aimed to do full justice to the richness and specificity of the interactions between the history of the sciences and history of the book, whilst providing a reasonably comprehensive coverage of this interdisciplinary terrain. Accordingly each section of the volume combines chapters of relatively broad scope with chapters tackling specific works and episodes in depth.

Periodisation is a tricky business. The narrative conventions and readers' expectations of synthetic works like the present one demand it. But, at least in history of the sciences and history of the book, it is potentially misleading. It may be inaccurately suggestive of radical discontinuities: between labile manuscript and stable print, between a culture of patronage and privileges and one of commerce and competition, between the authority of books and that of specialist journals in the sciences. Further, it tends to obscure the diversity in the timescales of significant developments – consider the contrast highlighted by Johns in relation to reading between the long duration of traditional commonplaces about healthy daily regimen and the rapid turnover of physiological theories purporting to underpin and explain them.

The first section, 'Triumphs of the book', runs roughly from the Carolingian revival of learning to the end of the Baroque. Among the earlier triumphs which figure here are the preservation of classical learning and the installation of university teaching; later ones include the formation of new genres, new philosophies and new technologies, along with the protracted and uneven achievement for certain kinds of printed books of unprecedented levels of stability and authority. At the same time there are certain distinctive continuities, notably in the didactic realm: thus throughout the period book cataloguing, annotating, commenting, and the forming of commonplace books played major roles in the practice and maintenance of the sciences.

The second section, 'Learned and conversable reading', covers a 'long eighteenth century', roughly 1688 to 1815. It looks at the changes and continuities in the uses of books at the traditional learned sites, that is, cabinets, libraries and universities. It is widely claimed, following Jürgen Habermas, that this period saw the emergence of a 'public sphere' associated with the formation of a bourgeois society and the commercialisation of the book market.[12] Aspects of Habermas' thesis, notably an idealistic characterisation of a public sphere of rational and disinterested discussion, and failure to do justice to earlier and later domains of public debate, have been criticised;[13] however, as our contributors demonstrate, there were formed in this period important new sites of debate of the sciences, from academies and journals to salons, coffee-houses and public displays.

The third section, 'Publication in the age of science', covers the first half of the nineteenth century, when science itself was formed as an alliance of disciplines. Innovations in production, distribution and consumption of books, themselves viewed at the time as prime examples of scientific progress, made possible the mass production of books. The contributors focus on the varied roles of books in the establishment of the new science in its educational, national and international settings.

The 1840s and '50s provide a natural end-point for the volume. In these decades the scale of production of books showed a remarkable increase;[14] by then science, vigorously promoted by mass-produced textbooks and treatises, was consolidated and institutionalised as an educational and cultural formation; and specialist journals were well on the way to taking over from treatises as the primary loci of scientific authority and vehicles of professional scientific innovation.

The Owl of Minerva flies at dusk. In the 1980s and 1990s both science and books became problematic. Where once science was widely perceived and promoted as an autonomous and disinterested master discipline united by a single methodology, historians, sociologists and philosophers now tend to argue for the irreducible multiplicity of the methods of the sciences, and for the inextricability of pure science from its technological and socio-political engagements. Books too are losing their privileges and integrity, and in this the sciences are at the cutting edge: in all their long-standing central roles in the sciences – in teaching, in research, in the delegation and control of agents, and in the establishment of public images – books are on the way out. With the waning of these functions, books, like science, are losing their obviousness. Many, the editors included, are becoming nostalgic for the handiness and heftiness of books; and many are coming to miss the reassuring authority of science. But for all of us as historians there is a brighter side. The problematisation of science and its media surely makes us alert to issues concerning the history of the book in relation to the sciences, or at least more alert than were historians in bygone days when science and its books were familiar fixtures. And we may hope that historical studies, including those we present here, will make us more critical and sensitive in reflecting on and coping with the anxieties and predicaments generated by the fragmentation of science and the electronic revolution.

Notes

1. E. Darwin, *The Temple of Nature* (London, 1803), Canto IV, lines 269–72. Thanks to Elly Bell for drawing our attention to this passage.
2. G. Sarton, 'Synthetic bibliography', *Isis*, 3 (1921): 159–70.
3. H. Butterfield, *The Origins of Modern Science, 1300–1800* (London, 1949), p. 39.
4. J. Kepler, *De stella nova* (Prague, 1606), p. 188.
5. E. Eisenstein, *The Printing Press as an Agent of Change* (Cambridge, 1979). For a critique of Eisenstein's claims see A. Grafton, 'The importance of being printed', *Journal of Interdisciplinary History*, 11 (1980): 265–86.
6. B. Latour, *Science in Action* (Milton Keynes, 1987), ch. 6.
7. On genres as shared systems of expectations, see E. Hirsch, *Validity in Interpretation* (New Haven, 1967), ch. 3; on genres as covenants, see D. Damrosch, *The Narrative Covenant: Transformation of Genre in the Formation of Biblical Literature* (San Francisco, 1987).
8. See especially G. Genette, *Paratexts: Thresholds of Interpretation* (1987), trans. J. E. Lewin (Cambridge, 1997).
9. R. Darnton, 'What is the history of books?' *Daedalus* (Summer 1982): 65–83; reprinted in *The Kiss of Lamourette. Reflections in Cultural History* (New York, 1990), pp. 107–35, at p. 111.
10. J. Secord, *Victorian Sensation* (Chicago, 2000), Introduction.
11. See W. T. Stearn, 'Introduction', in *Carl Linnaeus' Species Plantarum. A Facsimile of the First Edition*, vol. I, pp. 1–176, at p. 115.
12. J. Habermas, *The Structural Transformation of the Public Sphere: An Inquiry into a Category of Bourgeois Society* [1962], trans. of 2nd edn. T. Burger (Cambridge, MA, 1989).
13. For telling comments see N. Fraser, 'Rethinking the public sphere: a contribution to the critique of actually existing democracy', in C. Calhoun (ed.), *Habermas and the Public Sphere* (Cambridge, MA, 1992), pp. 109–42; and J. Raymond, 'The newspaper, public opinion, and the public sphere in the seventeenth century', *Prose Studies*, 21 (1998): 109–40.
14. See S. Eliot, 'Some trends in British book production, 1800–1919', in J. O. Jordan and R. L. Patten, *Literature in the Marketplace: Nineteenth-Century British Publishing and Reading Practices* (Cambridge, 1995), pp. 19–43; R. Wittmann, *Geschichte des deutschen Buchhandels. Ein Überblick* (Munich, 1991); C. Bellarger, J. Godechot, P. Guiral and F. Trevrou (eds.), *Histoire générale de la presse française*, vol. II (Paris, 1969).

I Triumphs of the book

Book-wheel, from Agostino Ramelli, *Le diverse et artificiose machine* ... (Paris, 1588). (Courtesy of the Whipple Library, Cambridge)

1 Books and sciences before print

The role of texts in the transmission of scientific knowledge and discovery has to be considered in relation to a number of issues. Firstly, what do these texts indicate about the degree to which empirical knowledge and experience were transmitted orally? Secondly, to what degree could the authority of texts be thought to outweigh personal experience and observation? Thirdly, the historian of science is bound by the texts that have survived, but can nevertheless be alert to the implications in them concerning how they may have been read and understood, how they may have been used in teaching, how they were produced and how they were disseminated. Contrary to the assumption that there was a total revolution in the dissemination of ideas as a consequence of the advent of printing,[1] moreover, it needs to be stressed that the printers were able to exploit well-established modes for the dissemination of ideas, habits of book ownership, markets and patterns of distribution in the book trade.

Book production and distribution in the middle ages

We need to consider first of all, therefore, how books were made in the middle ages until printing with movable type transformed the technology for putting words on the page.

Before paper making was introduced from the Arab world into Spain and Sicily in the tenth and eleventh centuries and was produced at a commercial level in western Europe in the thirteenth century, the material for the pages of books was animal skin. The commonest skins were those of sheep, goat and cow, though in principle any animal skin, even human, can be made into parchment. In the central and later middle ages, especially in England with the economic importance of rabbit breeding for felt and meat, it is likely that rabbit skins provided the wherewithal for cheap books as well.

The skins, best treated when they are fresh, were soaked in a lime solution to loosen the fat and hair or wool. They were squeezed, scraped down and washed and stretched on a frame to dry. Once dry they would be scraped to the requisite thinness with a very

Figure 1.1 Illustration redrawn from the ninth-century Frankish manuscript Düsseldorf, Universitätsbibliothek MS A. 14, fol. 119v, showing Titus writing a letter to St Paul. The scribe's desk holds knife, pricking implement, pumice stone or sponge for erasing and ink pot and the scribe is writing with a quill. (Courtesy of Cambridge University Press)

sharp half-moon-shaped implement. The scraping is a highly skilled process, requiring strength as well as accuracy. Too much force would simply cut holes in the skin and varying pressure would result in an uneven surface. In principle, thereafter the parchment would be ready for use, to be trimmed to page sizes, pricked in the margins to provide a guide for ruling, the writing lines and space ruled and the text written by the scribe (Figure 1.1). Sometimes the surface might be treated by being rubbed with chalk. It is likely that there were parchment makers and suppliers whose business it was to prepare sheets for any individual or institution needing new writing materials. The palimpsest evidence from the middle ages, moreover, which is most abundant in periods of high book production such as the ninth, twelfth and fifteenth centuries, suggests that there was also a market in second-hand parchment that had had the original text washed off or otherwise erased. It has often been assumed that palimpsested texts are a consequence of negative judgements about their contents made by the scribes who replaced the texts with something else. They may, however, simply be using 'recycled' parchment supplied from elsewhere.

With parchment supplied the scribes made their own ink, largely based on oak gall apples though sometimes carbon based (mixed with vinegar), and pens made from either goose feathers or the common reed, *Phragmites communis* (found in marshlands or on river banks all over Europe). The office of scribe came to carry moral weight. Writers decribing the process stress that it is a labour and a virtue. It could be counted as a manual labour within the conventions of the monastic rule. It could be counted as a virtue and creative endeavour when considering achievements. A famous story, illustrated in the Munich codex, Bayerische Staatsbibliothek Clm 13031, fol. 1r, from Prüfening c. 1160–5, is of Swicher the monastic scribe. On his deathbed, the scribe's sins were weighed in the balance with the books he had written. One extra word from his pen made all the difference, the demons were vanquished thereby and Swicher went safely to heaven.

Scriptoria

There are any number of possible contexts for writing and writing materials to have been used. Thus a local official might have a secretary who needed to record the local judicial transactions presided over by the official, keep estate records and tax lists, and write the occasional business letter (though some of this might also be done on wax tablets, on wooden sticks and tablets or pieces of horn, bark or bone). A religious institution might have a writing office for its record keeping. From the seventh century onwards we can document what are known as scriptoria, that is, groups of scribes working in one institution, usually but not necessarily attached to a monastery or cathedral, who wrote the books needed both by their own institution and in some cases for export to other centres without their own sufficiently trained scribes. The scribal discipline of such a scriptorium was co-ordinated and there emerges the phenomenon of a 'house style' of script, that is, the script of a particular place that was distinctive in its letter forms and shapes, ductus, style, format and scribal conventions. All scribes of that place were trained to write the same kind of script so that the product of that particular scriptorium can be instantly recognised. Scriptoria such as Luxeuil, Jarrow and Wearmouth have been identified in late seventh-century Gaul and England,[2] and there was a proliferation of scriptoria in western Europe from the eighth century onwards associated with what is known as the Carolingian renaissance. All these scriptoria wrote in a script known as caroline minuscule (see, for example, Figure 1.2) which had evolved slowly and at different rates on the Continent from the basis of the Roman system of scripts. Such evolution was subject to regional variations, dependent in the first place on the way in which the script might have developed from the Roman script system and the choice made of different letters forms in the various regions. The scripts written in Gaul, Spain and the

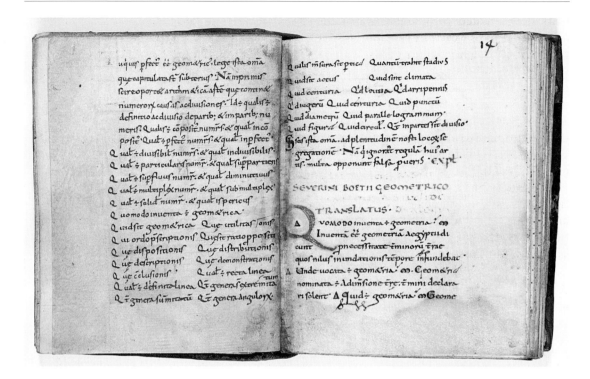

Figure 1.2 Cambridge, Trinity College Library, MS R.15.14, fols. 13v–14r, the Carolingian gromatic and geometrical compilation, showing the attribution of the text on geometry to Boethius. The script is caroline minuscule of the later ninth or early tenth century. (Courtesy of the Master and Fellows of Trinity College, Cambridge)

British Isles, all based on the Roman script system, differed greatly. Secondly, local idiosyncrasies, often manifest in the predilection for particular ligatures, abbreviations or letter forms, could be retained in a script which might otherwise appear uniform. What this process of script development means for the history of culture and knowledge is that script can act as an identifier for a region or precise place of origin. Because of this, we can draw maps of intellectual concentration and exchanges which are entirely dependent on the evidence provided by the letter forms in the manuscripts.[3]

Schools and libraries

Many centres concentrated on producing books to accommodate the needs of their schools and to fill the chests in their own libraries.[4] Others, however, made distinctive efforts to disseminate copies of particular texts. These could be of local authors, as can be seen in the production of the works of Bede at Jarrow-Wearmouth in response to demands for them from centres on the Continent, or of Hraban Maur of Fulda's commentaries. We can estimate the effectiveness of the distribution of texts from the evidence of extant library catalogues and manuscripts from all over Europe. They could also be copies of authorised and recommended texts in relation to a particular policy, such as the biblical, liturgical and legal texts disseminated from the Carolingian royal court.[5] A scribe could present his work to a patron (whether his abbot or bishop), or to a saint, or to Christ, just as an author could. The process for the publication and distribution of new texts was just the same as that for older classical and patristic works; in part it depended on networks of personal acquaintance and patronage, such as is mirrored in the scholars' correspondence of the middle ages. Books could also be supplied to other monasteries and churches on commission, or a centre could request the books of another in order to make copies of them.[6]

The letters of Lupus and Gerbert, for example, raise the questions not only of the availability of books containing scientific knowlege from antiquity of which some indication will be given below, but also how both Lupus and Gerbert, and their contemporaries and successors, gained their knowledge of the works and authors that they should know. As we shall see, such information was partly to be gained from ancient compilations, quotations and references. A further source was Cassiodorus' *Institutiones*. Of major importance, however, were the bibliographical guides and library catalogues compiled in the ninth century.

Lupus of Ferrières in writing to Einhard refers to a list of the books belonging to Einhard that he had seen.[7] The extant library catalogues provide us not only with clear evidence of the outcome of such networks of exchange and production; it is also clear that in themselves the library catalogues of the ninth century defined the canon of required knowledge and served as lists of the books which a library should possess.[8] As such they were used as bibliograpical handbooks. Libraries in medieval Europe thereafter built on the foundations laid in the Carolingian period.[9] Library catalogues in the ninth century were circulated so that other centres could know a possible source of exemplars for copying. Indeed, we can establish from the medieval library catalogues something like an inter-library loan system. The Dominicans, for example, operated a highly sophisticated system of inter-library loans in the later middle ages as well. These same library catalogues, moreover, provide us with a continuous line of evidence concerning science books in medieval libraries, especially works on computus, astronomy and geometry, which more often than not are incorporated into the lists of school books.

In the course of the twelfth and thirteenth centuries and in conjunction with the establishment of the universities of Paris, Bologna, Montpellier, Oxford and Cambridge, the production of books became a matter increasingly of commercial stationers, with professional scribes supplying texts for the students according to an organised system of *pecia*. Particular stationers were licensed by the universities to hire out portions of a book (*pecia*) in turn to a scribe (whether the student himself or someone a student was paying to do the work). This scribe would then copy the text, return it to the stationer and hire the next portion until the whole book was copied. The great majority of medieval university books were used in the Faculties of Civil and Canon Law and Theology; except for the University of Paris the *pecia* system did not apply in the Faculty of Arts, where the teaching was *viva voce*.[10] The supply of texts and the relationship between author, stationer and manuscript have to be extrapolated from the evidence for the copying of the works of scholars such as Peter Lombard and Aquinas. They can then be surmised for the production of texts in the *artes* and for the teaching of medicine.

On the one hand, moreover, such 'textbook' production has to be seen against an increasingly diverse framework of educational institutions. On the other hand, the efforts on the part of the authorities in these institutions as well as in the church and in the sphere of secular government to exert control over production and distribution, as and when they thought it necessary, added tensions and directed the selection of texts and the development of particular subjects.[11] Secular stationers and booksellers played an increasingly active role in the production and distribution of new and second-hand books as a speculative venture rather than simply on commission. It was on all these circulation and distribution patterns, both private and commercial, that the printers were able to draw. The new technology speeded up production and many more copies of established texts were made available to a market and to a reading public that was ready for them.

Yet printers also continued to make available, as the scribes and stationers of the middle ages had done before them, not only new authors and texts, but also the older stalwarts of the medieval school and university curricula in natural philosophy, mathematics, astronomy and medicine. Added to these the scientific works of classical antiquity were also among the early incunables, whether in the original Latin or translated into Latin from Greek and Arabic.[12] These classical texts had been preserved and studied throughout the middle ages. The contexts of survival and use in the middle ages of this third category of science books, therefore, merit detailed consideration.

The survival and transmission of scientific knowledge

The knowledge and exposition of the natural world in writing is associated with the literate society of the Greeks and their conquerors and heirs the Romans. It was transmitted to medieval western Europe and the Arabic world. The last named, in particular, preserved in Arabic translations from the fourth century onwards a great deal of Greek scientific thought which in due course was translated from the Arabic into Latin in medieval Spain and Sicily and disseminated further within western Europe. Commentaries and notes were also contributed by every generation of scholars and these preserve the memory and ideas of some thinkers which would otherwise have been lost. The authority of the book as a vehicle of knowledge, moreover, was further enhanced with the conversion of the Roman empire to Christianity.[13] The book, and writing, were firmly associated with the word and law of God, just as they were within the Hebrew tradition on which the Christians drew. Although the ultimate goal of medieval education was a fuller understanding of God and the Christian faith, most subjects could be accommodated as contributing to that goal. Study of the natural world was study of God's creation: 'God and nature were two aspects of the same truth.'[14]

Throughout the intellectual history of Europe, moreover, it is self-evident that the organisation of thought, the development of logic, deductive reasoning and argument always held a place in the educational curriculum, even if they were studied in the context of rhetoric and dialectic in the ancient world and most commonly as elements of theology in the middle ages.[15] Such training of the mind taught intellectuals to follow lines of deductive reasoning that could be and were applied to natural phenomena as much as they were to questions of doctrine or philosophy. The careers of scholars such as Robert Grosseteste in the thirteenth century and William of Ockham and the Oxford 'Mertonians' in the fourteenth century exemplify this obvious point. Robert Grosseteste lectured in theology at the University of Oxford and became Bishop of Lincoln, but his writings on mathematics, his explanations of the different phenomena of light, his work on the *Posterior Analytics* of Aristotle, and his emphasis on the primacy of 'experiment' (personal experience) over authority had a profound influence. William of Ockham and the 'Mertonians' in the fourteenth century made many connections between the study of logic, physics and theology and introduced many new approaches to science, especially in the link between mathematics and natural philosophy.[16]

Averroes' Commentary on Aristotle's *Posterior Analytics* became known in Latin in 1230, but Robert Grosseteste had already written his commentary between 1228 and 1230 on the Latin version supplied by James of Venice. The reception in Europe of many of the works of Aristotle and other works of science in the thirteenth century, and their rapid dissemination throughout the universities, however, should not obscure the various texts by which a knowledge of ancient science, notably geometry, mathematics and astronomy, had been preserved and transmitted in the preceding centuries. In this respect the Frankish scholars of the Carolingian period in Europe, especially those of monasteries in Picardy (Corbie, Rheims and St Amand), the Loire valley (Tours, Fleury, Auxerre, Orléans, Ferrières), the Rhine-Main area (Lorsch and Fulda) and the Lake Constance region (Reichenau and St Gallen) in the ninth century, made arguably the most fundamental contribution of all.[17] It was largely due to the deliberate efforts of the Carolingian scholars that the bulk of Latin classical learning and literature in all its richness survived at all. With only a few exceptions the earliest surviving manuscript of most ancient Latin texts, whether literary or scientific, is a ninth-century copy from one of these monasteries. Even Apicius' *De re coquinaria* goes back only as far as two ninth-century manuscripts, one from Tours and written for the Frankish King Charles the Bald (840–77) (grandson of Charlemagne) and the other from Fulda.

Geometry

Although Boethius had translated Euclid's geometry in the sixth century, for example, and ninth-century copies of this are known, the form in which Euclid's geometry was most widely disseminated was in a special late eighth- or early ninth-century compilation, possibly made at Corbie, namely the *Ars geometriae* in five books. This incorporated, alongside some of the so-called ancient gromatic texts or *Agrimensores* on surveying, the work of Hyginus (called 'Gromaticus' because of the subject matter of his principal writings). The latter remained, with Boethius' partial translation, the only source for Euclid's geometry in Europe until the twelfth century. The *Ars geometriae* tended also to go under Boethius' name, as it does in the tenth-century copy from northern France now in Cambridge Trinity College Library R.15.14 (Figures 1.2 and 1.3).[18] In addition there was a polished compilation of technical treatises on land surveying available in the ninth century, based most probably on some fifth- or sixth-century set of texts, as well as a host of individual compilations on surveying. These transmitted fragments of ancient geometrical and arithmetical knowledge such as the *De die natali* of Censorinus. A further important practical application of geometry and arithmetic was in the technical sphere,[19] exemplified by the work of Vitruvius on architecture and building, whose earliest surviving manuscripts, as well as those of Faventinus' epitome of Vitruvius, are also of ninth-century date.[20]

Astronomy

For astronomy (Figures 1.4 and 1.5) early medieval scholars depended on the *Phaenomena* and *Prognostica* of Aratus, a popular

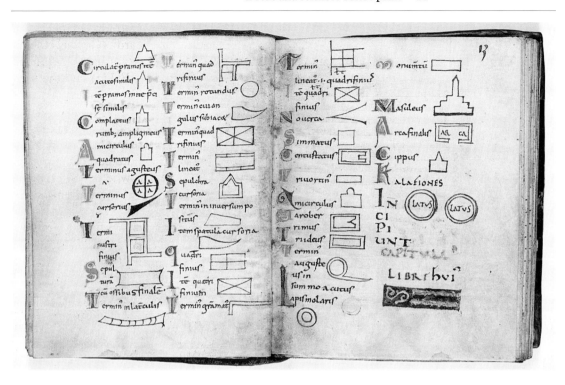

Figure 1.3 Cambridge, Trinity College Library MS R.15.14, fols. 12v–13r, a section of the *Nomina lapidum finalium et archarum positiones* with figures. (Courtesy of the Master and Fellows of Trinity College, Cambridge)

non-mathematical exposition of the chief features of the celestial globe. It was transmitted in the Latin verse adaptations by Cicero, Germanicus and Rufus Festus Avienus, and in translations known as the Aratus Latinus and the 'revised Aratus Latinus'. In the various ninth-century manuscripts of these works, scholia based on Hyginus' *Astronomica* and *Fabulae* were sometimes added or substituted. These included the myths and star catalogues for each constellation. In some manuscripts such as Leiden, Bibliotheek der Rijksuniversiteit Voss. lat. Q 79, some knowledge of the Ptolemaic star catalogue is indicated.[21] The variety of illustration in the many astronomical books that survive, moreover, reflects late antique astrological illustrations. Indeed, the imagery apparently preserved by Carolingian scribes and artists is a forceful reminder of the way in which pictures may also transmit knowledge, and how imagery, notably that of the constellations and signs of the zodiac, can itself hint at a complex ancient tradition in addition to that of the texts.[22]

Arithmetic

Astronomy was linked with early medieval arithmetical study. This was known as *computus*. Computistic calculations, particularly of the date of Easter in the Christian year, required astronomical knowledge as well as mathematical agility. In the sixth century, Dionysius Exiguus, in his *Libellus de cyclo magno*, introduced calculation

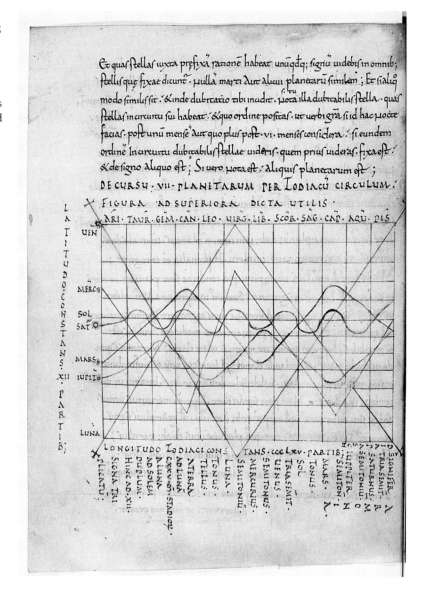

Figure 1.4 Cambridge, Trinity College Library MS R.15.32, a Carolingian astronomical collection containing Hyginus and Cicero's version of the *Aratea* and Abbo of Fleury's works on astronomy, copied probably at Winchester in the late tenth or early eleventh century, fol. 6v from Abbo's *De cursu planetarum per Zodiacum circulum*. (Courtesy of the Master and Fellows of Trinity College, Cambridge)

according to the Christian era, starting from the year of the Incarnation. He thereby inculcated a sense of the annual Christian cycle in relation to the movements of sun, moon and stars. Many tracts on *computus*, comprising short arithmetical and astronomical discussions and tables, were compiled which followed Dionysius Exiguus. Bede's *De ratione temporum* (On the reckoning of time),[23] for example, was one of many influential texts. Carolingian scholars, moreover, independently devised what has been described as an 'official *computus*', which it has been suggested was the outcome of a colloquium of astronomers and computists held in 809 under the

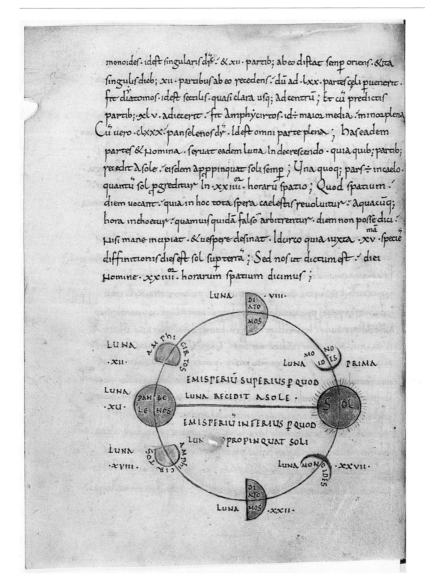

Figure 1.5 Cambridge, Trinity College Library MS R.15.32, fol. 7v from the *cursus lunae*. (Courtesy of the Master and Fellows of Trinity College, Cambridge)

auspices of the Carolingian ruler Charlemagne. It is extant in the compilations of 810 known as the '*3-book-*' and '*7-book-computus*', both linked with the royal court. The primary focus of these two compilations was also on the calculation of time. In addition there were Carolingian 'encyclopaedias of time' compiled in 793 and 809.[24]

How closely *computus* and astronomy were associated in the ninth century is apparent from such compilations as the scrap book of Walafrid Strabo of Reichenau. Walafrid (d. 849) studied at Fulda from 827 to 829, acted as tutor to the young prince, later King Charles the Bald (grandson of Charlemagne) at Aachen from 829 and thereafter was a renowned scholar at Reichenau and abbot from 839. His scrap book, now in St Gallen (Stiftsbibliothek MS 878)[25] was compiled continuously between 825 and his death. Into it Walafrid copied material on Latin word usage and grammar, as well as on the use of the calendar, medicine, natural history and science. One major treatise he incorporated was the *Computus* of his teacher Hraban Maur of Fulda. In it, Hraban discussed the concept of number, how to use numerals, and time reckoning. Another compilation from Fulda (Oxford, Bodleian MS Canon. misc. 353) from this period adds a sequence of questions and answers indicating how students at Fulda were taught calculation in relation to the movements of the sun and the moon. It also contains material on cosmology.[26]

Cosmology

In early medieval discussions of cosmology, most notably the eighth-century *Cosmographia* of Aethicus Ister (who had borrowed material from Pomponius Mela's *De chorographia*), the spherical shape of the world and of the universe are clearly assumed. Aethicus, however, incurred criticism from his contemporaries for venturing to suggest that the Antipodes might be inhabited.[27] The *Tusculan Disputations* of Cicero, apparently dispersed from the Carolingian royal court, was another source of cosmological ideas. It was a school text conveying some sense of Platonic cosmology, though fuller notions of Plato's ideas were derived from his *Timaeus*, the Latin translations of which by Cicero (incomplete) and Calcidius (with accompanying commentary) are extant in ninth- and tenth-century Frankish manuscripts. Thus the Carolingians ensured the transmission of Plato's ideas into the eleventh and twelfth centuries.[28]

Much ancient scientific knowledge, of course, was transmitted in Pliny's *Natural History*. All the pre-Carolingian copies extant are fragments only, and even in the Carolingian period the text has to be reconstructed from the bits and pieces used by the Carolingians for computistical and astronomical purposes, including the 3-book- and 7-book-*computus* referred to above. Charlemagne himself took an interest. Alcuin, the English scholar from York who spent time at the royal court before retiring to the abbey of Tours, wrote as follows:

A traveller came at speed with a sheet of questions from your Majesty . . . urging a weak-witted old man to examine the heavens when he has not yet learned the principles of earthly things, to expound the erratic courses of the planets when he is far from able to recognize the plants in the ground . . . What can be said more clearly on the agreement of the paths of the sun and the moon through the signs of the zodiac than what such an investigator as

Master Bede has left us in his writing? What research can be more penetrating than the publications on the stars by that dedicated enquirer into nature, Plinius Secundus? . . . I dare not give random answers to such deep questions so I ask you in your kindness to send me the first books of the aforesaid Pliny.[29]

Much of the content of Pliny, moreover, was incorporated into the other great repository of ancient learning, namely the seventh-century encyclopaedia or *Etymologiae* of Isidore of Seville. Isidore was also much indebted to the writings of Solinus, Pomponius Mela and a great deal else.[30] The *Etymologiae* was one of the most widely disseminated and well-known works of the entire middle ages; nearly every medieval library had a copy.[31]

Medicine

For the study of medicine, important texts which transmitted some of the teaching of Galen and of Scribonius Largus included the *De medicamentis* of Marcellus, extant in two ninth-century manuscripts, and the translation of Galen by Constantinus. The *Liber medicinalis* of Quintus Serenus, a verse treatise in sixty-four chapters, was also very popular in the ninth and tenth centuries, though some scholars, ignoring the evidence of the medical compilations of the Carolingian period, think it was read as poetry rather than as a medical manual. A great many medical miscellanies survive, including some eighty-two manuscripts from the ninth century alone.[32] They comprise medical recipes, explanations, discussion of diseases, citations and extracts from such authorities as Hippocrates, Galen, Oribasius, Dioscorides or Soranus, even though in most cases full versions of these authors were not known in Europe before the twelfth century. Such compilations, with their very distinctive mixture of home-grown treatments and scientific commentary, continued to be produced, and were further augmented by the new treatises and translations emanating from Salerno in the eleventh and twelfth centuries and also from Montpellier in the thirteenth century, both of which were centres for the study of medicine. Contreni has argued, however, that it is from the ninth century that medicine was established on new educational principles as part of the curriculum; he has been able to document members of the intellectual elite who pursued medical studies alongside their other interests.[33]

The communication of ideas

In focusing on the survival and influence of ancient scientific texts, as well as the dissemination of new texts, it is important to remember that the process of such transmission depended on both casual and formal methods and on oral and written communication. Letters and books, teaching, disputation and conversations, text production and libraries: all these enabled ideas to be circulated, disseminated and handed down through generations. The pattern of formal and informal, oral and written, transmission of knowledge established in the early middle ages has remained substantially the same ever since, despite dramatic technological changes.

Lupus of Ferrières (c. 805–62), one of a group of Frankish scholars in the constellation of the monasteries of Fleury, Auxerre, Ferrières and Orléans in the Loire valley in the middle of the ninth century, provides a straightforward example of both informal and formal methods for the exchange of ideas. He wrote many letters to friends discussing scholarship, taught pupils in his own monastery, and borrowed books from other libraries. He wrote to Archbishop Orsmar of Tours (837–46), for example, urging him to obtain the Commentaries of Boethius on the *Topica* of Cicero in the library of the abbey of St Martin. He had obviously seen it for he knew it was written on papyrus (and was therefore possibly as old as the sixth century). He promised to take great care of it but did not want Orsmar to tell anyone that it was he, Lupus, who had borrowed it. Rather, he suggested Orsmar should tell Abbot Amalric that Orsmar wished to send it to certain relatives of his who were eager to have it.[34]

Similarly, Gerbert of Rheims (c. 940–1003) (later Pope Sylvester II), famed for his teaching of mathematics and astronomy at Rheims in the tenth century, passed on the learning he had apparently acquired in Spain. There it is supposed that he had come into contact with some Greco-Arabic teaching in mathematics, geometry and astronomy. He provides an important indication of how knowledge could be transmitted. He says in a letter to his friend Constantine of Fleury, written c. 980, that it has been some years since he had a book or any practice in explaining the rules of the abacus, so instead he offers some rules repeated from memory which he now puts into writing once more. He writes as follows:

Here in this letter diligent researcher, you now have the rational method, briefly expressed in words it is true but extensive in meaning, for the multiplication and division of the columns [of the abacus] with actual numbers resulting from measurements determined by the inclination and erection of the geometrical radius, as well as for comparing with true fidelity the theoretical and actual measurement of the sky and of the earth.[35]

In his history of France, moreover, written at the end of the tenth century, Richer describes Gerbert's practical demonstration of the use of an abacus, remembered from his own time at the school of Rheims. Richer comments that the use of the abacus enabled Gerbert to manipulate the 'thousand characters' (that is Arabic numbers) Gerbert had had made out of horn. These indicated the multiplication or division of each number when shifted about in the twenty-seven parts of the abacus (described as a board of suitable dimensions, to be made by the shield maker). Its length was divided into twenty-seven parts (columns) on which Gerbert arranged the symbols, nine in number, signifying all numbers. Gerbert did the manipulation with such speed in dividing and multiplying large numbers that, in view of their very great size, they could be shown (seen) rather than grasped mentally by words. Richer thus also describes Gerbert's use of Arabic numerals though he would appear still at this stage to have lacked the nought.[36]

Gerbert therefore used a practical demonstration of the abacus in his classes. Richer's reference to Gerbert's letter to Constantine as a *liber*, and the rules there provided, on the other hand, indicate the process by which what Gerbert had once learnt was transformed into a textbook. It is clear that the letter to Constantine was a format chosen for the transmission of these rules but it did not remain a personal letter. Rather, as Richer makes clear, it was made generally available and circulated as a useful text.[37] Yet there are also indications in the text itself of an awareness of the difficulty of transferring from one medium of explanation to another. Gerbert has already referred to a book or practice and offers to Constantine only 'certain rules repeated from memory'. Gerbert taught, therefore, by both oral and written means. Generally, of course, we only know about oral communication in antiquity and the middle ages from descriptions of it in writing or from texts thought to be written versions of something first known orally. Poetry in the vernaculars of early medieval Europe (Old English, Irish, Old High German, Latin) in particular is thought to have been composed first orally and subsequently written down. The process of written composition, however, is disputed. What has to be borne in mind is the possible impact on the literary form made by the translation from the oral to the written medium. The latter had inherited rules for written verse from classical antiquity. It is by no means a simple matter of oral forms subsequently written down. Genre, memory, the changes wrought by transmission over the centuries, all have potential impact. Further, the degree to which change can be deliberately introduced in relation to a different understanding or set of priorities may also play a role.

Certainly the process of teaching was usually oral, even if closely tied to a text with a master providing extempore or prepared commentary on it. Sometimes this commentary was itself incorporated into the text, though the lay-out of twelfth- and thirteenth-century manuscripts enables one to distinguish between original text and the different layers of commentary.[38] Other texts, treatises and letters record dialogues between students and masters, disputes and discussions taking place within a group, and teaching at institutions of learning. Thus Richer describes Gerbert's teaching and the books he used. Gerbert, therefore, clearly combined teaching from a set of standard texts inherited from antiquity, augmented by the commentaries of Boethius from the sixth century, with his own discussion of all these texts and his contraptions as visual aids for the understanding of astronomy and arithmetic.

According to Richer, Gerbert expounded dialectic with reference to the *Isagoge* of Porphyrius in the translation of Victorinus and Boethius' commentary on the *Isagoge*. He explained Aristotle's *Categories*, the *Perihermenias*, Cicero's *Topica* and used also Boethius' commentary in six books on Cicero's *Topica* (that is, the very same commentary Lupus requested in the papyrus late-antique copy from Tours). Because he feared his students would not master the art of oratory without a knowledge of poetry, Gerbert prepared his students for instruction in rhetoric by reading and commenting on Virgil, Statius, Terence, the satires of Juvenal, the poetry of Persius and Horace and the historian Lucan. When the student was familiar with these poets and their style they proceeded to the study of rhetoric and thereafter were assigned to a sophist in order to master the art of argument. Gerbert then instructed the students in mathematics, first of all in arithmetic and then in music and its structure. He devoted himself to imparting the principles of astronomy to the astonishment of all, with the aid of certain instruments, such as a round sphere made of wood, inclined obliquely and with two poles. With its help he discussed the idea of the horizon which divided the visible constellations from the invisible. He taught the students about the stars, the equinox, the division of the sphere by parallels and the place of the Arctic circle, the north and south poles and the hot regions. He had another special instrument to demonstrate this. He also constructed an ingenious armillary sphere to help the students understand the revolution of the planets and another to demonstrate the position of the constellations. He took no less pains over his instruction in geometry, for to help with this, as we learnt in his letter to Constantine, Gerbert had had an abacus made. Richer adds that Gerbert was an inspiring teacher and the number of his students grew by the day. His renown as a teacher spread from Gaul to Germany and across the Alps, even as far as the Adriatic.

When we look at John of Salisbury's account of his education in the twelfth century, it is very similar in its emphases. John studied in France at Chartres. He describes the teaching method of Bernard of Chartres[39] who

> taught the figures of grammar, the adornment of rhetoric, the quibbles of sophistries and where the subject of his own lesson had reference to other disciplines, these matters he brought out clearly, yet in such wise that he did not teach everything about each topic but in proportion to the capacity

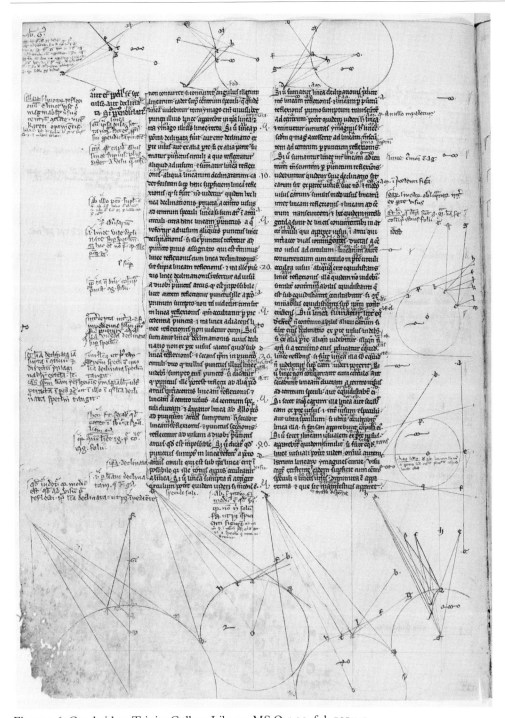

Figure 1.6 Cambridge, Trinity College Library MS O.5.30, fol. 121v, a thirteenth-century manuscript probably copied in England containing Alhacen's *Perspectiva*. It shows a late medieval reader's annotations and diagrams made in order to understand the text. (Courtesy of the Master and Fellows of Trinity College, Cambridge)

of his audience dispensed to them in time the due measure of the subject. Learning by rote was crucial, particularly the evening grammar drill. Poets and orators provided essential models but only to teach pupils to express their thoughts in their own language: borrowing of phrases was actively discouraged and imitation likewise.

History and poetry should be diligently read without the spur of compulsion . . . students were encouraged to read and remember.[40]

This might be all that there was to be said, were it not for the problem of books containing knowledge which either depended on practical empirical knowledge to make sense of them or seemed to contain knowledge imparted through the authority of texts which actually bore little relation to the reality revealed through observations. Examples of the former are the recipes for the making of artists' pigments or advice on the construction of an organ such as are to be found in Theophilus.[41] These texts, wholly inadequate in themselves as instruction manuals, nevertheless hint at an enormous fund of empirical knowledge of such subjects as acoustics, mathematics and chemistry passed on orally from craftsman to craftsman (Figure 1.6). Astronomy books from the early middle ages, moreover, contain diagrams of constellations which are far from faithful representations of the real stars in the heavens: one cannot learn accurate astronomy from the illustrations and diagrams. Similarly, it is rare for early medieval geographical texts to express knowledge in the form of direct observation.[42] More often they act as vehicles for ancient tradition and have to be understood as a representation of God's creation rather than an accurate depiction thereof. When Gerbert combined what he had been told with what he had read, therefore, and when Robert Grosseteste stressed the value of relating what was read to one's own empirical observation, they were opening up patterns of thought and writing which would have a direct and crucial bearing on the subsequent development of the sciences. In other words, what is recorded in writing may require oral or visual elucidation to be fully effective. Books therefore were and are by no means the only media for the dissemination of scientific ideas and knowledge though they were the primary means for their preservation.

Notes

I should like to thank Marina Frasca-Spada and Nick Jardine for their helpful criticisms.

1 E. Eisenstein, *The Printing Press as an Agent of Change: Communications and Cultural Transformations in Early-modern Europe* (Cambridge and New York, 1979) is the most obvious proponent of this misguided idea. See the criticisms by P. Needham in *Fine Print*, 6 (1980): 23–35; and A. Grafton, 'The importance of being printed', *Journal of Interdisciplinary History*, 11 (1980): 265–86.

2 M. B. Parkes, 'The scriptorium of Wearmouth-Jarrow', Jarrow Lecture 1982, reprinted in M. B. Parkes, *Scribes, Scripts and Readers. Studies in the Communication, Presentation and Dissemination of Medieval Texts* (London, 1991); and E. A Lowe, 'The "Script of Luxeuil": a title vindicated', in E. A. Lowe, *Palaeographical Papers 1907–1965*, ed. L. Bieler (Oxford, 1972), vol. II, pp. 389–98.

3 See in particular B. Bischoff, *Latin Palaeography. Antiquity and the Middle Ages* (Cambridge, 1990, from German 3rd edn of 1986); and R. McKitterick, 'Carolingian book production: some problems', *The Library*, 6th Series 12 (1990): 1–33, reprinted in R. McKitterick, *Books, Scribes and Learning in the Frankish Kingdoms, 6th–9th Centuries* (Aldershot, 1994), chapter XII.

4 B. Bischoff, 'Libraries and schools in the Carolingian revival of learning', in B. Bischoff, *Manuscripts and Libraries in the Age of Charlemagne*, trans. M. Gorman, Cambridge Studies in Palaeography and Codicology 1 (Cambridge, 1994), from the original German of 1972.

5 R. McKitterick, 'Unity and diversity in the Carolingian church', in R. Swanson (ed.), *Unity and Diversity in the Church*, Studies in Church History 32 (Oxford, 1996), pp. 59–82.

6 R. McKitterick, 'Script and book production', in R. McKitterick (ed.), *Carolingian Culture: Emulation and Innovation* (Cambridge, 1994), pp. 221–47.

7 Lupus of Ferrières, Ep. 1, *Correspondence*, ed. L. Levillain (Paris, 1964); English translation, G. W. Regenos, *The Letters of Lupus of Ferrières* (The Hague, 1966), no. 1.

8 I have argued this in detail in R. McKitterick, *The Carolingians and the Written Word* (Cambridge, 1989), pp. 165–210.

9 See, for example, A. J. Piper, 'The libraries of the monks of Durham', in M. B. Parkes and A. G. Watson (eds.), *Medieval Scribes, Manuscripts and Libraries. Essays Presented to N. R. Ker* (London, 1978), pp. 213–50.

10 G. Pollard, 'The *pecia* system in the medieval universities', in Parkes and Watson, *Medieval Scribes*, pp. 145–61.

11 On the development of the universities see H. De Ridder-Symoens (ed.), *A History of the University in Europe. I, Universities in the Middle Ages* (Cambridge, 1992); and the particularly valuable case study provided by medieval Oxford: J. I. Catto (ed.), *The History of the University of Oxford, I. The Early Oxford Schools* (Oxford, 1984); and J. I. Catto and T. A. R. Evans (eds.), *The History of the University of Oxford. II, Late Medieval Oxford* (Oxford, 1992).

12 See M. Stillwell, *The Awakening of Interest in Science During the First Century of Printing 1450–1550. An Annotated Checklist of First Editions Viewed from the Angle of Their Subject Content – Astronomy, Mathematics, Medicine, Natural Science, Physics, Technology* (Bibliographical Society of America, 1970), and M. Stillwell, *The Beginning of the World of Books. A Chronological Survey of the Texts Chosen for Printing During the First 20 Years of the Printing Art* (Bibliographical Society of America, 1972).

13 R. McKitterick, 'Essai sur les représentations de l'écrit dans les manuscrits carolingiens', in F. Dupuigrenet Desroussilles (ed.), *La*

Symbolique du livre dans l'art occidental du haut moyen âge à Rembrandt, Revue Française d'Histoire du Livre, 86–7 (1995), pp. 37–64.

14 J. D. North, 'Natural philosophy in late medieval Oxford', and 'Astronomy and mathematics', in Catto and Evans, *The History of the University of Oxford. II, Late Medieval Oxford*, pp. 65–174.

15 J. J. Contreni, 'The Carolingian Renaissance: education and literary culture'; D. Ganz, 'Theology and the organization of thought', in R. McKitterick (ed.), *The New Cambridge Medieval History. II c. 700–c. 900* (Cambridge, 1995), pp. 709–85; and J. Marenbon, 'Carolingian thought', in McKitterick, *Carolingian Culture*, pp. 171–92.

16 J. A. Weisheipl, 'Science in the thirteenth century' and 'Ockham and the Mertonians', in Catto , *The History of the University of Oxford. I, The Early Oxford Schools*, pp. 435–70 and 607–58.

17 L. D. Reynolds (ed.), *Texts and Transmission. A Survey of the Latin Classics* (Oxford, 1983).

18 M. Folkerts (ed.), *Boethius' Geometrie*, vol. II (Wiesbaden, 1970), esp. pp. 173–218.

19 B. Bischoff, 'Die Überlieferung der technischen Literatur', in B. Bischoff, *Mittelalterliche Studien*, vol. III (Stuttgart, 1981), pp. 277–97 and plates XIX–XXVII.

20 Reynolds, *Texts and Transmission*, pp. 440–5.

21 B. Bischoff, B. Eastwood, T. A.-P. Klein, F. Mütherich and P. F. J. Obbema, *Aratea. Kommentar zum Aratus des Germanicus MS. Voss. lat. Q. 79 Bibliotheek der Rijksuniversiteit Leiden* (Luzern, 1989). See also P. McGurk, 'Carolingian astrological manuscripts', in J. L. Nelson and M. T. Gibson (eds.), *Charles the Bald: Court and Kingdom*, BAR International Series 101 (Oxford, 1981), pp. 317–32.

22 B. Eastwood, 'The astronomies of Pliny, Martianus Capella and Isidore of Seville in the Carolingian world', and A. von Euw, 'Die künsterliche Gestaltung der astronomischen und computistischen Handschriften des Westens', in P. L. Butzer and D. Lohrmann (eds.), *Science in Western and Eastern Civilization in Carolingian Times* (Basel, Boston and Berlin, 1993), pp. 161–81 and 251–72. See also the illustrations from Leiden, Bibliotheek der Rijksuniversiteit, Voss. lat. Q 79 and Madrid, Biblioteca Nacional Cod. 3307, both codices of the ninth century, in F. Mütherich and J. E. Gaehde, *Carolingian Painting* (London, 1977), Plates 18, 19 and 27.

23 L. W. Jones (ed.), *Bedae Opera de Temporibus*, Medieval Academy of America (Cambridge, Mass., 1943), and Bede, *The Reckoning of Time*, trans. F. Wallis (Liverpool, 1999).

24 A. Borst, 'Alcuin und die Enzyklopädie von 809', in Butzer and Lohrmann, *Science in Western and Eastern Civilization*, pp. 53–78; and A. Borst, *Die karolingische Kalendarreform*, Monumenta Germaniae Historica Schriften 46 (Hanover, 1998). On Pliny see Reynolds, *Texts and Transmission*, pp. 307–16.

25 B. Bischoff, 'Eine Sammelhandschrift Walahfrid Strabos (cod. Sangall. 878)', in B. Bischoff, *Mittelalterliche Studien*, vol. II (Stuttgart, 1967), pp. 34–51.

26 W. Stevens, 'Fulda scribes at work: Bodleian Library Manuscript Canonici Miscellaneous 353', *Bibliothek und Wissenschaft*, 8 (1972):

287–317; and 'A ninth-century manuscript from Fulda: Canonici Miscellaneous 353', *The Bodleian Library Record*, 9 (Oxford, 1973), both reprinted in W. Stevens, *Cycles of Time and Scientific Learning in Medieval Europe* (Aldershot, 1995), chapters VI and VII; see also chapters IX and XI.

27 H. Löwe, 'Ein literarischer Widersacher des Bonifatius. Virgil von Salzburg und die Kosmographie des Aethicus Ister', *Akademie der Wissenschaften und der Literatur in Mainz, Abhandlungen der Geistes- und Sozialwissenschaftlichen Klasse* (1951): 899–988.

28 R. McKitterick, 'Knowledge of Plato's *Timaeus* in the ninth century: the implications of Valenciennes Bibliothèque Municipale MS 292', in H. J. Westra (ed.), *From Athens to Chartres. Neoplatonism and Medieval Thought* (Leiden, 1992), pp. 85–95; and A. Somfai, 'The transmission and reception of Plato's *Timaeus* and Calcidius's *Commentary* during the Carolingian Renaissance', unpublished Ph.D. dissertation, University of Cambridge (1998).

29 Alcuin, Epistola 155, ed. E. Dümmler, *Monumenta Germaniae Historica Epistolae Karolini Aevi*, vol. IV (Hanover, 1902); English trans. S. Allott, *Alcuin of York* (York, 1974), no. 76.

30 *Isidori Hispalensis Episcopi Etymologiarum sive Originum Libri XX*, ed. W. M. Lindsay (Oxford, 1911).

31 See J. Fontaine, *Isidore de Séville et la Culture Classique dans l'Espagne Wisigothique* (Paris, 1959 + suppl. 1983); C. H. Beeson, *Isidor-Studien* (Munich, 1912); E. Brehaut, *An Encyclopedist of the Dark Ages, Isidore of Seville* (New York, 1912); J. Hillgarth, 'The position of Isidorian studies: a critical review of the literature 1936–1975', *Studi medievali*, 24 (1983): 817–905.

32 A. Beccaria, *I codici di medicina del periodo pre-salernitano secoli IX, X e XI* (Rome, 1956), and E. Wickersheimer, *Les Manuscrits latins de médecine du haut moyen âge dans les bibliothèques de France* (Paris, 1966).

33 J. J. Contreni, 'Masters and medicine in northern France during the reign of Charles the Bald', in Nelson and Gibson, *Charles the Bald: Court and Kingdom*, 2nd edn (Aldershot, 1990), pp. 267–82.

34 Lupus of Ferrières, Ep. 53, ed. L. Levillain, *Loup de Ferrières, Correspondence*, 2 vols. (Paris, 1964); English trans. G. W. Regenos, *The Letters of Lupus of Ferrières* (The Hague, 1966), no. 53.

35 Gerbert, ed. N. M. Bubnov, *Gerberti opera mathematica 972–1003* (Berlin, 1899), pp. 6–8; English translation H. P. Lattin, *The Letters of Gerbert with his Papal Privileges as Silvester II* (New York, 1961), no. 7.

36 R. Latouche (ed.), *Richer, Histoire de France*, 2 vols. (Paris, 1967), vol. II, pp. 55–65.

37 See Bubnov, *Gerberti opera mathematica*, pp. 8–22.

38 See C. F. R. De Hamel, *Glossed Books of the Bible and the Origins of the Paris Book Trade* (Woodbridge, 1984); and H.-J. Martin and J. Vezin (eds.), *Mise en page et mise en texte du livre manuscrit* (Paris, 1990).

39 See R. Southern, *Scholastic Humanism and the Unification of Europe*, vol. I (Oxford, 1995).

40 John of Salisbury, *Metalogicion*, ed. C. C. J. Webb (Oxford, 1929); English trans. C. H. Haskins, *The Renaissance of the Twelfth Century* (Cambridge, Mass., 1928), pp. 135–6 and 372–4.

41 Theophilus, *De diversis artibus*, ed. and trans. C. R. Dodwell, *Theophilus. The Various Arts,* Oxford Medieval Texts (Oxford, 1986).
42 N. Lozovsky, 'Carolingian geographical tradition: was it geography?', *Early Medieval Europe,* 5 (1996): 25–44.

Further reading

W. Berschin, *Greek Letters and the Latin Middle Ages from Jerome to Nicholas of Cusa* (Washington, D.C., 1988)

G. Constable, *Letters and Letter Collections,* Typologie des sources du moyen âge occidental, fasc. 17 (Turnholt, 1972)

J. Contreni, 'The Carolingian school: letters from the classroom', in J. Contreni, *Carolingian Learning, Masters and Manuscripts* (Aldershot, 1992), ch. XI.

Lucien Febvre and Henri-Jean Martin, *The Coming of the Book. The Impact of Printing 1450–1800,* 1st edn (Paris, 1958), trans. D. Gerard, ed. G. Newell-Smith and D. Wootton (London, 1976).

D. Gutas, *Greek Thought, Arabic Culture. The Graeco-Arabic Translation Movement in Baghdad and Early 'Abbasid Society (2nd–4th/8th–10th Centuries)* (London and New York, 1998)

H. I. Marrou, *A History of Education in Antiquity,* trans. G. Lamb (Madison WI, 1956)

B. Musallam, 'The ordering of Muslim societies', in *Cambridge Illustrated History of the Islamic World* (Cambridge, 1996), pp. 164–207

L. D. Reynolds and N. G. Wilson, *Scribes and Scholars. A Guide to the Transmission of Greek and Latin Literature,* 3rd edn (Oxford, 1991)

F. Robinson, 'Knowledge, its transmission and the making of Muslim societies', in *Cambridge Illustrated History of the Islamic World* (Cambridge, 1996), pp. 208–49

C. S. Smith, *Mappae Clavicula: A Little Key to the World of Medieval Techniques* (Philadelphia, 1974)

D. V. Thompson, *Materials and Techniques of Medieval Painting* (New York, 1956)

P. Williams, *The Organ in Western Culture, 750–1250* (Cambridge, 1993), 'Organs and written technology', pp. 235–313

2 Printing the world

The traffic in maps

In 1795 one of the more intriguing examples of sixteenth-century printed geography was unearthed in the Venetian archives. The find consisted of six blocks made from pear wood, which made up a detailed map of the world (Figure 2.1). Constructed on the novel cordiform, or 'heart-shaped' geographical projection, this map, and the blocks from which it was composed, was curious not because of its design, but because it purported to be a Turkish map of the world dated 1560, and designed by one Hajji Ahmed, a native of Tunis, who had undertaken to print a map of the world, designed entirely with the use of Arabic script, for the benefit of his fellow Muslims.[1] The title of the map, printed in Arabic, can be translated as 'The Representation of the Whole World Depicted in its Entirety', and the legend which explains the creation of the map, also written in Arabic, announces:

In the name of God, the Merciful, the Compassionate: O ye wise and O ye learned, the blessings of God be upon you! Be it known unto you that I . . . Hajji Ahmed from the city of Tunis . . . became, through the decree of revolving destiny, a captive in Europe [Firengistan]. There I was bought by one of the Frankish lords, a good and learned man, so that I never lacked freedom to perform my religious duties or failed to fulfil them according to the rule and prescription of Islam; and thanks to the learning which I had acquired the people here treated me with all honour and respect.

Now the people of these countries have drawn and produced this representation of the world according to the teaching of the philosophers of old, Plato, Socrates, Abu'l-Fida and the great Lokman, and have in this map written down and communicated fully, according to the demands of science and logical arrangement, the facts concerning the Heavens and the surface of the Earth, in order that those who peruse it, of high and low degree, may draw great benefit from it.

I therefore, on seeing this really excellent and important work, and realising that it was of value and essential to all the Moslems and their rulers, translated it systematically from the language and script of the Franks into the Moslem script; and they undertook to grant me my manumission as the reward of my labour. But I swear by the Mighty and Gracious God that the troubles and trials that I underwent before bringing it to this form are

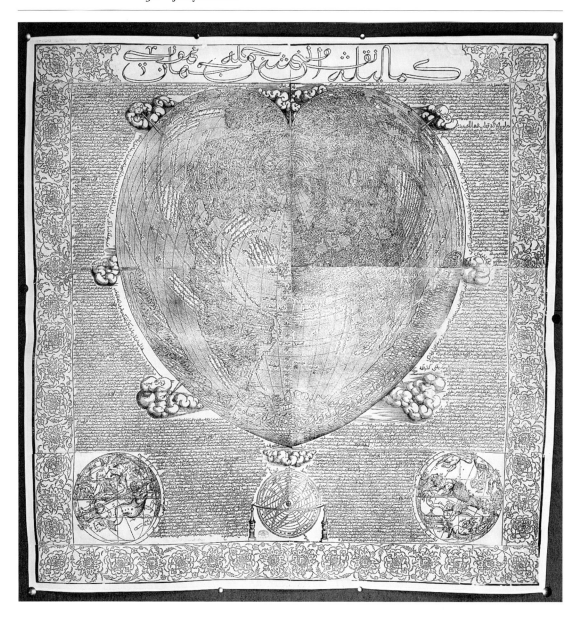

Figure 2.1 Cordiform world map, attr. Hajji Ahmed, Venice, c. 1560. (By permission of the British Library)

beyond description. However, praise be to God, Who has granted us understanding and solicitude for others, for by means of this valuable work I have become the instrument for benefiting all the Moslems.[2]

The legend is worth quoting at such length to stress the aggressive nature which Hajji Ahmed adopts in his barely veiled attempt to sell the map; he appeals to his fellow Muslims to buy his map not only because of its geographical accuracy and worldly benefits, but also because of the heavenly reward in store for those who assist in the manumission of their enslaved brother.

But on closer inspection it becomes clear that Hajji Ahmed's map is not quite what it seems. The map, its projection and much of its nomenclature are clearly indebted to an earlier, French map which adopted the cordiform projection, designed by the French geographer Oronce Fine, and first printed in 1534.[3] Even more significantly, the map's Arabic is littered with mistakes; towns are given their Latinised titles, and sources are drawn exclusively from European rather than Islamic sources. It can only be concluded that the map was not in fact designed by Hajji Ahmed, but was the creation of a particularly daring and imaginative Venetian publisher. Consultation of the Venetian archives bears this out, and reveals the truth behind the implausible figure of Hajji Ahmed. In 1568 the Riformatori dello Studio di Padova licensed the printing of an 'Arabic' map of the world by one 'Cagi Acmet', and translated by two scholars referred to as 'Membre' and 'Cambi'.[4] In the same year the Venetian printer Marc'Antonio Giustinian was granted a monopoly on the printing and selling of the map, presumably working in collaboration with Michel Membre, a distinguished linguist fluent in Arabic, Persian and Turkish. In 1545 Giustinian established a Hebrew press in Venice, and from 1548 he held a monopoly on Hebrew printing in Venice. Giustinian appears to have used his experience of producing type for Hebrew printing to create blocks using Arabic script to assemble the Hajji Ahmed map. He seems to have thought that a map masquerading as a genuine Islamic map of the world would make money in the Muslim world. Hajji Ahmed was therefore a clever fabrication, an attempt to encourage Muslims to buy a map whose profits would line the pockets of the Venetian printing industry, rather than benefit Allah. Ultimately the enterprise failed. Shortly after the printing of the map Giustinian was accused of contravening the Venetian Press Laws, and some of his printed material was confiscated by the authorities. The reappearance of the blocks for the Hajji Ahmed map in the Venetian archives over a hundred years later suggests that the blocks were seized due to the sensitive nature of Venetian–Ottoman relations at the time.

The production of the Hajji Ahmed map says a great deal about the nature of printed geography in the sixteenth century. First and foremost it reveals the commercial imperatives which drove the production of much printed material within the period. Giustinian appears to have created a map for a highly specific, primarily Turkish audience, aggressively marketed through the legend which accompanies the map. This was not an esoteric, intellectually disinterested printing venture, but a bold and enterprising attempt to cash in on a relatively unexploited market of consumers. A large print run of say 1,000 copies of the map could be sold for a higher price in the Ottoman dominions than a comparable map would fetch on the European market.

But alongside its commercial contexts, the fact that Giustinian chose to print a map rather than a book is also highly significant in terms of the development of early print culture. William H. Ivins has argued that the ability of the printing press to produce an 'exactly repeatable pictorial statement'[5] created a communications revolution in its own right. To standardise and reproduce a unique image via the printing press enables a profound change in visual consciousness, which could allow images such as maps to have a much more far-reaching impact upon a potential audience than the circulation of manuscript maps. Whilst Ivins' work has been subject to revision and criticism, there is little doubt that the ability to print and reproduce standardised geographical images created a definable shift in apprehensions of the terrestrial world throughout the sixteenth century. This was after all a period which saw the creation of the first recognisably modern atlas of the world, the *Theatrum orbis terrarum*, produced by the Flemish geographer Abraham Ortelius in Antwerp in 1570, and which had gone through forty-two different editions by 1612. The *Theatrum* was also swiftly followed by Gerard Mercator's first printed *Atlas*, published in 1595. The Hajji Ahmed map was just one example of the massive outpouring of maps which emerged from the printing presses of sixteenth-century Europe, recording the contours of a rapidly changing world, and unique as a prized commercial and intellectual object only in so far as it offered itself as having a particularly unique selling point: it was a map which combined the innovative techniques of the European printing press with the local knowledges of an Arab savant.

If printing defined a new commercial ethos in relation to learning, as well as novel possibilities for the development of the predominantly visual discipline of geography, then it has also most famously been seen to create the conditions for the development of a community of tightly knit scholars, a 'republic of letters'[6] whereby humanism and printing, the scholar and the printer, worked side by side not only to solve the logistical problems of printing, but also to redefine the intellectual world of early modern Europe. In many respects, the Hajji Ahmed map can be seen as an example of this union of learning and technology. Michel Membre, one of the translators of the map's legend, worked as an interpreter for perhaps the most influential of all sixteenth-century Italian geographers, Giovanni Battista Ramusio, the chancellor of Venice's Council of Ten and compiler of the *Navigazioni e viaggi*, published in Venice in 1550. Ramusio had himself gleaned much of his information on Ottoman geography from the French Orientalist scholar Guillaume Postel, who had spent several years in the Ottoman empire. Both men were part of the wider network of scholars, printers, travellers and merchants cited by Ortelius in 1570 in the foreword to his *Theatrum*. The Hajji Ahmed map was thus just another instance of the unification of technological innovation, entrepreneurial vision and humanist learning.

This perception of the production of the map fits into standard notions of the development of printing and geography in the sixteenth century; but as more cautious and perceptive scholars of print culture have pointed out, it is notoriously difficult to sustain grand claims as to the radical, innovative and even revolutionary significance of early print history. Such claims can often also have distinctly ideological and political overtones, and I want to suggest that this is indeed the case if Giustinian's Hajji Ahmed map is made to fit so seamlessly into a neat history of early printed geography. What accompanies many of the suppositions concerning print culture outlined above is that printed maps such as the Hajji Ahmed map represent an image of a confident, technically innovative and intellectually superior European identity over and against a technologically ignorant, intellectually backward and easily duped Islamic (and predominantly Turkish) culture. The fact that the Ottoman empire did not adopt the printing press until the 1720s only adds weight to an ideologically charged argument that sees the Ottomans as the technological, intellectual and cultural antithesis to Europe, a belief which Edward Said has argued contributed to Europe's sense of its innate superiority over 'The Orient'.[7] The adoption of print culture therefore becomes a decisive marker of cultural enlightenment, something which the Ottomans were seen as decidedly lacking. I would like to suggest that a closer inspection of the origins of printed geography fails to sustain this ideologically weighted perception of the Ottomans, and to do this it is necessary to examine the maps which predate Giustinian's Hajji Ahmed map by nearly a hundred years, and in particular the early printed editions of the figure most often viewed as the founder of the discipline of geography, the Alexandrian scholar Ptolemy.

Ptolemy goes to press

Ptolemy's *Geographia* first emerged in Alexandria in the second century AD. The founding encyclopaedia of the discipline of geography, it not only defined the term itself, but also listed descriptions and locations of over 8,000 known locations in the classical world. As such it came to be seen as the first and last word on the subject of geography. By the late 1300s, Greek manuscript copies of Ptolemy's *Geographia* (also referred to in the period as *Cosmographia*) started to arrive in Italy from Byzantium as part of the recovery of classical learning which defined so much of the intellectual activity of the period. Already a prized manuscript, the *Geographia* was one of the most obvious texts to undergo the transmission into print in the second half of the fifteenth century. The first printed text of the *Geographia* was published in Bologna in 1477, containing Ptolemy's written text as well as twenty-six maps, including a map of the known world, produced by the use of the rel-

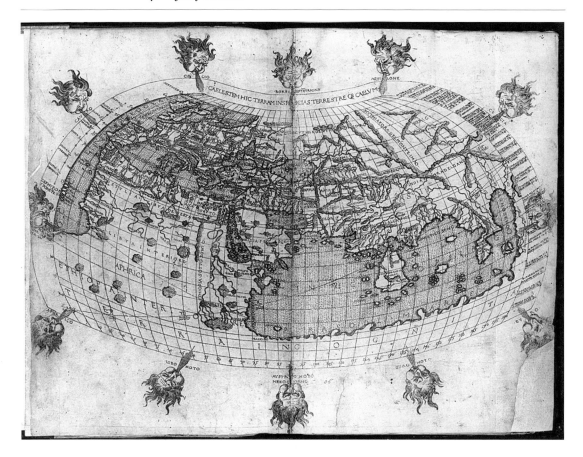

Figure 2.2 World map, Francesco Berlinghieri, Florence, 1482. (By permission of the British Library)

atively experimental medium of copperplate engraving. The first printed atlas in the history of European cartography, the Bologna edition ran to 500 copies, a remarkably confident investment in the *Geographia*, and in the anticipated public eager and able to buy such a book. By 1487 no fewer than six incunable editions of Ptolemy's *Geographia* had been printed in Rome (1478), Florence (1482), Ulm (1482 and 1486), as well as Bologna. These atlases contained at least twenty-six maps, including a world map followed by regional maps of Europe, Africa and Asia. The world map reflected the coordinates of the Ptolemaic world picture, as can be seen in the first map printed in the 1482 Florence edition (Figure 2.2). The influence of Ptolemy's *Geographia* can be gauged by the calculation that of 222 recorded maps printed prior to 1500, half were directly based on Ptolemy's writings.[8]

Opinion remains divided on the intellectually innovative nature of these early printed editions of Ptolemy's *Geographia*. Many historians of cartography have pointed out that, just at the point at which Castilian and Portuguese maritime expansion began to transform the world picture established by Ptolemy's *Geographia*, these incunabula acted as a brake on the dissemination of such information, rather than a catalyst. The investments made in compiling a printed text combining over 100 pages of written material, with between twenty-six and thirty-two separately designed maps, was significant. Printers did not want to have to consistently update and redesign expensively engraved plates or woodcut blocks in line with the most recent discoveries. Sixteenth-century editions of Ptolemy began to look increasingly out of step with the more accurate manuscript maps and portolans, or sea-charts recording Castilian discoveries in the New World and Portuguese incursions beyond the Cape of Good Hope.

Nevertheless, practical considerations were also crucial in defining the shape of these early printed atlases. The early German Ptolemy atlases were printed using the more established technique of woodcut blocks (of the type used by Giustinian in his printing of the Hajji Ahmed map). However, the early Italian atlases used the more sophisticated and expensive medium of copperplate engraving. This technique allowed for greater accuracy and precision in the representation of territory, as well as providing greater scope for adding or correcting specific details than that provided by the woodcut. However, printers could often combine different types of printing techniques in the one atlas, as in the case of the elegant maps contained in the 1478 Rome Ptolemy, whose letters and numerals were in fact punched from a set of dies, rather than being engraved by hand.[9] This use of punched lettering on engraved maps is symptomatic of the conceptual difficulties with which these early printers struggled in an attempt to create a vocabulary of geographical representation which united the visual dimension of an atlas with its written component. This distinction between the printed word and image is graphically made when considering the requirements made of the print type itself. Once the type was set for, say, an atlas, and the required number of copies printed, the type was then broken up and prepared for the next job; but blocks or engraved plates of maps could not be dispersed, as unique objects in their own right. As a result printers often solved this problem by printing the letter press, or text, on one side of a large number of sheets, whilst printing the maps on the other side in a much smaller number required for one particular issue. Imprints of the maps could therefore be easily taken again at a later point, whereas the type for the text could not be so easily reassembled. As a result many early sixteenth-century atlases find maps on the other side of a sheet of text often in different states (having been corrected or expanded), and sometimes even in a different order.[10] In this respect the commercial exigencies of the printing shop could often condition the ways in which printed geography books and atlases developed through their various editions.

As a result, the innovations of the printing press were often counterbalanced by cautiousness regarding the appearance of the final product. The earliest printed maps were made to look as close to their manuscript forebears as possible. This can be explained not only because of the ways in which scribes shifted from the reproduction of manuscripts to the creation of printed texts; patrons were also keen to compare and contrast the relative practical and aesthetic merits of manuscripts over and against printed texts (in many cases comparing the same text in manuscript with its printed version). As a result the shift from both manuscript to print, like the shift from the older Ptolemaic image of the world to a more recognisably 'modern' representation of the terrestrial world, incorporating the maritime discoveries of both the Castilian and Portuguese voyages of the late fifteenth and early sixteenth centuries, was a gradual and painstaking procedure. What was particularly significant in the Ptolemaic atlases was the formal dimension of the maps produced on the basis of Ptolemy's calculations. The geometrical projections contained within Ptolemy's treatise envisaged a completely new dimension of terrestrial space decidedly at odds with the more sacred perception of space defined within the medieval tradition of *mappae-mundi*. As a result the global and regional maps printed in line with Ptolemy's calculations on the size and shape of the *oikoumene* were plotted across a predetermined geometrical grid of lines of latitude and longitude informed by the principles of abstract geometry. Within such a grid printers and geographers were able to plot a proliferation of locations across an empty, continuous and potentially infinite concept of terrestrial space. As a result subsequent printed editions of the *Geographia* gradually incorporated new geographical information within the Ptolemaic world picture. The 1482 Florence edition contained four additional maps to the standard running order of maps included in the *Geographia*, which drew on more contemporary geographical knowledge. By 1507 Martin Waldseemüller printed an edition of Ptolemy in St Dié, which incorporated knowledge of both Columbus' voyages and the Portuguese voyages via the Cape of Good Hope which effectively shattered the integrity of the Ptolemaic world picture. Nevertheless, Ptolemy's text remained the primary point of departure where such information was gradually assimilated. It would take another sixty years before Ortelius' *Theatrum* had the confidence to offer a universal compilation of global and regional maps which did not rely upon the guiding template of Ptolemy's *Geographia*. It is significant that, whilst Ortelius was fulsome in his praise for Ptolemy, maps based on the *Geographia* were placed in an appendix to the *Theatrum*, defining a decisive and irrevocable shift in the nature of early modern printed geography.

What the preceding analysis has been concerned to stress is the extent to which the impact of printing, particularly within the field of geography, did not necessarily create a decisive and irrevocable 'shift' or 'break' within the culture of the period. With the benefit of historical hindsight it is tempting to impose retrospectively this interpretation of a decisive intellectual and technological 'break' on to the development of late fifteenth-century printed geography, viewing the shift from manuscript to print culture as symptomatic of a wider transformation of European culture itself. This tradition, exemplified most recently in John Hale's book *The Civilization of Europe in the Renaissance* (1994), sees the transmission of classical texts from Byzantium to Italy as a prelude to the assimilation of such texts into a renewed European intellectual heritage through the medium of print. The resulting self-conscious awareness of Europe as an intellectual and geographical entity in its own right is, according to critics like Hale, defined over and against cultures perceived to be at best indifferent or at worst hostile to the humane and civilising dimension of this investment in learning and scholarly enquiry. Most significantly, it is the Ottoman empire to the east of Europe which comes to be seen as the antithesis to this renewed pursuit of learning, personified in the failure of the Ottomans to adopt printing as a defining symbol of this spirit of intellectual enquiry. In looking at one final example from the early history of printed geography, I want to contest this interpretation, as not only an ideologically weighted interpretation which orientalises the Ottomans as the antithetical 'other' through which early European consciousness identifies itself, but also as a profound misrecognition of the cultural and intellectual exchanges which took place between Italy and the Ottoman empire via the medium of printed geography.

East meets West

The example I choose to emphasise this point is Francesco Berlinghieri's *Geographia*, printed in Florence in 1482. Berlinghieri's edition of Ptolemy has often been seen as one of the most innovative of the early printed editions of the *Geographia*, as its text was printed in Italian, and it also added for the very first time several modern maps to the usual running order of Ptolemy's text. But what is even more remarkable about Berlinghieri's text was that it was initially dedicated to none other than the Ottoman Sultan Mehmed the Conqueror. The Topkapi Saray Library in Istanbul still possesses a printed copy of Berlinghieri's text, with a manuscript letter dedicated to Mehmed, which reads:

To Mehmed of the Ottomans, illustrious prince and lord of the throne of God, emperor and merciful lord of all Asia and Greece, I dedicate this work.[11]

The rest of the manuscript letter goes on to address Mehmed's son, Bayezid II. In his letter Berlinghieri explains that he had intended to dedicate his *Geographia* to Mehmed, but that when the Sultan died in May 1481, as the edition was going through the press, he decided to dedicate the *Geographia* to the Duke of Urbino, Federigo da Montefeltro. Unfortunately for Berlinghieri, by the time the corrected type for the edition had been finally completed in late 1482, Federigo had also died. Berlinghieri explained that he was therefore returning to his initial attempt to dedicate the text to an Ottoman sultan, namely the recently crowned Bayezid. The Topkapi text bears this story out, as the printed heading addressed to Federigo is obliterated by a blue and gold painted dedication addressed to Mehmed.

In his excellent account of the complicated printing history of this most remarkable edition, R. A. Skelton has convincingly argued that it was Ottoman, rather than Italian, patronage which seems to have concerned Berlinghieri as he put his text through the printing press. As the text was going to press, Berlinghieri appears at the last moment to have changed the type to include the dedicatory preface to Federigo; following his death, Berlinghieri hastily added his manuscript letters and dedications to Bayezid in an attempt to establish his patronage of the atlas. However, the connection between Berlinghieri's *Geographia* and the Ottoman court did not end there. Mehmed's death sparked a struggle for succession between his sons Bayezid and the younger Cem. After being defeated in battle in 1482, Cem fled to Rome, where he was wooed as a possible ally by the courts of Europe. In the summer of 1484 Paolo da Colle, an agent working on behalf of Lorenzo de' Medici, and who had already delivered the presentation copy of Berlinghieri's *Geographia* to Bayezid in Istanbul in 1482, was sent to Savoy, where Cem was living. Da Colle carried with him a copy of Berlinghieri's *Geographia*, dedicated to Cem, and identical to the text da Colle had presented to Bayezid eighteen months earlier, down to the dedicatory hand-written letters written by Berlinghieri in both copies. The only change was a hurriedly added dedicatory preface in manuscript addressed to Cem, extolling his princely attributes in terms almost identical to those addressed to Bayezid and Federigo.[12]

This may at first seem to be an extraordinarily anomalous attempt by a Florentine scholar to dedicate a classical text, written and printed in Italian, to not one but two Ottoman sultans. But in fact as recent studies have argued, Mehmed the Conqueror would have been seen as a perfectly logical and suitably powerful and cultivated patron for classical texts of the stature of Ptolemy's *Geographia*. Following the fall of Constantinople in 1453, Mehmed embarked on an ambitious programme of public rebuilding, which included the involvement of European artists of the calibre of

Gentile Bellini, Constanzo de Ferrara and Antonio Filarete. Despite subsequent European claims to the contrary, Mehmed also appears to have seen himself as the inheritor, rather than terminator, of the Byzantine cultural tradition. As well as encouraging the shift of Islamic astronomical enquiry from Samarkand to Constantinople, Mehmed also built up an extensive personal library under the supervision of several Greek scholars. These included a range of Islamic, Latin and Greek maps. Rather than appearing antithetical to the scholarly pursuits of Byzantine culture, Mehmed seemed to be vying for the mantle of its inheritor alongside the courts of Italy, and nowhere more so than in the field of maps and geography. As well as receiving a series of contemporary manuscript maps from Italy, Mehmed was also responsible for commissioning an Islamic copy of Ptolemy's *Geographia*, from the Greek scholar Georgius Amirutzes. According to a contemporary chronicle:

Having read the works of the renowned geographer Ptolemy and perused the diagrams which explained these studies scientifically, the Sultan found these maps to be in disarray and difficult to construe. Therefore he charged the philosopher Amirutzes with the task of drawing a new clearer and more comprehensible map. Amirutzes accepted with pleasure, and worked with meticulous care. After spending the summer months in study and research, he arranged the sections in scientific order. He marked the rivers, islands, mountains, cities and other features. He laid down rules for distance and scale, and having completed his studies he presented the Sultan and those engaged in scholarship and science with a work of great benefit. Amirutzes wrote the names of the regions and cities in the Arabic script, and for this purpose engaged the help of his son, who was master of both Arabic and Greek.[13]

Amirutzes' manuscript map, completed in 1465, and still held in the Ayasofya Library in Istanbul, complemented Greek manuscript copies of the *Geographia* which were also held in the Sultan's library. It was not only Ptolemy's *Geographia* to which Mehmed's patronage was associated. In 1451 Amirutzes' colleague Georgius Trapezuntius (George of Trebizond) composed a translation and commentary in Latin of Ptolemy's other great work, the thirteen books known as the *Almagest*. Complementing the *Geographia*, the *Almagest* dealt with every aspect of theoretical mathematical astronomy. In 1465 Trapezuntius travelled to Constantinople, where he stayed with Amirutzes and dedicated his translation and part of the commentary on the *Almagest* to Mehmed.[14]

As a result, by 1482, less than a year after Mehmed's death, the Ottoman court was directly associated with manuscript copies of Ptolemy's *Geographia* in Arabic, Greek and Italian, as well as possessing a printed copy of the text, along with Trapezuntius' dedication of the *Almagest*, scholarly and intellectual credentials which few Italian patrons could match. Rather than appearing antithetical to the classical culture of the Byzantine empire, Mehmed and his scholarly entourage emerge as self-styled heirs to this intellectual tradition. It is therefore no surprise that scholars like Berlinghieri and Trapezuntius attempted to associate their geographical and astronomical texts with Ottoman power and patronage. I would suggest that the anxiety directed towards the Ottomans, both within contemporary accounts and subsequent historiography, is based on an anxiety regarding them as highly successful *competitors* in the pursuit of classical geographical learning, rather than its barbaric opponents.

In the aftermath of Mehmed's reign a vigorous tradition of geographical and cosmographical enquiry flourished, from Ottoman sea-charts of the Mediterranean to the astronomical studies which continued to influence the development of European astronomy.[15] However, these traditions were exclusively transmitted through the medium of scribal rather than print culture. This gradual separation between the print culture of Christian western Europe and the manuscript culture of the Ottoman empire led to increasingly divergent political and cultural perceptions between the two cultures. As historians of the impact of printing upon European culture have observed, print 'paved the way for the more deliberate purification and codification of all major European languages'.[16] As a result, print culture created the conditions for the creation of what Benedict Anderson has called an 'imagined community' of national consciousness, where the pervasive dissemination of print culture creates a sense of identification amongst the vernacular speakers defined as 'belonging' to a particularly nationality.[17] Manuscript production is unable to provide this sense of cultural and political 'belonging', by nature not only of its smaller level of reproducibility and circulation, but also in terms of its relationship to patronage and political authority. Manuscript cultures tend to disseminate outwards from a privileged centre of production, and, unlike print culture, their authority and influence wane the further away they move from such a centre. This situation affects not only the status of the manuscript but also the form of political authority which builds its influence upon the power to disseminate its ideas through different media. As a result, it became increasingly convenient for European thought to dismiss Ottoman learning as atavistic and limited in its refusal to embrace print.

This denigration of Ottoman learning, and specifically its geography, can be glimpsed in the shift observed between the dedication of Berlinghieri's *Geographia*, to the creation of the Hajji Ahmed map less than a hundred years later. Here the Ottomans are seen as a lucrative but intellectually limited market for a printed map such as that produced by Giustinian. The possibilities for a much more intolerant discourse of Orientalist denunciation can be glimpsed in this later map, which establishes the growing political and cultural differences which increasingly defined the relationship between western Europe and the Ottoman empire. Histories of early modern Europe have invariably marginalised the place of Ottoman culture, partly as a result of its failure to embrace the technological and intellectual innovation of the printing press. But as I hope to have shown through this brief analysis of Ottoman involvement in both manuscript and early printed maps and atlases, historians should be wary of ideologically mobilising the adoption or rejection of the printing press as a sign of cultural atavism.

Notes

1 See V. L. Menage, 'The map of Hajji Ahmed and its makers', *Bulletin of the School of Oriental and African Studies*, 21 (1958): 291–314.
2 Ibid., pp. 296–7.
3 See R. W. Shirley, *The Mapping of the World: Early Printed Maps, 1472–1700* (London, 1983), pp. 118–19.
4 Menage, 'Hajji Ahmed', p. 308.
5 William H. Ivins, *Prints and Visual Communications* (New York, 1969).
6 See Elizabeth Eisenstein, *The Printing Press as an Agent of Change*, 2 vols. (Cambridge, 1979), vol. I, pp. 136–58.
7 See Edward Said, *Orientalism* (London, 1978).
8 On Byzantine texts and their transmission, see O. A. W. Dilke, 'Cartography in the Byzantine empire', in J. B. Harley and David Woodward (eds.), *The History of Cartography: Cartography in Prehistoric, Ancient and Mediaeval Europe and the Mediterranean*, vol. I (Chicago, 1987), pp. 258–75. On early printed Ptolemy atlases and their influence, see Tony Campbell, *The Earliest Printed Maps, 1472–1500* (London, 1987), pp. 122–38.
9 See Campbell, *Earliest Printed Maps*, p. 126.
10 See R. A. Skelton, 'The early map printer and his problems', *The Penrose Annual*, 57 (1964): 171–87.
11 Quoted in Kemal Özdemir, *Ottoman Nautical Charts* (Istanbul, 1992), p. 52.
12 See R. A. Skelton (ed.), *Francesco Berlinghieri: Geographia*, p. vii. Cem was eventually defeated and Bayezid ruled until his death in 1512.
13 Quoted in Özdemir, *Ottoman Nautical Charts*, p. 52.
14 See Noel Swerdlow, 'The recovery of the exact sciences of antiquity: mathematics, astronomy, geography', in Anthony Grafton (ed.), *Rome Reborn: The Vatican Library and Renaissance Culture* (Yale, 1993), pp. 125–67.
15 On local forms of Ottoman mapping, see my *Trading Territories: Mapping the Early Modern World* (London, 1997), ch. 3.
16 Eisenstein, *Printing Press*, vol. I, p. 117.
17 See Benedict Anderson, *Imagined Communities: Reflections on the Origins and Spread of Nationalism* (London, 1991), pp. 6–46.

Further reading

J. Brotton, *Trading Territories: Mapping the Early Modern World* (London, 1997)

T. Campbell, *The Earliest Printed Maps, 1472–1500* (London, 1987)

E. Eisenstein, *The Printing Press as an Agent of Change*, 2 vols. (Cambridge, 1979)

J. B. Harley and David Woodward, *The History of Cartography* (Chicago, 1987–)

W. H. Ivins, *Prints and Visual Communications* (New York, 1969)

J. Raby, 'East and west in Mehmed the Conqueror's library', *Bulletin du Bibliophile*, 3 (1987): 297–321

R.W. Shirley, *The Mapping of the World: Early Printed Maps, 1472–1700* (London, 1983)

R. A. Skelton, 'The early map printer and his problems', *The Penrose Annual*, 57 (1964): 171–84

R. A. Skelton (ed.) *Francesco Berlinghieri: Geographia* (Amsterdam, 1966)

N. Swerdlow, 'The recovery of the exact sciences of antiquity: mathematics, astronomy, geography', in A. Grafton (ed.), *Rome Reborn: The Vatican Library and Renaissance Culture* (Yale, 1993), pp. 125–67

ANTHONY GRAFTON

3 Geniture collections, origins and uses of a genre

At some point in the 1550s, an unidentified reader worked his way through the collection of five treatises and one hundred genitures that Girolamo Cardano had published in 1547. Like many, if not most, sixteenth-century scholars, this one read pen in hand, transforming the blank margins of his book into a running commentary on it. This reflective, critical kind of reading often formed only one stage in a larger set of learned practices, one which involved systematic, large-scale recording and recycling of passages from the text. After filling the margins of a printed book with summaries and clarifications, the reader might enter extracts from what he had read in notebooks, arranging them under one of the many schemes of subject headings that were widely used. At a still later stage, he might reuse them in his own compositions, fitting the old pieces into new mosaics. We do not know if that happened in this case.[1]

We do know, however, that this reader was both well connected and well informed. Like many makers and users of genitures, he seems to have had medical training, and may well have been a physician himself. When Cardano, discussing the human life-span, remarked that it basically came to a close at the age of ninety-eight, after which one lived on 'like a plant, and no change takes place in his way of life until he dies', the reader recalled that he and the well-known Piedmontese physician Giovanni Argenterio had once seen a man 'who was in his hundred and first year, and whom we judged to be barely in his sixties'.[2] When Cardano insisted that the art of astrology did exist, even though no adequate account of it had yet been given, the reader remarked that 'Vesalius thinks the same' – evidence that he had at least read, and possibly knew, the most innovative medical man of the time.[3]

Cardano's anonymous reader also had some familiarity with astronomical techniques and tools. He listed the positions of the sun, the moon and the planets for 1 November 1552.[4] When he read Cardano's statement that if the stars that lay outside the body of Scorpio and in the foot of Ophiulcus 'are with the lord of the ascendant, and that is with the sun, the body of the person in question will be publicly torn', he clearly felt a shock of recognition. For, as he admitted, 'that constellation is in my ascendant, and with the lord of the ascendant, precisely'. But he consoled himself with the reflection that 'still, the lord of the ascendant is not with the sun'.[5] Evidently, then, this reader did not encounter astronomy and astrology for the first time as he read Cardano's work. He could not only read the clipped, technical prose of Cardano's explanations of astrological principle, but also evaluate the quantitative information displayed in the one hundred appended genitures – charts of the positions of the sun, moon, planets and the ascending node of the moon's path, or dragon's head, laid out on the zodiac, which was divided into twelve segments or houses.[6]

In the early stages of his encounter with Cardano's complex, demanding book, the anonymous medical man clearly found it instructive. As he traced his path through the printed text, he marked it in a highly traditional way, using procedures developed over the centuries by readers of manuscripts. He left a trail of underlinings, asterisks, one-word references and short content summaries which transformed the uniform, black-and-white product into a kind of embodied memory of its content.[7] Returning to the work years later, he would never be at a loss to find Cardano's efforts to establish the value and the limitations of astrology, his references to Copernicus and Ptolemy, or his statement that the movement of the line of apsides and the great conjunctions, which genitures normally did not record, determined the births of great men. The reader took a similarly intense interest in Cardano's technical doctrines – for example, his belief, a controversial one, that one should divide a geniture into twelve equal houses, rather than into unequal segments that more nearly approximated the lengths of the divisions of the zodiac as seen from a given point on Earth. And he made clear that he thought this arduous game of reading eminently worth the candle: 'all this', he exhorted himself at one point, 'must be learned by heart'.[8] Though the reader's remarks on the actual genitures are sparse, moreover, he seems to have paid close attention not only to Cardano's explicit comments but also to the planetary positions, and the astrologically significant geometrical relations between planets, that they recorded. Beside the chart of Henry Cornelius Agrippa, author of a celebrated work on natural magic, he listed no fewer than eleven examples of planets in trine, quadrature and sextile to one another – that is, planets that were 120, 90 or 60 degrees apart in the geniture.[9] These geometrical relations belonged to the astrologically significant ones known as aspects. Each planet theoretically exercised a particular influence: Saturn, for example, was notoriously connected with the melancholy humour. But their angular relations with other planets modified these effects, often radically, and the aspects which appeared in a geniture served as a standard code for evaluating their influence in a specific geniture. In tabulating them, this scholar showed that he could read a geniture as a modern reader decodes a graph, identifying and listing its significant quantitative contents.

In this case as in many others, the informed reader did not grant his assent to everything he encountered. As he tested Cardano's assertions against his other readings and experiences, he found a number of them wanting. Sometimes – as in the case of the healthy old man he and Argenterio encountered – he cited instances of experience that contradicted Cardano. As so often in the sixteenth century, he did not distinguish in any explicit way between specific experiences like meeting the old man – experiences which he himself witnessed – and more general statements which he had turned up in books. At one point Cardano contradicted Haly, the medieval Arabic commentator on Ptolemy's *Tetrabiblos*, the most rigorous ancient textbook of astrology: 'Do not believe Haly when he says that new teeth come into being when the bones are naturally at their driest, as Galen shows.' The reader contradicted Cardano in his turn, but left the source he relied on unmentioned: 'Still, this has been observed.'[10] The reader's language was not, in other words, entirely governed by the grammar of assent.

Over time, however, the reader clearly identified one set of grounds for tension in Cardano's own text. In his short treatises on how to draw up and interpret natal charts, Cardano formulated general aphorisms about the effects of particular planets, and the relations between planets, on individuals. In his genitures, by contrast, he identified the particular planets and configurations of planets which had shaped individual lives. Each part of the text was designed to complement the other. But as the reader read theory and practice against one another, he encountered what he saw as contradictions and errors of fact and argument. In his commentary on Agrippa's geniture, for example, Cardano noted that the magus 'after becoming a jurisconsult, reached the high office of royal secretary: this is shown by Jupiter, which is in quadrature to the sun'. Cardano went on to analyse the positions of Jupiter and the sun in detail. But he failed to convince the reader, who declared that Cardano had simply devised the astrological principle he cited after the event, pretending to predict something that he was retrospectively explaining: 'You say this because it happened that way.'[11]

A much more serious problem emerged from the reader's encounter with Cardano's geniture for the fifteenth-century Florentine poet and scholar Angelo Poliziano. One of Cardano's introductory texts laid out the natures of the planets, describing Saturn as 'the significator of death', Jupiter as 'the significator of religion and wealth', and Mars as 'the significator of violent infirmities, such as wounds and bruises'.[12] In Poliziano's geniture, as the reader noted, Mars was in opposition to Saturn and to the ascendant – normally the position of the 'significator', which exercised a determining influence on the individual in question.[13] And yet, as he also remarked, Poliziano had died a natural death. Cardano had not called attention to this apparent collision between his principles and these facts. But contradiction mattered in the sixteenth century – and not only in the formal realm of logic. Many commentators and theorists of interpretations believed what the Lutheran theologian Matthias Flaccius Illyricus formally stated in his treatise on the interpretation of Scripture. Every text had, or should have, a single basic purpose and argument; as an anatomy laid bare the bones which gave order and structure to the human body, so an interpretation should lay bare the goals and arguments which did the same for any written work.[14] Signs of contradiction, accordingly, undermined a writer's authority. 'It is curious', the reader observed in the margin, addressing himself to the author as if to a partner in discussion, 'that in accordance with your principles, he did not die by a violent illness.'[15] The more closely the reader scanned Cardano's principles and evidence, in other words, the looser he found the connections among them.

Eventually the reader's patience evaporated. In a characteristically sharp comment on his competitors, Cardano reflected that 'those who set down together things which are unconnected, want to give the impression that they have something profound to teach, though they are really absurd and empty'. 'Here', replied the reader, 'you do a splendid job of criticising yourself.'[16] In this case at least, a set of genitures – and the form of astrology that they were intended to embody and teach – lost credibility when studied by someone who applied the normal sixteenth-century protocols of learned reading in a systematic way. Anatomised, the text failed to reveal the strong muscles and hard skeleton of coherent argument that held true classics together. As we will see, though this reader followed a normal path, he arrived at unusual conclusions. Cardano and other collectors of genitures had many readers in the sixteenth and seventeenth centuries, most of whom showed only the fascinated interest with which this one began his work.

Genitures and their collectors

Genitures – charts like the ones Cardano assembled and printed – were nothing new in the sixteenth century. They were born, in fact, with astrology itself, in ancient Mesopotamia, where the earliest preserved geniture was drawn up in the fifth century BC.[17] Hellenistic and imperial astrologers composed genitures for both historical characters and living clients, for some of whom they even created 'luxury horoscopes' with detailed commentaries.[18] Some of these circulated fairly widely. When the Roman antiquarian Varro tried to find precise dates for the life of Romulus, the legendary founder of Rome, he asked Tarrutius, a diviner and astrologer, to infer Romulus' chart from the events of his life. Tarrutius did so, and his results were quoted and discussed by the astrological poet Manilius, the biographer Plutarch, the antiquary Censorinus and – much later – the Byzantine scholar-bureaucrat John Lydus, clear evidence that the geniture reached a wide public.[19]

Ptolemy, the second-century astronomer whose *Almagest* provided the systematic treatment of the methods and results of classical astronomy which remained standard down to the sixteenth century, also provided, in his *Tetrabiblos*, a detailed manual on how to create and interpret genitures. And though the *Tetrabiblos* did not include any genitures to serve as examples of the principles it stated, Ptolemy's contemporary Vettius Valens composed an astrological work, entitled *Anthologies*, which contained some 130 of them. In both this and later astrological works, composed in Rome, Byzantium and in the Islamic world, the charts of legendary heroes and real philosophers, emperors and cities circulated, providing astrologers with sources and models. Many of these posed problems to their readers. Some – like Tarrutius' geniture for Romulus – were garbled in the course of transmission. Others – like those presented in the *Mathesis* of the Roman astrologer Firmicus Maternus (fourth century), who offered a *thema mundi*, a chart for the Creation itself, as well as others for Plato and some of Homer's heroes – were compiled arbitrarily, so that the planetary positions they recorded did not match those of any real historical date.[20]

In the fifteenth and early sixteenth centuries, moreover, many scholars drew up and collected genitures of great men and women. Some of the tattered, bearded prophets who wandered the streets and squares of Italy in the years around 1500, proclaiming that Antichrist had already arrived and that the Messiah would soon appear and engage him in a final combat, insisted that only the Bible or divine inspiration, not the movements of the planets, could predict the imminent transformation of the world. Many others, however, saw planetary conjunctions as predicting the future – including a second universal flood, which was set to take place in 1524 and the expectation of which caused widespread panic. They collected the genitures of the great, using them too as keys to the kingdom of the future.[21] Some of the dozens of prognosticators and almanac-writers whose leaflets predicting Turkish invasions and bad weather flooded the book-market studied the genitures of kings and princes to predict their policies and actions.[22] So, on a higher social and technical level, did the royal and princely astrologers, whose tasks included drawing up genitures for their lords, their lords' children, and – no less important – their lords' rivals, allies and enemies. Doctors, many of whom thought that astrology should govern much or all of medical practice, drew up genitures as well as elections and interrogations – charts designed to identify what a particular moment, enterprise, weather system or disease had in store for a client. The English royal physicians of the later middle ages, as Hilary Carey has shown, compiled elaborate, systematic case books.[23]

By the first half of the sixeenth century, in other words, genitures were widely collected, and many of them entered fairly wide circulation. Like many other segments of the system of communications in early modern Europe, the exchange system among astrologers used handwriting, rather than printing, as its normal means of transmission.[24] The natal charts of powerful rulers and popular writers reached astrological practitioners who did not belong to the social or intellectual elite. Even the authors of almanacs and pamphlet-sized predictions – who, as we will see, included Cardano – seem to have had collections of royal genitures, which they relied on when they issued their predictions about dynasties and battles.[25] Cardano's career as a doctor, by contrast, gave him access to other sets of charts, compiled not to predict the political and military moves of the great but to explain the bowel movements and illnesses of ordinary patients. In a later work, in which Cardano used the genitures of less celebrated clients to confirm his astrological principles, he identified one of his sources as the astrological notebook or notebooks of the celebrated Milanese physician Bonaventura Castiglione.[26] Like other visual and partly visual media, genitures crossed social and cultural lines; like religious images, moreover, they rested on principles known to poor and rich, barely literate and highly educated readers.

Genitures circulated, moreover, for several reasons. Canny astrologers told their powerful clients what they wanted to hear about their own futures – as Cardano's more prestigious rival, Luca Gaurico, did when he promised Charles V's brother Ferdinand that he would conquer the Turks and become more powerful than the Emperor himself.[27] Such documents naturally interested a ruler's enemies as well as his loyal courtiers. Equally charged with political interest were the numerous genitures artfully designed and put into circulation to further a particular interpretation of a given individual's career. In the course of his journey through the Holy Roman Empire in 1532, Gaurico impressed Philipp Melanchthon and others with his Latinity and his astrological expertise. Consulted about Luther's geniture, he accepted one of the dates then circulating as Martin Luther's birthday, setting the event in October 1484. Gaurico chose this date – which Luther himself apparently accepted, though he was by no means certain of the year – because it connected the Protestant Reformer's appearance both to an ominous great conjunction of Saturn and Jupiter and to Johannes

Lichtenberger's widely circulated prediction that an evil prophet would appear in 1484. Both Protestant astronomers like Erasmus Reinhold and Catholics like Cardano had examined and copied out Gaurico's nativity for Luther long before it first appeared in print in 1552. Cardano, in fact, published his own geniture for Luther – which set the Reformer's birth in 1483 and attacked Gaurico's geniture as false – in 1547, five years before Gaurico's own work was published. Within the interlocking circles of the astrological community, genitures transferred information rapidly and efficiently.[28] Humanists and other non-specialists, moreover, also studied genitures – often equipped by their astrological advisers with long explanatory texts.

In the period between 1538 and 1552, however, astrologers and entrepreneurs collaborated to open up this subterranean realm. The 'private science' of astrologers and medical men became at least partly public, as the first printed geniture collections exposed data, principles and technical debates of many kinds to the scrutiny of an increasingly large public. Cardano himself played a central role in this process. In the 1530s, his medical career hindered by the illegitimate birth which made him ineligible to join the Milanese College of Physicians, Cardano found himself confined to a small town outside Milan. Desperate to attract the attention of patrons, he resorted to his astrological skills as a way of making money and reaching the public. His first publication was a *Pronostico*, an astrological booklet in Italian, which appeared in 1534 or 1535, and which used the stars to predict everything from the future history of the Catholic Church and the European states for the next several decades to the weather for the next several years. Very soon, as Ian Maclean has shown, Cardano became ashamed of this publication, aimed at an uneducated public and cast as much in the language of popular prophecy as that of technical astrology. He did not go on with almanac-writing, suppressed any reference to the *Pronostico* in his many later lists of his own works, and turned to more erudite genres. Cardano's turn to learned astrology led to his first publication of genitures.[29]

In 1538, Cardano brought out two short treatises in Latin on astronomy and astrology. To these he appended a collection of ten genitures: five of them for scholars, Petrarch, George of Trebizond, Francesco Filelfo, Fabio Cardano and Gualterio Corbetta, and five of them for rulers, Pope Paul III, Charles V, Francis I, Suleiman I and Ludovico Sforza. From a modern standpoint, the individuals in question hardly seem of comparable importance. Even if a Milanese like Cardano might naturally see his erudite fellow-countryman Corbetta and the last of the independent dukes of Milan, Ludovico Sforza, as great men, it seems curious that Cardano also included a chart for his own father – by his own account a man of great and curious learning, especially skilled in the arts of magic, but also one whose only substantial publication was an edition of Archbishop Peckham's *Perspectiva*, with a short preface. Cardano himself justified his choice of genitures in a preface, which sketched the qualities he and others thought a geniture collection should have:

I have not added any of these genitures without careful thought, or for a trivial reason. For every one of them had some remarkable quality. The brilliant virtues and successes of these rulers lit up the world, so much so that histories for centuries to come will record them. True, the last of them ended in disaster, with the exile of [Ludovico Sforza's] sons, his loss of his kingdom, his betrayal, and the imprisonment in which he met his wretched death. But all of them have one thing in common: I had a solid knowledge of their deeds and their qualities.[30]

A geniture collection, evidently, should rest on solid knowledge of the facts: Cardano took evident pleasure in attesting that one of his ten subjects, Gualterio Corbetta, had given him his own geniture, and in pointing out that numerous sources confirmed the dates he gave of Petrarch's, Filelfo's and George's births.

Cardano also believed that the charts of private individuals – like his humanists – had a special value for the astrologer. The stars, he argued, genuinely determined the fates of such men: their careers corresponded precisely to their genitures. For kings and princes, he could not claim as much. Many astrologers emancipated them entirely from planetary influence. Since late antiquity, some astrologers had followed the lead of the fourth-century writer Firmicus Maternus, who avoided political risk by exempting emperors from the control of the stars, on the grounds that as gods, they were above the planets.[31] Cardano insisted that true Christians, 'who cultivate true piety and observe in the heavens nothing superstitious, but only see them as natural causes', could not follow this lead. 'Just as all men are troubled, without exception, by heat, cold, and pain, so we cannot deny that the bodies and minds of rulers are violently changed and gently pulled to good and evil effect.' But he also admitted that the larger fate of each kingdom did much to shape that of its ruler. In any case, astrologers found it hard to discuss royal genitures as honestly as they could analyse the lives of private men like Petrarch.[32] A geniture collection, in other words, had to contain the charts of ordinary men and women as well as those of the princes of state and church. Otherwise it would not directly embody and confirm astrological principles.

But geniture collections, even as Cardano presented them, were not meant to be read only for their scientific qualities. Even his ordinary subjects had to be, like the children in Lake Woebegone, above average. Cardano underlined that each of his first ten genitures had 'some remarkable quality' – 'aliquid admiratione dignum'.[33] A geniture collection, in other words, should exhibit the workings of unalterable law, should reveal how the stars in their regular motions determined the fates of mankind. But it should also surprise the reader, offering startling tidbits of news and gossip as well as confirmation of technical astrological axioms.

By the middle decades of the sixteenth century, printers who took a serious interest in the natural sciences had realised that a public existed for many forms of astrological publication. The Nuremberg publisher Johannes Petreius, best known for having printed both Copernicus' *De revolutionibus* and Cardano's *Ars magna*, scoured libraries in his own city and elsewhere for unpublished astrological texts, ancient and modern. The work of Antonius de Montulmo, which he printed in 1540, included, in its preface, a first announcement that he planned to publish the *De revolutionibus* of Copernicus, and an appeal to his readers to provide him with further new offerings.[34]

Petreius' preface was not simply a piece of advertising for his scholarly ways, moreover. A few years later he asked the astronomer Erasmus Reinhold to come to Nuremberg and write the sort of astrological text that he had always wanted to publish. Petreius offered Reinhold time off from his duties as Dean at the University of Wittenberg, where he had to spend much of his time urging his students not to clamber over the fortifications and impede the garrison's operations, and from his arduous work as an astronomer, which required him both to direct the teams of young calculators that produced his famous *Prutenic Tables* of astronomy and to draw up and explicate the genitures of important political and military figures for his patron, Albrecht of Hohenzollern. In Nuremberg, Petreius would put a nice 'little room and study' at Reinhold's disposal, for as long as he needed it. All he had to do was to produce a little work on how to draw up a geniture. Petreius made clear that he wanted a how-to book, one that defined such mysterious terms as 'house' and 'angle' in terms a layman could understand and that showed precisely how an astrologer took data from tables and entered them in a form to create the sort of genitures that clients wanted. Petreius discouraged Reinhold from providing worked examples, genitures already drawn up for clients. For, as he pointed out, 'I think this collection ought to be saleable, and there are plenty of those things already in print.'[35]

Petreius knew what he was talking about. For he himself had turned Cardano's first set of genitures from a one-off curiosity to the start of a vogue. Failing to find publishers who could take on and distribute his scientific works in Milan or elsewhere in Italy, Cardano had taken out a privilege for thirty-eight distinct texts. He included this in a publication that appeared in Nuremberg in 1541. One of Petreius' advisers, Georg Joachim Rheticus, noticed the list, and they contacted Cardano to express an interest in his work.[36] Within a year or two, a massive stream of works by Cardano welled up in Nuremberg and began to saturate northern book markets. In 1543, Petreius printed a collection of 67 genitures compiled by Cardano, which accompanied a reprint of the mathematician's two treatises on astronomy and astrology. In 1547 Petreius brought out a second, enlarged edition, this one with 100 genitures, equipped with commentaries and a longer series of five introductory treatises. Both of these editions contained not only Cardano's materials, but further genitures supplied by Georg Joachim Rheticus, a disciple of Copernicus who worked, at times, with Petreius.[37] Where Cardano's 1538 collection had expressed warm admiration for the counter-reforming pope Paul III, these two new ones showed considerable sympathy for reforming Catholic humanists like Erasmus – and even some admiration for Martin Luther. Evidently Cardano and his publisher edited his work to make it more appealing to northern – and Protestant – tastes. Cardano's collection, in its final form, was thus not the result of individual effort, but of an intensive collaboration between a Catholic savant, a Protestant printer and the printer's advisers. When Petreius referred to geniture collections as a genre already well represented in the contemporary book market, in short, he described his own accomplishment as a publisher. Without his entrepreneurship, the genre of the geniture collection might never have been launched.

Genitures and their readers

Cardano's collection rapidly found readers. Janus Cornarius – a Basel physician and a prolific editor of texts in his own right – bought his copy of the 1543 edition in Marburg, in the year of its appearance.[38] Erasmus Reinhold took the same work as one of the central sources for his own manuscript collection of genitures, which he evidently compiled around 1545.[39] Philipp Melanchthon, Fridericus Staphylus and others used their copies not only to study the principles of astrology, but also to compile further genitures which complemented the published ones. Melanchthon entered the charts of his own four children as well as those of a number of princes in the blank spaces and end-papers of his copy.[40] By the 1550s, when the English statesman and political theorist Sir Thomas Smith, caught by a sudden passion for astrology, spent months perfecting his mastery of technique, it was natural for him to choose Cardano's work, in the 1547 edition, as one of his basic texts, and to fill its margins with notes indicating how carefully he had worked through it.[41] And Cardano's work was far from the only one to attract such interest: a copy of Albohali's work on the judgement of nativities, now in Paris, contains fifty-two nativities written in.[42] Some of these – like those for Cardano and his father – clearly come from Cardano's work, but others seem to have been added by the compiler, a scholar apparently active in Frankfurt.

As Cardano's rivals noted the interest with which his work was read, moreover, they too began to publish genitures with commentaries. The Nuremberg astronomer Johannes Schöner brought out in 1545 an elaborate work on astrology, which he illustrated with multiple, detailed analyses of the geniture of the Holy Roman Emperor Maximilian I.[43] In 1552, Cardano's chief rival, Luca Gaurico, responded to Cardano's work by publishing his own geniture collection, the *Tractatus astrologicus*.[44] Cardano replied by equipping his own edition of Ptolemy's *Tetrabiblos*, the chief ancient textbook of the art, with a new collection of twelve elaborate genitures, including his own (which he had already studied in a less detailed way in 1543 and 1547). He analysed these in as much detail, and from as many points of view, as Schöner studied the Emperor's chart – and took the opportunity to get in some digs at Gaurico as he did so.[45] Geniture collections had become not just a source of information, but a site of scholarly contests for supremacy. In the course of the sixteenth century, new collections appeared, as astrologers like Johannes Garcaeus and Rudolph Goclenius equipped their own textbooks with worked examples. Collections gradually swelled: Gaurico's *Tractatus* of 1552 included around 160 genitures, but that of Garcaeus, published in 1576, had more than 400.[46]

Geniture collections differed on many points, from the method of dividing the houses used by their compiler to the space they devoted to women. Cardano, who used the system of equal houses and considered even his own mother's chart, which he discussed in his commentary on his own geniture, worthy of little remark, 'because she was a woman', stood at one end of two spectrums: most astrologers used the more sophisticated and arduous methods of house division associated with Alcabitius and Regiomontanus, which required the use of tables and computation, and many showed more interest in the fates of women. Erasmus Reinhold, for example, included a whole section of women's genitures in his manuscript notebook.

Both the published texts and the comments left by their readers, however, indicate certain common elements. One kind of information that all geniture collections provided took the form of intimate details about the health, character, intellect, fortunes, births and deaths of prominent scholars, artists and rulers on the one hand and notorious monsters, criminals and lunatics on the other. In the dedicatory letter to his collection of 1543, Cardano promised to show readers

all the different forms of death, by poison, by lightning, by water, by public condemnation, by iron, by accident, by disease: and after long, short or middling periods; also the various forms of birth that yield twins, monsters, posthumous children, bastards, and those in the course of whose birth the mother dies; and then of the forms of character, timid, bold, prudent, stupid, possessed, deceptive, simple, heretics, thieves, robbers, pederasts, sodomites, whores, adulterers; and also, with regard to the disciplines of the rhetoricians, the jurisconsults, and the philosophers, and those who will become the greatest physicians and diviners, and famous craftsmen, and also those who will become despisers of the virtues. I have also followed out the different incidents of life, explaining what sort of men kill their wives, suffer exile, prison and continual ill health, convert from one religion to another or pass from the highest position to a low status, or, on the other hand, from a low fortune to kingship or power.[47]

Readers evidently agreed. Cornarius, for example, used Cardano's work as a biographical compendium about the great and the good. He entered the death dates for Francis I and Henry VIII a few years after buying the treatise, and took special interest in Cardano's hyperbolic commentary on the chart of Charles V, in whose time, the astrologer suggested, civilisation might have reached its final peak. But he also took an interest in the curious story of the chiromancer and astrologer Bartolomeo Cocles, who put on armour, knowing that he was in danger on a certain day, only to be felled by the blow of a club. 'Mira res', wrote Cornarius, showing that he did see Cardano's book, as its preface indicated, as a virtual chamber of wonders.[48]

Other readers, who corrected and supplemented Cardano's work, still clearly saw it as exemplary. Melanchthon added to his own collection alternative charts for several of the individuals treated by Cardano: Paul III, Poliziano, Charles V and Luther. But he read them, and the genitures offered by Cardano himself, exactly as Cardano had, as clear demonstrations of the validity of such rules as that the conjunction of the moon, Mercury and Venus in Libra had made his daughter produce a series of female children, or that the presence of benevolent planets in Pisces at the birth of Charles V had tempered the influence of the malevolent ones.[49] For Melanchthon, in other words – as for most early readers – Cardano's collection seemed to support the validity of the general principles about each planet's character and influence that he offered in his introductory treatises. Ever fascinated by the patterns of world history, Melanchthon took careful notes on the passages in which Cardano offered, in line with a tradition that went back to the Islamic astrology of the middle ages, horoscopes for the rise of new religions.[50] Even the most explosive of Cardano's genitures – the one for Jesus which he included in his commentary on Ptolemy's *Tetrabiblos* in 1555, at a time when most readers had apparently forgotten that medieval and earlier Renaissance astrologers had often done the same – was studied and copied without evident distress by some readers whose Christian faith we have no reason to doubt.[51]

Above all, geniture collections stimulated readers by what they had to leave out. In interpreting any geniture, astrologers could potentially have referred to thousands of factors: fixed stars, constellations, degrees of the zodiac, parts of houses, all had astrological significance, as well as the sun, the moon, the planets, the head and tail of the dragon (the ascending and descending nodes of the moon's path). Every geniture collector briefly interpreted the data, calling attention to those factors which, in his judgement, would determine the character and fate of the individual. Genitures for children summed up their potential strengths and weaknesses in advance, for the use of parents, guardians and teachers. Genitures for adults explained their victories and defeats, diseases and recoveries. In no case, however, did the astrologer disclose exactly how he had known which of the thousands of potentially relevant factors were most important.[52] Cardano – always frank to a fault – admitted that his vital form of judgement belonged not to the formal protocols of astrology, but to the spiritual gifts with which a practitioner began and the tacit knowledge he acquired over time. Like the supple, multivalent attention to details that a jeweller needed to assess stones, the astrologer's judgement, he argued, was too 'sublime' to be expressed in words.[53] The geniture collection presumably supplied a virtual equivalent to the actual experience that a practitioner like Cardano acquired as he supplied genitures for kings and scholars. Reading genitures, accordingly, not only informed the memory but sharpened the judgement, in ways no other form of learning the art could. This sense above all must explain the enthusiasm with which so many readers – many of whom were not astrologers or medical men by profession – worked their way through geniture collections, seeking that delicious but indefinable high prudence that Gabriel Harvey scented in both Gaurico and Cardano.

Astrology was anything but uncontested in the sixteenth century. Astrologers, in the first place, often disagreed about everything from the date of a particular birth to the identity of the most significant planetary configurations in a given chart. Geniture collections embodied these polemics, sometimes implicitly – as when Cardano insisted that his own chart for Martin Luther was correct, knowing that informed readers would already have seen the unpublished geniture drawn up by Luca Gaurico that he was criticising. Gaurico – a curial insider and a favourite of Pope Paul III, who took a strong interest in astrology – replied in the same idiom, tacitly attacking Cardano for his use of equal houses, his computation of planetary positions 'only in whole degrees', and his explanatory principles – like his view that those who had the planets above the horizon at the moment of their birth would live longer than those who had them below it. Informed readers noted these polemics and immediately identified their object. The English humanist Gabriel Harvey was something of a connoisseur of astronomical and astrological literature: he entered long and well-informed discussions of the mathematical practitioners of late sixteenth-century London into his copy of Guarico's book. It is revealing that he also compared Gaurico and Cardano systematically (Figure 3.1).[54] Harvey noted and understood Gaurico's references to Cardano's technical shortcomings. But he also entered in the margins of his Gaurico references to Cardano's later efforts at self-defence. He admitted that Gaurico's willingness 'to oppose his own astrological experience to the ingenious observations of Ptolemy' impressed him. But so did Cardano's 'marvellous nativity of Christ'.[55]

Harvey's reading of Cardano and Gaurico brings out another aspect of their work. Both of them – like most of their colleagues in the period – saw astrology as offering not absolutely valid predictions, a determinist reading of the individual's character and future, but rather as setting the general possibilities within which the individual would have to act – and as offering useful information about flaws of character and weaknesses of temperament for which the client and his parents or guardians might have to correct. Such information was of obvious political interest and utility – especially when offered by astrologers who themselves gave every sign of having rich political experience. Harvey noted that both Gaurico and Cardano could help the reader who hoped to hone his political skills. Gaurico, after all, had obtained his 'vast experience' and 'authority' in 'the court of a prudent pope'.[56] But Cardano's rich collection of lives – so Harvey also reflected – was, like Diogenes Laertius' *Lives of the Philosophers*, Philostratus' and Eunapius' *Lives of the Sophists*, and Paolo Giovio's modern *Elogia*, just the sort of book that a clever aspiring intellectual, or 'polytechnus', should read.[57] The political activities of astrologers in Renaissance Europe have remained largely unstudied by historians who have tended to see the pragmatic counsels of Machiavelli as the key to understanding political decision-making. But Barbara Bauer and others who have begun to study the relations between astrologers and rulers have shown that many members of the ruling elites took astrology seriously as a source of political counsel – sometimes more seriously than history itself.[58] In this respect, Harvey and other readers of geniture collections understood the political ramifications of astrology very well indeed.

Astrology itself, however – as we saw at the beginning of this

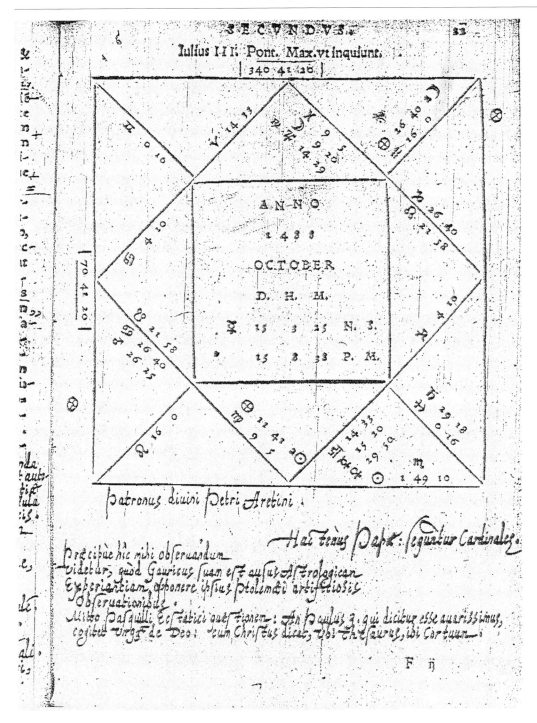

Figure 3.1 Gabriel Harvey compares genitures: in this case, those drawn up for Pope Julius II by Luca Gaurico in his *Tractatus astrologicus* (Venice, 1552) and Girolamo Cardano in his earlier collection of genitures. Harvey's note is in his copy of Gaurico's book (Oxford, Bodleian Library).

essay – was not uncontested intellectual ground. At the end of the fifteenth century, Pico della Mirandola had devoted his last, posthumously published book to an unsparing critique of the art. Though earlier writers like Cicero, Augustine and Nicole Oresme had attacked astrology, none of them knew the subject and its literature as well as Pico did. After his book appeared, both astrologers and their critics had to take account of his work, which argued that astrology lacked any empirical foundation.[59] Cardano – at least in retrospect – saw his work as supplying this lack. More than once he compared astrology to medicine, to the disadvantage of the former. The founder of ancient medicine, Hippocrates, had both written general treatises and supplied, in the *Epidemics*, the sober case histories from which he derived his theory of the temperaments and his classification of diseases. Ptolemy, by contrast, had left only a systematic treatise, not the rich case books from which he must have derived it. Cardano's geniture collection could thus be seen as a sort of equivalent to the *Epidemics* – a strictly empirical report on the data.[60]

Some readers, as we have already seen – probably a majority – used their copies of Cardano in exactly this way. But others – like the one from whose notes we began – read more critically. In the later years of the sixteenth century, critics of astrology began to produce their own geniture collections, designed to show that the empirical evidence actually disproved the principles of astrology. Sixtus ab Hemminga, whose *Astrologia ratione et experientia refutata* appeared in 1583, devoted much of the work to reviewing the nativities drawn up by the imperial astrologer Cyprian Leowitz, which, he showed, generally did not accurately predict the outcome of their subjects' lives. Pierre Gassendi, half a century later, anatomised the genitures of Nostradamus with equal severity and greater wit.[61] In cases like these, the doubts which assailed Cardano's anonymous contemporary reader swelled into a condemnation of the art as a whole. The problems which the geniture collectors did not explain – for example, that of determining which of the dozens of angular relations among planets, stars, houses, signs and ascendants registered in any given geniture actually determined the character of its object – seemed insoluble to some of those who actually tested the supposed empirical foundations of the art by the criteria of rigour and consistency characteristic of the age of the Mechanical Philosophy.

Yet geniture collections continued to be read, and to be made, deep into the seventeenth century. John Aubrey, who had read both the renaissance astrologers and their critics, admitted that 'we haue not that Science yet perfect'. But if he acknowledged that Cardano and Gaurico had done their work of collection imperfectly, he still saw their method as the only valid one: 'The way to make it perfect is to gett a supellex of true Genitures.'⁶² When interpreting the genitures that he collected, he still applied the general principles which Cardano, Schöner and others had stated. His work yielded not only a massive collection of genitures with commentaries, but also one of the seventeenth century's masterpieces of character analysis and English prose, the *Brief Lives* – though modern editions generally conceal the fact that this magnificently quirky collective study of characters and temperaments is also the last of the great renaissance geniture collections.⁶³

The geniture collections had many faults. Few published charts and commentaries compared in quantitative precision or qualitative profundity to the manuscript genitures prepared by astrologers for wealthy individual clients. Like the 1496 edition of Regiomontanus and Peurbach's *Epitome of the Almagest*, which gave the public an imperfect version of that great work, the printed editions of Gaurico and Cardano offered only a rough and approximate version of astrological practice to the reader. But the printed text of Regiomontanus and Peurbach served a vital purpose for Copernicus and the many other astronomers who learned the subject from it. Similarly, the printed and manuscript geniture collections that formed a central part of so many learned libraries also served vital purposes for contemporary readers, informing and stimulating them in a variety of ways. Though largely forgotten except by historians of astrology and a few other scholars who have used them for biographical purposes, they form a significant, if not a major, part of the everyday life of scientific and scholarly practice in the sixteenth and seventeenth centuries.

Notes

1 For the larger context, see A. Blair, 'Humanist methods in natural philosophy: the commonplace book', *Journal of the History of Ideas*, 53 (1992): 541–51, and q.v., ch. 4; A. Moss, *Printed Commonplace-Books and the Structuring of Renaissance Thought* (Oxford, 1996). For a case in point, see A. Grafton, *Commerce with the Classics* (Ann Arbor, 1997), ch. 4.
2 G. Cardano, *Libelli quinque* (Nuremberg, 1547), Munich, Bayerische Staatsbibliothek 4° Astr. U 35ᵃ, fol. 51 recto.
3 Ibid., sig. A ij verso.
4 Ibid., fol. 44 recto.
5 Ibid., fol. 53 verso.
6 For the history and forms of genitures, see J. North, *Horoscopes and History* (London, 1986); and J. C. Eade, *The Forgotten Sky: A Guide to Astrology in English Literature* (Oxford, 1984).
7 M. Carruthers, *The Book of Memory: A Study of Memory in Medieval Culture* (Cambridge, 1990).
8 Cardano, *Libelli quinque*, fol. 24 verso.

9 Ibid., fol. 163 verso.
10 Ibid., fol. 51 verso; for Cardano and Haly see G. Ernst, *Religione, ragione e natura: ricerche su Tommaso Campalla e il tardo Rinascimento* (Milan, 1991).
11 Cardano, *Libelli quinque*, fol. 163 verso.
12 Ibid., fol. 24 verso.
13 See Eade, *Forgotten Sky*.
14 M. Flacius Illyricus, *De ratione cognoscendi sacras literas*, ed. and trans. L. Geldsetzer (Düsseldorf, 1968), pp. 90–3, 98–101; R. Sdzuj, *Historische Studien zur Interpretationsmethodologie der Frühen Neuzeit* (Würzburg, 1997).
15 Cardano, *Libelli quinque*, fol. 178 verso.
16 Ibid., fol. 242 recto.
17 A. Sachs, 'Babylonian horoscopes', *Journal of Cuneiform Studies*, 6 (1982): 49–75; F. Rochberg-Halton, 'Babylonian horoscopes and their sources', *Orientalia*, 58 (1989): 102–23; D. Pingree, *From Astral Omens to Astrology* (Rome, 1997).
18 L. Taub, 'The rehabilitation of wretched subjects', *Early Science and Medicine*, 2 (1997): 74–87.
19 P. Brind'Amour, *Le Calendrier romain* (Ottawa, 1983), pp. 240–9; A. Grafton and N. Swerdlow, 'Technical chronology and astrological history in Varro, Censorinus and others', *Classical Quarterly*, NS 35 (1985): 454–65.
20 See esp. E. S. Kennedy and D. Pingree, *The Astrological History of Masah'allah* (Cambridge, Mass., 1971); North, *Horoscopes and History*.
21 See e.g. the rich studies collected in *Prophetic Rome in the High Renaissance Period*, ed. M. Reeves (Oxford, 1992).
22 See in general L. Thorndike, *History of Magic and Experimental Science*, 8 vols. (New York, 1923–58), V, chs. x–xv. More recent studies include P. Zambelli (ed.), *'Astrologi hallucinati': Stars and the End of the World in Luther's Time* (Berlin and New York, 1986); O. Niccoli, *Prophecy and the People in Renaissance Italy*, trans. L. G. Cochrane (Princeton, 1990); R. Westman, 'Copernicus and the prognosticators: the Bologna period, 1496–1500', *Universitas: Newsletter of the International Center for the History of Universities and Science, University of Bologna*, December 1993: 1–5.
23 H. M. Carey, *Courting Disaster: Astrology at the English Court and University in the Later Middle Ages* (New York, 1992).
24 Cf. H. Love, *Scribal Publication in Seventeenth-Century England* (Oxford, 1993).
25 G. Cardano, *Pronostico* (Venice, 1534 or 1535), sig. [A vi] verso; O. Cane Gaurico, *Prognosticon anni 1537* (Rome, 1537), sig. [A iii] verso.
26 Cardano, *Libelli quinque*, fol. 232 verso.
27 Vienna, Oesterreichische Nationalbibliothek, MS 7433, fols. 2 verso – 3 recto, 10 recto. Cf. G. Minois, *Geschichte der Zukunft* (Düsseldorf and Zürich, 1998), p. 409.
28 See A. Warburg, *Heidnisch-antike Weissagung in Wort und Bild zu Luthers Zeiten, Gesammelte Schriften: Studienausgabe*, ed. H. Bredekamp et al., vol. I, pt. 2 (Berlin, 1998), pp. 199–303. Cf. now R. Staats,

'Luthers Geburtstag 1484 und das Geburtsjahr der evangelischen Kirche 1519', *Bibliothek und Wissenschaft*, 18 (1984): 61–84; R. Briggs, *Apocalypse and Gnosis: Apocalypticism in the Wake of the Lutheran Reformation* (Stanford, 1988).

29 See N. Siraisi, *The Clock and the Mirror: Girolamo Cardano and Renaissance Medicine* (Princeton, 1997); I. Maclean, 'Cardano and his publishers, 1534–1663', in E. Kessler (ed.), *Girolamo Cardano: Philosoph, Naturforscher, Arzt* (Wiesbaden, 1994).
30 Cardano, *Libelli duo* (Milan, 1538), fol. E ij recto.
31 M. T. Fögen, *Die Enteignung der Wahrsager: Studien zum kaiserlichen Wissensmonopol in der Spätantike* (Frankfurt, 1993).
32 Cardano, *Libelli duo*, sig. E ij recto.
33 Ibid.
34 See N. Swerdlow, 'Annals of scientific publishing: Johannes Petreius's Letter to Rheticus', *Isis*, 83 (1992): 270–4.
35 Berlin-Dahlem, Geheimes Preussisches Staatsarchiv, HBA A4 223.
36 See Maclean, 'Cardano and his publishers'.
37 P. McNair, 'Poliziano's horoscope', in C. H. Clough (ed.), *Cultural Aspects of the Italian Renaissance: Essays in Honour of Paul Oskar Kristeller* (Manchester and New York, 1976), pp. 262–75.
38 Houghton Library, Harvard, *IC5.C1782.543d.
39 Leipzig, Universitätsbibliothek, MS Staatsbibliothek 935.
40 Vienna, Österreichische Nationalbibliothek, 72 J 35 (72 X 5 is a second annotated copy of the same work); Houghton Library, Harvard, *GC5.C7906.540a.
41 Oxford, Bodleian Library, MS Ashmole 1557.
42 Paris, Bibliothèque Nationale, Rés. V.1300; see Thorndike, *History of Magic*, VI, p. 105.
43 J. Schöner, *De iudiciis nativitatum libri tres* (Nuremberg, 1545).
44 L. Gaurico, *Tractatus astrologicus* (Venice, 1552).
45 Ptolemy, *Tetrabiblos*, ed. G. Cardano (Lyons, 1555).
46 Thorndike, *History of Magic*, VI, 104.
47 Cardano, *Libelli duo* (Nuremberg, 1543), ep. ded., A iij verso–[A iiij recto].
48 Houghton Library, Harvard, *IC5.C1782.543d, fol. P ij verso.
49 Vienna, Österreichische Nationalbibliothek, 72 J 35, sig. M ij verso and rear end-paper.
50 Ibid., sig. N recto; for the background cf. Ernst and J. North, 'Astrology and the fortunes of churches', *Centaurus*, 24 (1980): 181–211.
51 See e.g. Herzog August Bibliothek, Wolfenbüttel, 35.2 Astron. For the background see W. Shumaker, *Renaissance Curiosa* (Binghamton, 1982), including a translation of and commentary on Cardano's geniture; L. Smoller, *History, Prophecy and the Stars: The Christian Astrology of Pierre d'Ailly* (Princeton, 1994); P. Zambelli, *Una reincarnazione di Pico ai tempi di Pomponazzi* (Milan, 1994).
52 John Dee did try to set the art on a quantitative basis: see *John Dee on Astronomy = Propaideumata aphoristica (1558 and 1568)*, ed. and trans. W. Shumaker, intro. J. Heilbron (Berkeley, 1979).
53 Cardano, *Libelli quinque*, fol. 122 recto–verso.

54 Oxford, Bodleian Library, 4° Rawl. 61.
55 Ibid., fols. 22 recto, 122 verso.
56 Ibid., fol. 21 verso.
57 Ibid., fol. 57 verso.
58 B. Bauer, 'Die Rolle des Hofastrologen und Hofmathematicus als fürstlicher Berater', in A. Buck (ed.), *Höfischer Humanismus* (Weinheim, 1989), pp. 93–117.
59 See esp. P. Zambelli, *L'ambigua natura della magia* (Milan, 1991; new edn, Venice, 1996). There is a modern edition of Pico's text with translation and commentary by E. Garin, *Disputationes adversus astrologiam divinatricem* (Florence, 1946–52).
60 For Cardano's Hippocratism see Siraisi, *The Clock and the Mirror*, ch. 6. For his portrayal of Hippocrates as his model, see esp. *Libelli quinque*, fol. 257 verso (cf. 215 verso).
61 Thorndike, *History of Magic*, VI, pp. 193–6; P. Brind'Amour, *Nostradamus Astrophile: les astres et l'astrologie dans la vie et l'oeuvre de Nostradamus* (Ottawa and Paris, 1993).
62 Oxford, Bodleian Library, MS Aubrey 6, fol. 12 verso = J. Aubrey, *Brief Lives*, ed. A. Clark, 2 vols. (Oxford, 1898), I.9.
63 See the classic study by M. Hunter, *John Aubrey and the Realm of Learning* (London, 1975).

Further reading

T. Barton, *Ancient Astrology* (London, 1994)
J. C. Eade, *The Forgotten Sky: A Guide to Astrology in English Literature* (Oxford, 1984)
A. Geneva, *Astrology and the Seventeenth-Century Mind: William Lilly and the Language of the Stars* (Manchester and New York, 1995)
A. Grafton, *Cardano's Cosmos: The Worlds and Works of a Renaissance Astrologer* (Cambridge, Mass., forthcoming)
J. North, *Horoscopes and History* (London, 1986)
L. Smoller, *History, Prophecy and the Stars: The Christian Astrology of Pierre d'Ailly* (Princeton, 1994)
K. V. Thomas, *Religion and the Decline of Magic* (London, 1971)

4 Annotating and indexing natural philosophy

Coping with information overload

Reading, compiling and commenting on texts long constituted one of the central practices of natural philosophy, from antiquity down to at least the late seventeenth century. From Pliny to Ulisse Aldrovandi to Johann Jonston, from Aristotle to Roger Bacon to the professors of philosophy in seventeenth-century universities, to do natural philosophy was in large part to gather, sort and critique causal explanations, reports of observations and philosophical opinions proffered by one's predecessors; the texts that were generated in the process would in turn fuel the discussions of future generations of natural philosophers. In the Renaissance this self-perpetuating cycle of textual commentary faced a massive increase in the range and number of relevant books to be read. The humanist programme of recovering lost ancient works made available for the first time in a millennium a number of works on natural topics, including those of Dioscorides, Lucretius and Archimedes, among others.[1] At the same time, travel to the new world as well as to exotic parts of the 'old world' yielded new accounts of flora and fauna and human customs. Finally, the new technology of printing made more readily accessible in a wide range of editions – from the bulky *editiones principes* and *opera omnia* to vernacular translations or cheap school editions of single works – not only this new material, but also the well-known ancient and medieval authorities still held in high esteem. Printing also fuelled the composition of works by an ever-increasing number of modern authors, many of whom would not have had the university or courtly connections to reach any significant diffusion in manuscript.

The result of these converging factors, from the sixteenth century on, was an over-abundance of works to be read and referred to in the cycle of textual commentary across all fields of study, including natural philosophy.[2] By the second half of the seventeenth century the sense of a crisis due to information overload had reached such proportions that printing, long praised as a 'divine' invention,[3] had to be defended against the charge of bringing on a new era of barbarity. In 1685 Adrien Baillet prefaced his collection of critical book reviews with this justification:

We have reason to fear that the multitude of books which grows every day in a prodigious fashion will make the following centuries fall into a state as barbarous as that of the centuries that followed the fall of the roman empire. Unless we try to prevent this danger by separating those books which we must throw or leave in oblivion from those which one should save and, within the latter, between the parts that are useful and those which are not.[4]

In response to Baillet, the German professor Daniel Georg Morhof, whose most famous work was a massive guide to learning, the *Polyhistor*, which reached some 2,000 pages in its final form, acknowledged the complaint that typographers produced too many useless books but concluded: 'one should not think that this remarkable art should be condemned; for nothing is good with which some evil from the vice of men is not admixed'. Morhof considered the abundance of books to be inevitable, as 'the most wise Solomon already then said that there is no end to the writing of books'.[5]

Despite some early humanist resistance to printing (including a humanist censorship plan) and the institution of censorship by numerous church and state authorities, which often had a devastating impact on individual authors, attempts to stop or stem the tide of books flowing from the presses were limited in their impact in the long run, notably due to the multiplication of possible venues for publication.[6] Instead, scholars and teachers responded to the new realities of an over-abundance of books with a variety of tools. The book review appeared in the late seventeenth century not only in works like Baillet's but especially in numerous periodicals devoted in whole or in part to reviews.[7] Morhof favoured the bibliographical survey, adding a critical dimension to a genre which Conrad Gesner in the mid-sixteenth century had practised with as little selection as possible, explicitly leaving the exercise of judgement to the individual reader.[8] In his advice on constituting a library, Gabriel Naudé recommended the use of reference tools of various kinds – 'Common places, Dictionaries, Mixtures, several Lections [such as the miscellanies entitled *Lectiones antiquae* or *variae*], Collections of sentences and other like Repertories.' He explicitly countered attacks that might be levied against them: 'In earnest, for my part, I esteem these Collections extreamly profitable and necessary, considering [that] the brevity of our life and the multitude of things which we are now obliged to know, e're one can be reckoned amongst the number of learned men, do not permit us to do all of ourselves.'[9] Finally, at the most immediate level of coping with the books themselves, scholars and teachers on the one hand, and printers on the other responded by formalising methods of annotation and indexing so as to devise more effective finding devices and aids to memory.

Diligent readers had presumably always taken notes, although we are usually left with almost no clues as to the specific media and methods they used. Portable wax tablets, bits of parchment or slips of paper used for quick jottings are rarely preserved for our analysis; a few literary accounts survive, such as that of Pliny the Elder who was reportedly always ready to note down (or have a servant note down) a passage from a book or an observation.[10] The best preserved kinds of notes are those left by readers in the margins of manuscripts or books, although these can be hard to decipher, not only when they have been cropped in later rebinding, or written in small and difficult hands, but also when they rely on a system of symbols or initials devised by the annotator for a personal mnemonic use and therefore left without explanation.[11] Other useful sources include the scattered advice of pedagogues and scholars concerning note-taking, and the printed equivalents to personal notes, increasingly marketed from the sixteenth century on as shortcuts for harried readers – notably printed commonplace books, which offered quotations and examples from a wide range of authors sorted under thematic headings, and printed indexes which guided both first-time and repeat readers of a book to its themes and examples.

Indexing and commonplacing in manuscript and print

Humanist pedagogues formalised note-taking practices as aids to memory, adding to a well-developed medieval repertory of mnemonic techniques.[12] Their advice books recommended that schoolboys keep a notebook of commonplaces in which they copied out remarkable passages from their readings under appropriate headings (based, for example, on the topical or thematic content of the passage, or on its rhetorical utility), for later use in their compositions. Adult readers were advised to continue this practice and to accumulate notes from readings as well as lived experience, notably during travel.[13] Few commonplace books actually survive from the sixteenth or seventeenth centuries; those which do are generally devoted to rhetorical and moral rather than natural topics. Nonetheless one can witness quite clearly the first stage of this kind of note-taking in the marginal annotations often extant in books from this period, including books about nature. Perhaps the final, more laborious stage of copying the passages out into a notebook was not carried out as often as the pedagogues would have liked, and adult readers especially may have relied only on their annotations rather than a separate notebook to find their way back to an interesting passage. In any case, it is clear that the most common function of the marginal note in the early modern period was to indicate the topic or theme being treated in a passage, as one would in selecting a heading for a commonplace book, to make it possible to find and retrieve the passage later if desired. The annotations would thus serve as a kind of running index throughout the book.

At its most simple-minded an annotation of this kind would copy out into the margin a key word or expression found in the text itself to signal the topic under discussion. The practice was soon imitated in print: to relieve the reader of some of this task, printed editions often included printed marginal summaries (or 'manchettes') – these were distinct from the printed notes that provided a bibliographical reference as in a modern footnote.[14] After listing all the topics in the margins, a reader taking notes might then collect on the blank fly-leaves of his book the page references and topics of those passages of greatest interest to him, thus producing a personal and select index to the book. In a printed equivalent to this practice, then, the simplest kind of index offered ready-made in printed books consisted of an alphabetised collection of the marginal summaries, for example at the back of the book.[15] If performed strictly mechanically, however, this kind of indexing was often of little help in defining key words or concepts. A case in point is a short French treatise on earthquakes in which a one-page index collects the marginal summaries, but alphabetises them by the first word of the summary, although it is often only a conjunction ('que', 'si', etc) rather than a keyword; nonetheless, this index is so short that it can easily be read through in entirety, which is presumably how it was used.[16]

At its most sophisticated, on the other hand, the reader's work of assigning a heading to a passage of interest involved careful consideration of the many possible contexts in which the passage might be useful in the future and a personal decision as to the one or more headings under which to classify the passage. An analysis of Montaigne's reading notes indicates that he revised commonplace heading assignments he had already made and pursued unconventional themes in his choice of headings; his reading notes were so thoughtful that they enabled him to bring together diverse material in his *Essays* in surprising ways. Montaigne explicitly denied in his *Essays* that he had sewn together a patchwork of commonplaces although he acknowledged constant borrowing from other authors; his choice of passages to select and combine was so idiosyncratic and self-consciously novel as to preclude (in his mind at least) their being considered commonplaces.[17] Sophisticated printed indexes did not pursue the idiosyncratic in the way that Montaigne did, but instead offered multiple avenues to each passage, using cross-references and different thematic and topical headings to guide readers with varied interests to useful material.

For example, in a printed commonplace book, the *Theatrum humanae vitae* of Theodor Zwinger (in the posthumous and largest edition of 1604), one anecdote, like the *bon mot* of the Roman emperor Vitellius who upon smelling the odour of a corpse, declared that it was good to smell a dead enemy, but even better to smell a dead compatriot, is indexed under 'pleasure', 'smell', 'odour of corpse' and 'cruelty' in the *index titulorum*, under 'Vitellius Caesar' in the *index exemplorum* (an index of historical figures, in which the cross-references are most complete) and under 'enemy' and 'compatriot' in the *index rerum et verborum*. The anecdote is repeated at least four times in the nine-volume, thematically arranged work and a number of cross-references (although not all of them) are provided in the text and in the indexes.[18] While this system of multiple indexes may seem cumbersome, it was clearly designed to help readers approaching the work with different kinds of commonplace headings in mind to locate anecdotes and quotations of use to them; multiple indexes and index entries and cross-references all served to allow for the idiosyncrasy of any one choice of commonplace heading and to cater to many individual readers' interests.

Manuscript and printed commonplacing and indexing, as practised by readers in their annotations and by authors and/or printers in drawing up printed indexes, could have two rather opposite kinds of consequences. On the one hand, in an unsystematic distribution of headings with little cross-referencing, related material could easily become scattered under separate headings and different, even contradictory, conclusions could be drawn in different places without those contradictions being readily apparent. The method could thus harbour and tolerate considerable cognitive dissonance. On the other hand, sophisticated thematic indexing or commonplacing which cross-referenced and systematically compared material in related categories could bring material together in new ways, highlighting contradictions and interconnections and potentially yielding new insights. Indeed, one can recognise the outline of a method of commonplacing, applied to direct observation more than to bookish sources, in Francis Bacon's ideal of scientific investigation.[19] The first step in Bacon's *New Organon* is to collect and present to the understanding all known instances of a phenomenon, combined with instances where it is absent or present to varying degrees; the next step, induction, proceeds by the systematic confrontation of the material arranged in tables: the prerogative instances revealed in this confrontation help to reach flawless general principles. Bacon's ideal (although he never adequately carried it out) is a method of natural commonplaces derived from direct observation which abides by systematic guidelines for sorting gathered material and reaching higher levels of generalisation.[20]

Unsystematic commonplacing

In practice, commonplacing rarely yielded systematic results. Francis Bacon himself, in following his own precepts, left in his unfinished *Sylva sylvarum* a vast collection of 'facts' both bookish and directly observed, arranged under loose topical headings, in which experimentation and critical judgement coexist with what appear to us as credulity and obvious inconsistencies.[21] Nonetheless this work, usually published with the *New Atlantis*, was among the most widely reprinted of all Bacon's works. As an example of a finished work of natural philosophy displaying many of the same characteristics, I would cite the *Universae naturae theatrum* of Jean Bodin – an author best known for his political philosophy and for his *Method of History* in which he recommends commonplacing as a method of reading.[22] In the *Theatrum* Bodin discusses a vast array of natural historical 'facts', which he has accumulated mostly from his readings, but occasionally also from personal observations or second-hand reports. At times Bodin brings together disparate material to pursue a specific question, as if the material had been gathered under a commonplace heading (unfortunately neither annotations nor manuscripts of Bodin's survive). For example, the great heat of the (humid) summers of Muscovy, the common observation that fire burns hotter in wood than in straw and in metal than in wood, and the practice of the sauna in which the air is made humid in order to keep the heat better, lead Bodin to explain that 'the thickness of the air, excited by the vapour of the water, keeps the heat, while earlier it could not because of its fineness'.[23] Although Bodin never actually states the principle that denser things hold heat better than finer ones, he amasses an original array of material on that theme and lays the foundation for such a generalization.

On the other hand, Bodin's method of assigning material to topical and thematic headings is hardly systematic, so that at other times he reaches contradictory conclusions. For example, one careful contemporary reader, who diligently indexed the entire work by copying key words and expressions in the margin, points out that on p. 284 Bodin explains that grafted trees yield more and sweeter fruit because of the more abundant sap called up to repair the wound; but five pages earlier Bodin had explained that older trees yielded sweeter fruit precisely because they were less full of sap.[24] The reader rightly wonders whether it is the abundance or the absence of sap which causes sweetness in fruit. But Bodin never addresses that general principle; instead he provides contradictory explanations in the answers to two different questions, focused on grafted and on old trees respectively. Here Bodin's unsystematic choice of categories under which to group his material served to hide rather than uncover a thematic link. Commonplacing was a form of personal note-taking in view of gathering *copia* to supply a composition of one's own. It was thus primarily goal-oriented, often hastily performed, and as a result subject to the vicissitudes of the attention and interests of one reader-turned-author.

Increasingly systematic printed indexes

The printed index was born from the same initial motivation and note-taking practice as commonplacing (by flagging topics in the margin), to facilitate the retrieval of specific passages. In the course of the sixteenth century printed indexes became increasingly numerous, voluminous and systematic, offering a less idiosyncratic and more multi-purpose guide to a work than commonplace notes. The utility and power of indexes were acknowledged not only by readers who requested them and printers who supplied them, but also by censors who targeted them and authors who in indexing their own works used them as devices for highlighting features that might otherwise have remained hidden from view.[25] The index as we think of it – an alphabetical guide to the contents of a work – first appeared in the thirteenth century, as a work separate from the work being indexed, undertaken by a collectivity (e.g. of clerics) or an individual (notably, in some cases, funded by the pope) to facilitate access to a particularly important text or set of texts: such as the Bible, the texts of the Church fathers or Vincent of Beauvais' *Speculum historiale*.[26] By the fourteenth century, major works often included an index supplied by the author. In the sixteenth century separate indexes were the exception, available only for large, canonical works, such as the Bible, Pliny's *Natural histories* or the *Canon* of Avicenna.[27] But, judging from the sequence of signatures, indexes were often printed in a separate run, notably after the rest of the work had been paginated, and could be variously bound at the front or the back of a volume.[28]

The Latin term already existed in antiquity (designating the index finger or the tag bearing the title of a papyrus roll) and throughout the early modern period was never exclusively used in its modern meaning. Early indexes were often called 'tables' and conversely many 'indexes' (even in the sixteenth century) are what we would call tables of various kinds – lists of chapter headings or questions in order of their appearance in the book, or lists of authorities cited without reference to page numbers in the text. In addition the term continued to refer to other kinds of reference markers: among them, the pointing finger symbol used in manuscript and print to call attention to a passage (see Figures 4.1 and 4.2), or the hands on a calendrical or astronomical dial.[29] Indexes to manuscripts necessarily used layout-independent means of referring to

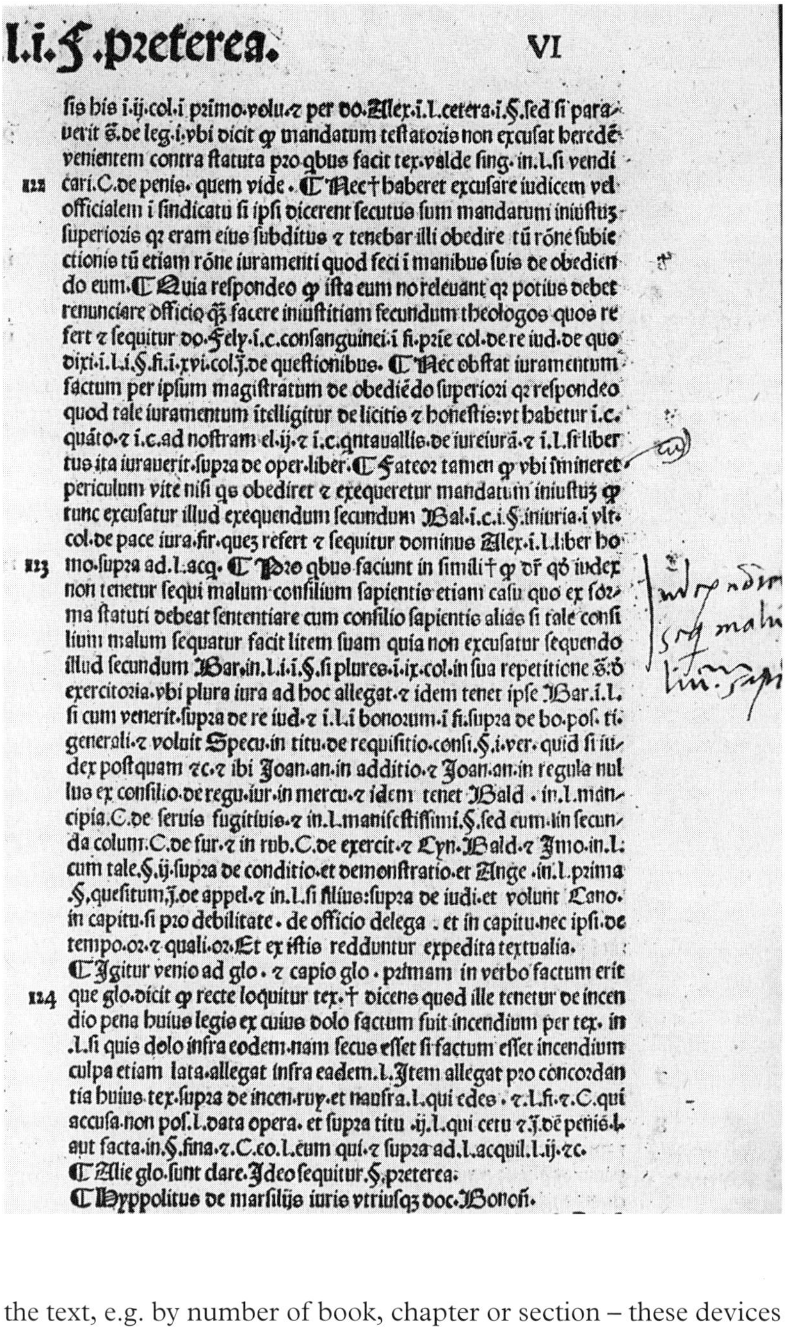

Figures 4.1 and 4.2 Pointing index fingers, either added by a reader alongside an annotation (Figure 4.1) or as part of the printed edition (Figure 4.2), used to highlight a subentry of the alphabetical index ('Repertorium') to the same work. From Hippolytus de Marsiliis, *Brassea* (Milan, 1522). (Reproduced by kind permission of the Houghton Library, Harvard University)

the text, e.g. by number of book, chapter or section – these devices appeared in the thirteenth century, as the Rouses have shown, at the same time as the first indexes, as part of the scholastic development of punctual consultation and precise referencing of texts. But as soon as printed texts came with foliation or pagination, and despite the added expense of modifying the references for every new edition of a work, most indexes provided page or folio numbers, often with

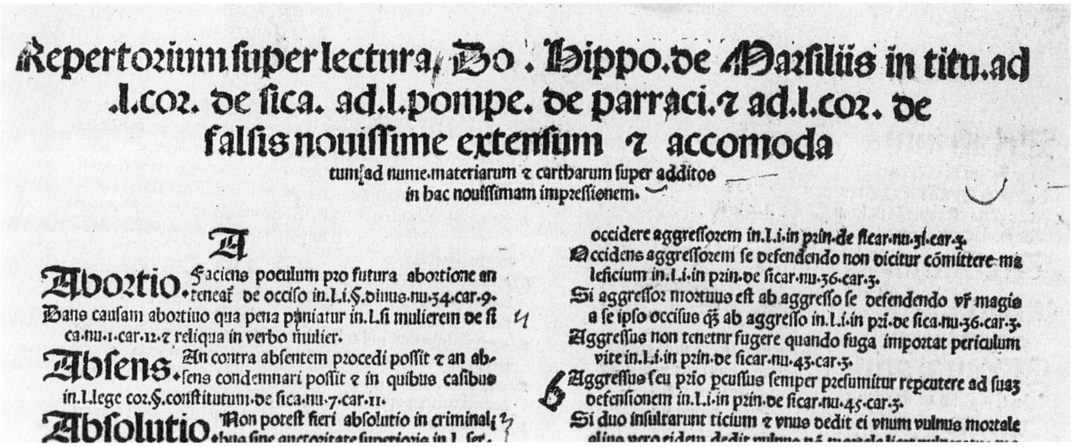

Repertorium super lectura, Bo. Hippo. de Marsiliis in titu. ad l. cor. de sica. ad l. pompe. de parraci. ꝛ ad l. cor. de falsis nouissime extensum ꝛ accomoda tum ad nume. materiarum ꝛ cartharum super additos in hac nouissimam impressionem.

A

Abortio. Faciens poculum pro futura abortione an teneat de occiso in. l. i. §. diuus. nu. 34. car. 9. Dans causam abortiuo qua pena puniatur in. l. si mulierem de sica. nu. 1. car. 12. ꝛ reliqua in verbo mulier.

Absens. An contra absentem procedi possit ꝛ an absens condemnari possit ꝛ in quibus casibus in. l. lege cor. §. constitutum. de sica. nu. 7. car. 11.

Absolutio. Non potest fieri absolutio in criminali ...

occidere aggressorem in. l. i. in prin. de sicar. nu. 31. car. 3.
Occidens aggressorem se defendendo non dicitur committere maleficium in. l. i. in prin. de sicar. nu. 36. car. 3.
Si aggressor mortuus est ab aggresso se defendendo vñ magis a se ipso occisus q̃ȝ ab aggresso in. l. i. in pri. de sica. nu. 36. car. 3.
Aggressus non tenetur fugere quando fuga importat periculum vite in. l. i. in prin. de sicar. nu. 43. car. 3.
Aggressus seu prio pcussus semper presumitur repcutere ad suaȝ defensionem in. l. i. in prin. de sicar. nu. 45. car. 3.
Si duo insultarunt ticium ꝛ vnus dedit ei vnum vulnus mortale ...

more specific references to a part of the page – to columns, the side of a folio (recto–verso), or to letters listed in the margins.

Figure 4.2

Printed indexes appeared as early as the 1480s, notably in herbaria which often provided an alphabetical index of plant names and an index of diseases to be treated, the latter usually classified from head to toe rather than alphabetically.[30] Early indexes are most striking today for their lack of strict alphabetisation. Medieval alphabetisation generally followed only the first two to three letters of a word and the sound rather than the spelling of a word (placing 'halcyon' under 'a' for example); early printed indexes often offered only a haphazard improvement on that practice. Starting in the second quarter of the sixteenth century some authors called for strict alphabetisation – most importantly Conrad Gesner, in his discussion of the significance and methods of indexing in 1548.[31] Early indexes also tended to neglect cross-referencing to alternative forms of an entry (e.g. an entry at 'ars imprimendi' but none at 'imprimendi ars')[32] and to list separately (and not always side by side) many different related occurrences of a concept or term (see Figure 4.3). As a result, while these indexes could lead to all kinds of interesting facts, they were not particularly reliable in providing a systematic mode of access to the text. Indeed one finds many indexes in early sixteenth-century editions, especially in miscellaneous genres where the text was self-consciously disordered and the alphabetical index constituted a primary mode of access to the text, bearing the annotations of dissatisfied readers. In extant copies of Caelius Rhodiginus' *Lectiones antiquae* or Erasmus' *Adages* contemporary readers corrected errors in page references, made additional entries, inserted at the proper place in the alphabetisation, or even made indexes of their own, of selected pages of the work, or of commonplace headings for example left in an arbitrary order in a printed 'index' of headings (in Erasmus' *Adages*).

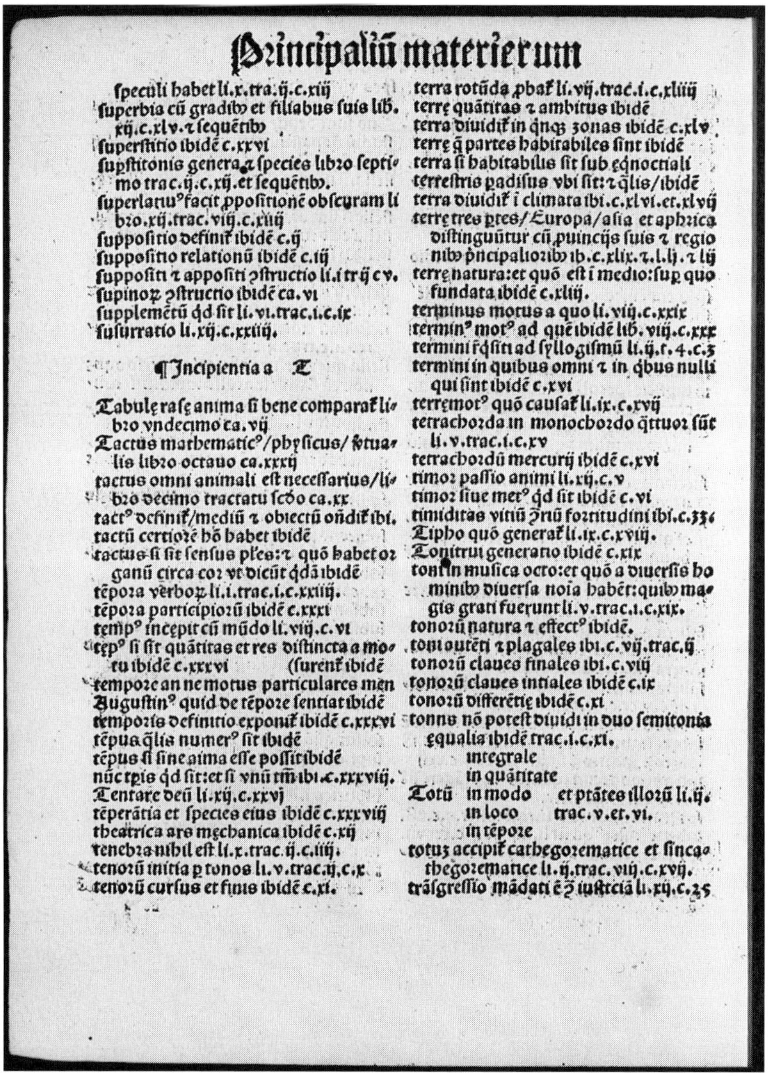

Figure 4.3 A page from the index to Gregor Reisch, *Margarita philosophica* (Strasburg, 1508; first published 1503, with the same index). Note the references to chapter rather than page or folio numbers (an uncommon practice in books with pagination or foliation like this one), the imperfect alphabetisation ('theatrica' between 'temperantia' and 'tenebra'), and the juxtaposition of many separate but related entries, e.g. for 'terra' (contrast with Figure 4.8 below) – as a result of the last two features, 'terraemotus' ends up separated from other 'terra' references by the 'terminus' entries. (Reproduced by kind permission of the Houghton Library, Harvard University)

Printers understood the demand of readers for indexes, even if they did not always satisfy it. Countless title-pages boast of a 'most complete' or 'augmented and corrected' index or indexes (see Figure 4.4). In some cases these boasts persisted on the title-page even when in the work itself the printer included an apology to the reader for the fact that the index was actually missing, notably for lack of time, or due to an outbreak of the plague, with the promise of including the index in the next edition.[33] Occasionally indexes were followed by a list of entries omitted.[34] These printers were aware that shortcomings in the index could be an irritant to readers and attempted to placate their audience in advance. Such statements indicate that the printer was often responsible for the index,

Annotating and indexing natural philosophy   79

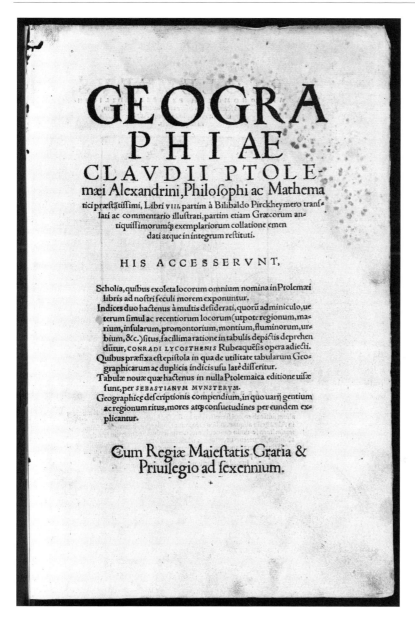

Figure 4.4 Title-page of Ptolemy's *Geography* (Basel, 1552), boasting of numerous additions including 'two indexes requested by many until now, through the use of which one can easily deduce the locations of sites both ancient and modern . . . added by the work of Conrad Lycosthenes'. (Reproduced by kind permission of the Houghton Library, Harvard University)

although we rarely know exactly who did the work. Only occasionally is the indexer identified, as in Figure 4.4. Account books reveal that printers employed their own correctors or typesetters for the task (which was paid separately) or on some occasions outside scholars.[35] In other cases authors no doubt supplied the index with the manuscript to be printed, as one letter accompanying a manuscript indicates.[36] Some indexes were preceded by a brief prefatory paragraph detailing the long nights and hard labour spent complet-

Figures 4.5 and 4.6 A comparison of the indexes to the German and Latin editions of Sebastian Münster's *Cosmographia*, both dated 1550. The Latin index is noticeably longer than the German and this disproportion remains even as both indexes become longer in subsequent editions. Comparison and illustrations from Frank Hieronymus, *1488 Petri – Schwabe 1988* (Basel: Schwabe, 1997), vol. I, pp. 624–5. (Reproduced by kind permission of Schwabe & Co., Basel; photograph by Ferranti-Dege, Cambridge)

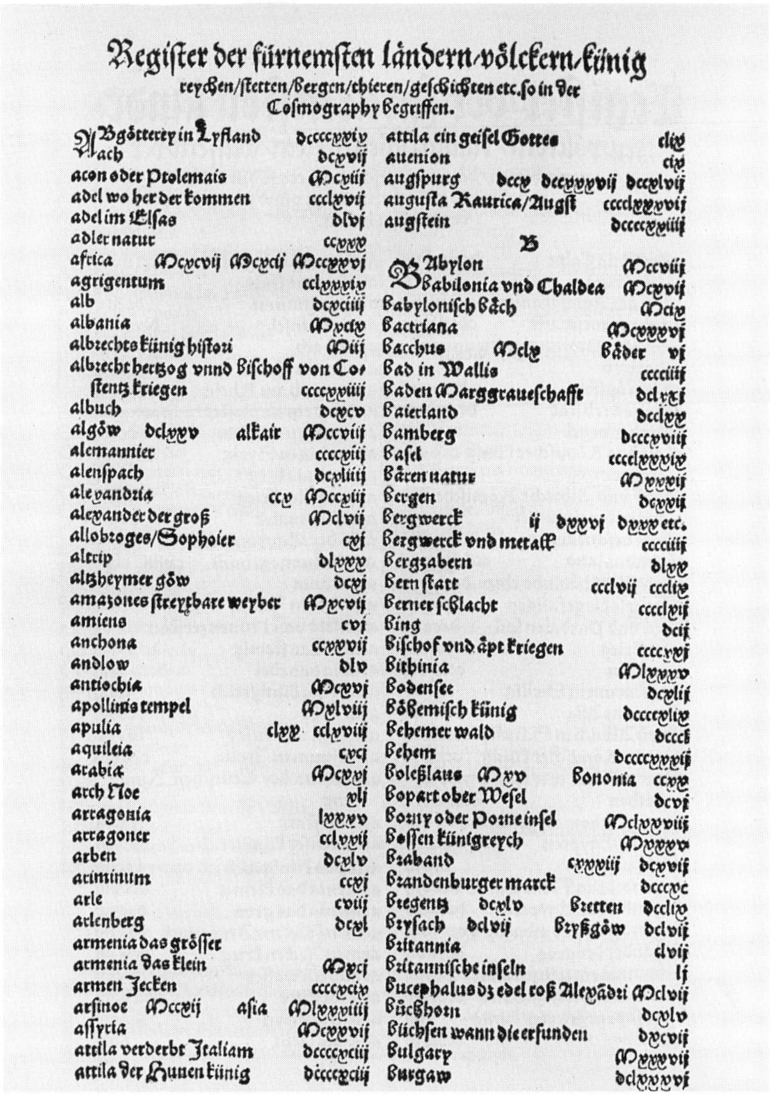

Register der fürnemsten ländern/völckern/künig
reychen/stetten/bergen/thieren/geschichten etc. so in der
Cosmography begriffen.

ing the index and explaining how to use it.[37] But we cannot always be sure who is speaking. The more prominent scholars, such as Erasmus, employed amanuenses who could have supplied much of the labour involved;[38] but they might also do the work themselves, as J. J. Scaliger did for a large collection of ancient inscriptions, not without complaining about the rigours of the task.[39]

The most important evidence that indexes were recognised to be useful and powerful tools was their growth: in the course of the sixteenth century, indexes consistently became more numerous, more copious and more systematic. From his database of 153 works of zoology published in the Renaissance, Laurent Pinon can show a steady rise in the percentage of books of zoology which are indexed,

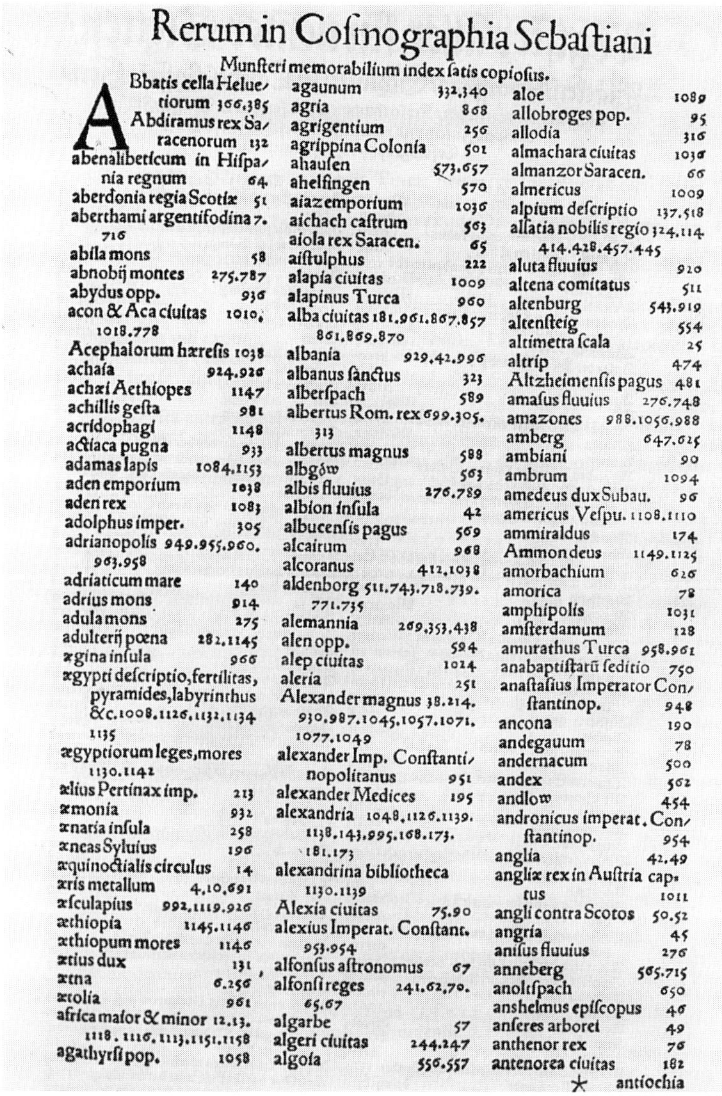

Rerum in Coſmographia Sebaſtiani
Munſteri memorabilium index ſatis copioſus.

ABbatis cella Heluetiorum 366.385
Abdiramus rex Saracenorum 132
abenalibericum in Hiſpania regnum 64
aberdonia regia Scotiæ 51
aberthami argentifodina 7.716
abila mons 58
abnobij montes 275.787
abydus opp. 936
acon & Aca ciuitas 1010.1018.778
Acephalorum hæreſis 1038
achaia 924.926
achæi Aethiopes 1147
achillis geſta 981
acridophagi 1148
actiaca pugna 933
adamas lapis 1084.1153
aden emporium 1038
aden rex 1083
adolphus imper. 305
adrianopolis 949.955.960.963.958
adriaticum mare 140
adrius mons 914
adula mons 275
adulterij pœna 282.1145
ægina inſula 966
ægypti deſcriptio, fertilitas, pyramides, labyrinthus &c. 1008.1126.1132.1134.1135
ægyptiorum leges, mores 1130.1142
ælius Pertinax imp. 213
æmonia 932
ænaria inſula 258
æneas Syluius 196
æquinoctialis circulus 14
æris metallum 4.10.691
æſculapius 992.1119.926
æthiopia 1145.1146
æthiopum mores 1146
ætius dux 131
ætna 6.256
ætolia 961
africa maior & minor 12.13.1118.1116.1113.1151.1158
agathyrſi pop. 1058
agaunum 332.340
agria 868
agrigentium 256
agrippina Colonia 501
ahauſen 573.657
ahelfingen 570
aiaz emporium 1036
aichach caſtrum 563
aiola rex Saracen. 65
aiſtulphus 222
alapia ciuitas 1009
alapinus Turca 960
alba ciuitas 181.961.867.857.861.869.870
albania 929.42.996
albanus ſanctus 323
alberſpach 589
albertus Rom. rex 699.305.313
albertus magnus 588
albgöw 563
albis fluuius 276.789
albion inſula 42
albucenſis pagus 569
alcairum 968
alcoranus 412.1038
aldenborg 511.743.718.739.771.735
alemannia 269.353.438
alen opp. 594
alep ciuitas 1014
aleria 251
Alexander magnus 38.214.930.987.1045.1057.1071.1077.1049
alexander Imp. Conſtantinopolitanus 951
alexander Medices 195
alexandria 1048.1126.1139.1138.143.995.168.173.181.173
alexandrina bibliotheca 1130.1139
Alexia ciuitas 75.90
alexius Imperat. Conſtant. 953.954
alfonſus aſtronomus 67
alfonſi reges 241.62.70.65.67
algarbe 57
algeri ciuitas 244.247
algoia 556.557
aloe 1089
allobroges pop. 95
allodia 316
almachara ciuitas 1030
almanzor Saracen. 66
almericus 1009
alpium deſcriptio 137.518
alſatia nobilis regio 324.114.414.428.457.445
aluta fluuius 920
altena comitatus 511
altenburg 543.919
altenſteig 554
altimetra ſcala 25
altrip 474
Altzheimenſis pagus 481
amaſus fluuius 276.748
amazones 988.1056.988
amberg 647.625
ambiani 90
ambrum 1094
amedeus dux Subau. 96
americus Veſpu. 1108.1110
ammiraldus 174
Ammon deus 1149.1125
amorbachium 626
amorica 78
amphipolis 1031
amſterdamum 128
amurathus Turca 958.961
anabaptiſtarū ſeditio 750
anaſtaſius Imperator Conſtantinop. 948
ancona 190
andegauum 78
andernacum 500
andex 562
andlow 454
andronicus imperat. Conſtantinop. 954
anglia 42.49
angliæ rex in Auſtria captus 1011
angli contra Scotos 50.52
angria 45
aniſus fluuius 276
anneberg 565.715
anolſpach 650
anshelmus epiſcopus 46
anſeres arborei 49
anthenor rex 76
antenorea ciuitas 182
antiochia

Figure 4.6

from 0 per cent of incunabula, to 31 per cent and 36 per cent of books published in 1501–30 and 1531–50, to 69 per cent of books published in 1551–70 and 1571–90 and 60 per cent of books published in 1591–1605.[40] Indexes clearly added to the expense of producing a book, and as a result tended to be longer in the Latin than in the vernacular (see Figures 4.5 and 4.6) and in the folio than in the octavo versions of the same work – in other words, in editions targeted to an audience generally less sensitive to price than to quality.[41] Nonetheless, by the second half of the sixteenth century indexes had spread to almost all genres of natural philosophy, from large and often disorganised compilations of material (e.g. the miscellanies of della Porta, Cardano or Scaliger; or the *Opera omnia* of ancient, medieval and recent authors, such as Pliny, Duns Scotus or Ficino) to shorter and more methodical genres such as academic textbooks and treatises (e.g. by Melanchthon, de Soto or Clemens Timpler). Indexes to the same work also grew steadily longer in successive editions; instead of covering only the main points under discussion, indexes increasingly included names and topics mentioned in passing or in digressions.

Finally, indexes became more systematic, due to a number of concurrent trends. By the end of the sixteenth century indexes were generally strictly alphabetised and more detailed, with multiple entries for one item and cross-references to related entries. Related entries were generally consolidated under one main entry divided into subheadings (as is common today). Whereas a number of early indexes consisted of separate sections by language or topic (see Figure 4.7), the trend towards a single topical index offered greater opportunities for bringing together disparately located but related material. The index could be used precisely to juxtapose divergent treatments of a topic, inviting readers to be aware of potential contradictions and comparisons. Thus in the *Encyclopedia* of Johann Heinrich Alsted (1630) the entry for 'terra' refers the reader in quick succession to a standard Aristotelian description of the Earth as stationary and heavy (under 'physica'), to geographical discussions of its divisions and parts, and suddenly to the question of whether the Earth moves (see Figure 4.8). From the main 'physics' section in the second volume the reader would never know that Alsted addresses the heated question of the Copernican hypothesis; indeed Alsted does so (with no cross-reference in the text) only at the end of the four-volume work, in a brief section of the miscellaneous 'farrago disciplinarum', entitled 'paradoxologia'. There he discusses the Copernican hypothesis (in one paragraph) with intelligence and respect, as a paradox the truth of which cannot easily be refuted. But Alsted breaks off rapidly, leaving to others to study the question in more depth.

In this case the index performs work that even an attentive reader might not have managed – to span some 4,000 folio pages in search of related material in disparate locations. For Alsted the mission of the *Encyclopedia* was a grand synthesis of all disciplines and schools of thought, and one of his main strategies of synthesis, as others have noted, was the simple juxtaposition of philosophical alternatives, as if the alternatives were not contradictory but could be reconciled by being brought together in one work.[42] However weak this strategy seems to us, it included as one of its elements the single, large alphabetical index at the end of the work, which alone made the reader aware of the presence of different treatments of a single topic. Readers were then left to their own devices in consult-

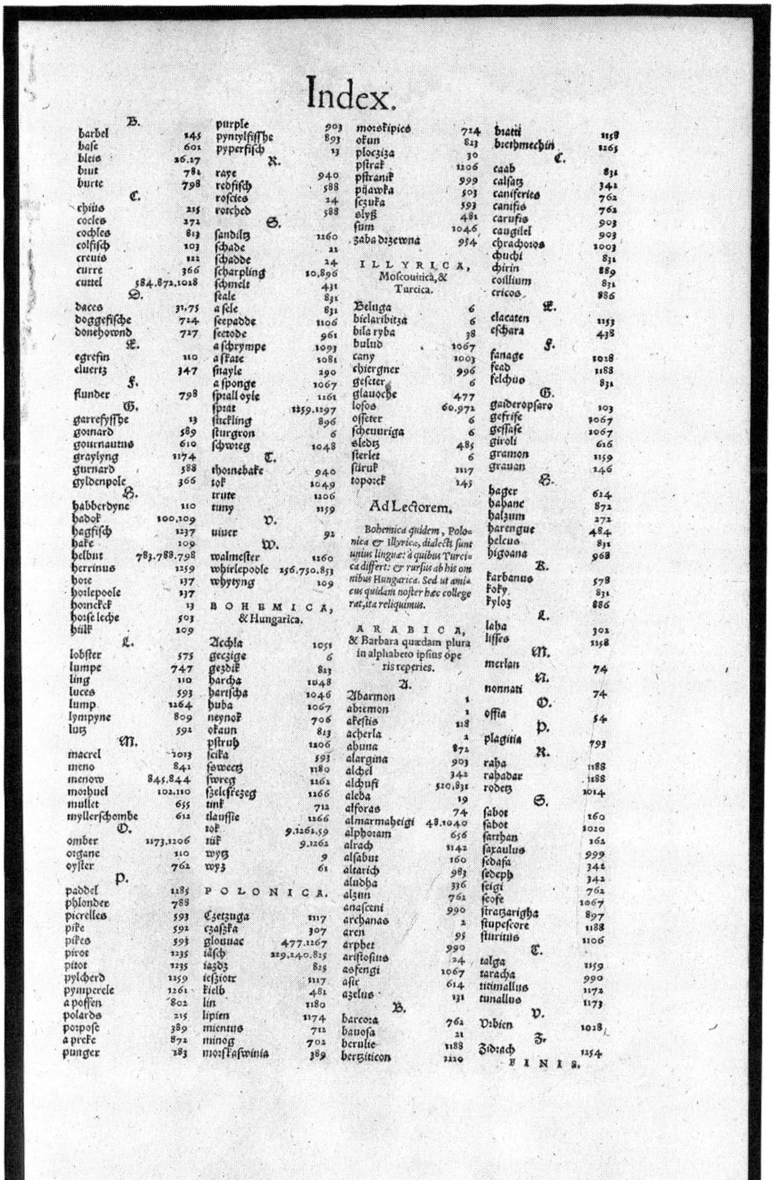

Figure 4.7 A page from the indexes in Conrad Gesner, *Historia animalium* (Zurich, 1551), showing the fragmentation by subject and language. This index covers fish only, as opposed to quadrupeds, birds or serpents. The page opens with English names, before moving on, in separately alphabetised lists, to names of species in Bohemian and Hungarian; Polish; Illyrian, Muscovite and Turkish (with an apology for merging three such different languages); and Arabic. (Reproduced by kind permission of the Houghton Library, Harvard University)

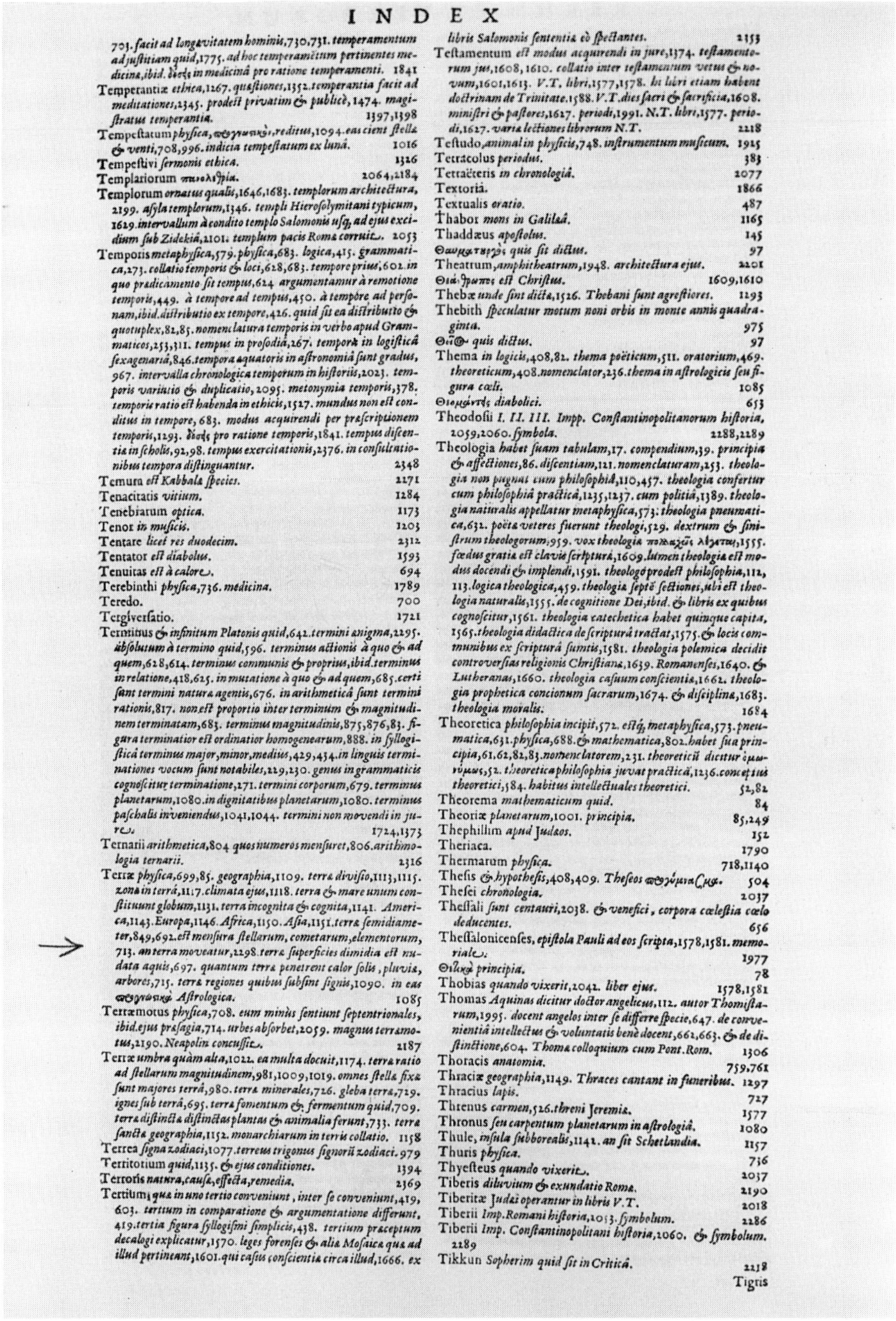

INDEX

703. *facit ad longævitatem hominis*, 730, 731. *temperamentum ad justitiam quid*, 1775. *ad hoc temperamẽtum pertinentes medicina, ibid.* δόσις *in medicinâ pro ratione temperamenti.* 1841

Temperantiæ *ethica*, 1267. *quæstiones*, 1352. *temperantia facit ad meditationes*, 2345. *prodest privatim & publicè*, 1474. *magistratus temperantia.* 1397, 1398

Tempestatum *physica*, προγνωστικὰ, *reditus*, 1094. *eas cient stellæ & venti*, 708, 996. *indicia tempestatum ex lunâ.* 1016

Tempestivi *sermonis ethica.* 1326

Templariorum πολιτεία. 2064, 2184

Templorum *ornatus qualis*, 1646, 1683. *templorum architectura*, 2199. *asyla templorum*, 1346. *templi Hierosolymitani typicum*, 1629. *intervallum à condito templo Salomonis usq; ad ejus excidium sub Zidekiâ*, 2101. *templum pacis Romæ corruit.* 2053

Temporis *metaphysica*, 579. *physica*, 683. *logica*, 415. *grammatica*, 273. *collatio temporis & loci*, 628, 683. *tempore prius*, 602. *in quo prædicamento sit tempus*, 624 *argumentamur à remotione temporis*, 449. *à tempore ad tempus*, 450. *à tempore ad personam, ibid. distributio ex tempore*, 426. *quid sit ea distributio & quotuplex*, 82, 85. *nomenclatura temporis in verbo apud Grammaticos*, 253, 311. *tempus in prosodia*, 267. *tempora in logisticâ sexagenariâ*, 846. *tempora æquatoris in astronomiâ sunt gradus*, 967. *intervalla chronologica temporum in historiis*, 2023. *temporis variatio & duplicatio*, 2095. *metonymia temporis*, 378. *temporis ratio est habenda in ethicis*, 1527. *mundus non est conditus in tempore*, 683. *modus acquirendi per præscriptionem temporis*, 1293. δόσις *pro ratione temporis*, 1841. *tempus discentiæ in scholis*, 92, 98. *tempus exercitationis*, 2376. *in consultationibus tempora distinguantur.* 2348

Temura *est Kabbalæ species.* 2171

Tenacitatis *vitium.* 1284

Tenebrarum *optica.* 1173

Tenor *in musicis.* 1203

Tentare *licet res duodecim.* 2312

Tentator *est diabolus.* 1593

Tenuitas *est à calore.* 694

Terebinthi *physica*, 736. *medicina.* 1789

Teredo. 700

Tergiversatio. 1721

Terminus & *infinitum Platonis quid*, 642. *termini ænigma*, 2295. *absolutum à termino quid*, 596. *terminus actionis à quo & ad quem*, 628, 614. *terminus communis & proprius, ibid. terminus in relatione*, 418, 625. *in mutatione à quo & ad quem*, 685. *certi sunt termini naturæ agentis*, 676. *in arithmeticâ sunt termini rationis*, 817. *non est proportio inter terminum & magnitudinem terminatam*, 683. *terminus magnitudinis*, 875, 876, 83. *figura terminatior est ordinatior homogenearum*, 888. *in syllogisticâ terminus major, minor, medius*, 429, 434. *in linguis terminationes vocum sunt notabiles*, 229, 230. *genus in grammaticis cognoscitur terminatione*, 271. *termini corporum*, 679. *terminus planetarum*, 1080. *in dignitatibus planetarum*, 1080. *terminus paschalis inveniendus*, 1041, 1044. *termini non movendi in jure.* 1724, 1373

Ternarii *arithmetica*, 804 *quos numeros mensuret*, 806. *arithmologia ternarii.* 2316

Terræ *physica*, 699, 85. *geographia*, 1109. *terræ divisio*, 1113, 1115. *zonæ in terrâ*, 1117. *climata ejus*, 1118. *terra & mare unum constituunt globum*, 1131. *terra incognita & cognita*, 1141. *America*, 1143. *Europa*, 1146. *Africa*, 1150. *Asia*, 1151. *terræ semidiameter*, 849, 692. *est mensura stellarum, cometarum, elementorum*, 713. *an terra moveatur*, 2298. *terræ superficies dimidia est nudata aquis*, 697. *quantum terræ penetrent calor solis, pluvia, arbores*, 715. *terræ regiones quibus subsint signis*, 1090. *in eas* προγνωστικὰ *Astrologica.* 1085

Terræmotus *physica*, 708. *eum minùs sentiunt septentrionales, ibid. ejus præsagia*, 714. *urbes absorbet*, 2059. *magnus terræmotus*, 2190. *Neapolin concussit.* 2187

Terræ *umbra quàm alta*, 1022. *ea multa docuit*, 1174. *terræ ratio ad stellarum magnitudinem*, 981, 1009, 1019. *omnes stellæ fixæ sunt majores terrâ*, 980. *terræ minerales*, 726. *gleba terræ*, 729. *ignes sub terrâ*, 695. *terræ fomentum & fermentum quid*, 709. *terræ distinctæ distinctas plantas & animalia ferunt*, 733. *terræ sanctæ geographia*, 1152. *monarchiarum in terris collatio.* 1158

Terrea *signa zodiaci*, 1077. *terreus trigonus signorũ zodiaci.* 979

Territorium *quid*, 1135. & *ejus conditiones.* 1394

Terroris *natura, causa, effecta, remedia.* 2369

Tertium; *quæ in uno tertio conveniunt, inter se conveniunt*, 419, 603. *tertium in comparatione & argumentatione differunt*, 419. *tertia figura syllogismi simplicis*, 438. *tertium præceptum decalogi explicatur*, 1570. *leges forenses & alia Mosaica quæ ad illud pertineant*, 1601. *qui casus conscientiæ circa illud*, 1666. *ex libris Salomonis sententiæ eò spectantes.* 2153

Testamentum *est modus acquirendi in jure*, 1374. *testamentorum jus*, 1608, 1610. *collatio inter testamentum vetus & novum*, 1601, 1613. *V.T. libri*, 1577, 1578. *hi libri etiam habent doctrinam de Trinitate.* 1588. *V.T. dies sacri & sacrificia*, 1608. *ministri & pastores*, 1627. *periodi*, 1991. *N.T. libri*, 1577. *periodi*, 1627. *variæ lectiones librorum N.T.* 2218

Testudo, *animal in physicis*, 748. *instrumentum musicum.* 1925

Tetracolus *periodus.* 383

Tetraëteris *in chronologiâ.* 2077

Textoria. 1866

Textualis *oratio.* 487

Thabor *mons in Galilæâ.* 1165

Thaddæus *apostolus.* 145

Θαυματουργὸς *quis sit dictus.* 97

Theatrum, *amphitheatrum*, 1948. *architectura ejus.* 2201

Θεάνθρωπος *est Christus.* 1609, 1610

Thebæ *unde sint dictæ*, 1526. *Thebani sunt agrestiores.* 1293

Thebith *speculatur motum noni orbis in monte annis quadraginta.* 975

Θεῖος *quis dictus.* 97

Thema *in logicis*, 408, 82. *thema poëticum*, 511. *oratorium*, 469. *theoreticum*, 408. *nomenclator*, 236. *thema in astrologicis seu figura cœli.* 1085

Θεομάντεις *diabolici.* 653

Theodosii *I. II. III. Impp. Constantinopolitanorum historia*, 2059, 2060. *symbola.* 2288, 2289

Theologia *habet suam tabulam*, 17. *compendium*, 39. *principia & affectiones*, 86. *discentiam*, 121. *nomenclaturam*, 253. *theologia non pugnat cum philosophiâ*, 110, 457. *theologia confertur cum philosophiâ practicâ*, 1235, 1237. *cum politiâ*, 1389. *theologia naturalis appellatur metaphysica*, 573. *theologia pneumatica*, 632. *poëtæ veteres fuerunt theologi*, 529. *dextrum & sinistrum theologorum*, 959. *vox theologiæ* πολλαχῶς λέγεται, 1555. *fœdus gratiæ est clavis scripturæ*, 1609. *lumen theologiæ est modus docendi & implendi*, 1591. *theologo prodest philosophia*, 112, 113. *logica theologica*, 459. *theologiæ septẽ sectiones, ubi est theologia naturalis*, 1555. *de cognitione Dei, ibid.* & *libris ex quibus cognoscitur*, 1561. *theologia catechetica habet quinque capita*, 1565. *theologia didactica de scripturâ tractat*, 1575. & *locis communibus ex scripturâ sumtis*, 1581. *theologia polemica decidit controversias religionis Christianæ*, 1639. *Romanenses*, 1640. & *Lutheranas*, 1660. *theologia casuum conscientiæ*, 1662. *theologia prophetica concionum sacrarum*, 1674. & *disciplina*, 1683. *theologia moralis.* 1684

Theoretica *philosophia incipit*, 572. *estq; metaphysica*, 573. *pneumatica*, 631. *physica*, 688. & *mathematica*, 802. *habet sua principia*, 61, 62, 82, 83. *nomenclatorem*, 231. *theoreticũ dicitur* ὁμωνύμως, 52. *theoretica philosophia juvat practicã*, 1236. *conceptus theoretici*, 584. *habitus intellectuales theoretici.* 52, 82

Theorema *mathematicum quid.* 84

Theoriæ *planetarum*, 1001. *principia.* 85, 249

Thephillim *apud Judæos.* 152

Theriaca. 1790

Thermarum *physica.* 718, 1140

Thesis & *hypothesis*, 408, 409. *Theseos* προγύμνασμα. 504

Thesei *chronologia.* 2037

Thessali *sunt centauri*, 2038. & *venefici, corpora cœlestia cœlo deducentes.* 656

Thessalonicenses, *epistola Pauli ad eos scripta*, 1578, 1581. *memoriale.* 1977

Θιλικὰ *principia.* 78

Thobias *quando vixerit*, 2042. *liber ejus.* 1578, 1581

Thomas *Aquinas dicitur doctor angelicus*, 112. *autor Thomistarum*, 1995. *docent angelos inter se differre specie*, 647. *de convenientiâ intellectus & voluntatis benè docent*, 662, 663. & *de distinctione*, 604. *Thomæ colloquium cum Pont. Rom.* 1306

Thoracis *anatomia.* 759, 761

Thraciæ *geographia*, 1149. *Thraces cantant in funeribus.* 1297

Thracius *lapis.* 727

Threnus *carmen*, 526. *threni Jeremiæ.* 1577

Thronus *seu carpentum planetarum in astrologiâ.* 1080

Thule, *insula subborealis*, 1141. *an sit Schetlandia.* 1157

Thuris *physica.* 736

Thyesteus *quando vixerit.* 2037

Tiberis *diluvium & exundatio Romæ.* 2190

Tiberitæ *Judæi operantur in libris V.T.* 2018

Tiberii *Imp. Romani historia*, 2053. *symbolum.* 2286

Tiberii *Imp. Constantinopolitani historia*, 2060. & *symbolum.* 2289

Tikkun *Sopherim quid sit in Criticâ.* 2218

Tigris

Figure 4.8 A page from the index to Johann Heinrich Alsted, *Encyclopedia* (Herborn, 1630; facsimile Stuttgart-Bad Cannstatt: Frommann-Holzboog, 1990), showing the grouping of subentries under major headings, including the reference to the motion of the earth ('an terra moveatur') under 'Terra'. Contrast with Figure 4.3. (Reproduced by kind permission of Frommann-Holzboog; photograph by Ferranti-Dege, Cambridge) ing, reconciling or choosing between alternatives which Alsted diligently reported and juxtaposed in the index.

During the 150 years preceding Alsted's *Encyclopedia*, the index had developed into an increasingly widespread and systematic tool for sorting and accessing the pieces of information in a text. Unlike commonplace indexing and note-taking which were practised by individuals with idiosyncratic and specific goals in mind, the printed index, clearly a selling point with readers, was designed to cater to as many different interests as possible. As a result indexes grew steadily longer and more detailed, less idiosyncratic and personal. Whereas personal commonplacing easily left contradictory material buried under different headings, systematic indexing brought to light potential contradictions which might otherwise have passed unnoticed. Although the tolerance for cognitive dissonance characteristic of much medieval and Renaissance natural philosophy[43] is still evident in Alsted's practice of synthesis as uncritical juxtaposition, Alsted's index guided a reader to the theoretical alternatives competing in 1630 in a way that personal note-taking might not have.

Commonplacing remained a central practice of bookish disciplines into the nineteenth century, prompting John Locke, for example, to publish in 1686 a 'new method of commonplaces' in which he explained how one could keep track of entries continued on non-consecutive pages of a notebook through the use of an alphabetical index at the front of the notebook.[44] In the meantime, however, natural philosophy relied less and less on bookish methods of research. Furthermore, from the eighteenth century on, commonplace books (such as those of George Berkeley or Thomas Jefferson) typically took the form of a diary of readings arranged in the order in which the texts were encountered rather than according to the systematic thematic order recommended by humanist pedagogues. The latter function had been effectively taken over by the printed index, which performed it more successfully.

Notes

In addition to the debts acknowledged in the notes I am grateful to Peter Burke, Anthony Grafton, Richard Yeo and the editors of the volume for helpful comments on earlier drafts. This research was partially funded by a Radcliffe Junior Faculty Fellowship at the Bunting Institute of Radcliffe College.

1 See Anthony Grafton, 'The availability of ancient works', in Charles Schmitt et al. (eds.), *The Cambridge History of Renaissance Philosophy* (Cambridge: Cambridge University Press, 1988), pp. 767–91.
2 On the impact of information overload in botany, see Brian W. Ogilvie, 'Encyclopaedism in renaissance botany: from historia to pinax', in Peter Binkley (ed.), *Pre-modern Encyclopaedic Texts. Proceedings of the*

Second COMERS Congress, Groningen, 1–4 July 1996 (Leiden: Brill, 1997), pp. 89–99.
3 See for example Rabelais, *Pantagruel*, ch. 8.
4 Adrien Baillet, *Jugements des savans sur les principaux ouvrages des auteurs* (Amsterdam: aux dépens de la Compagnie, 1725), 'avertissement de l'auteur', p. xi.
5 Daniel Georg Morhof, *Polyhistor*, ed. Johannes Moeller (Lubeck: sumtibus Petri Böckmanni, 1732), I, IV, 2 #6 and I, I, 22 #23.
6 For examples of some early opposition to printing, see Martin Lowry, *The World of Aldus Manutius: Business and Scholarship in Renaissance Venice* (Oxford: Basil Blackwell, 1979), pp. 26–41. On censorship, see John Monfasani, 'The first call for press censorship: Niccolò Perotti, Giovanni Andrea Bussi, Antonio Moreto and the editing of Pliny's *Natural History*', *Renaissance Quarterly*, 41 (1988): 1–43; Martin Davies, 'Making sense of Pliny in the Quattrocento', *Renaissance Studies*, 9 (1995): pp. 240–257; and Luigi Firpo, 'The flowering and withering of speculative philosophy – Italian philosophy and the Counter Reformation: the condemnation of Francesco Patrizi', in E. Cochrane (ed.), *The Late Italian Renaissance 1525–1630* (New York: Macmillan, 1970), pp. 266–84.
7 See, for example, the 'accounts of books' present in the *Philosophical Transactions*.
8 Conrad Gesner, *Bibliotheca universalis* (Zurich: Froschauer, 1545), sig. *3v.
9 Gabriel Naudé, *Advis pour dresser une bibliothèque* (1627), trans. John Evelyn as *Instructions concerning erecting of a library* (1661) (repr. Cambridge: Houghton, Mifflin and Co., 1903), pp. 59–61.
10 Pliny the Younger, *Epistles*, III, 5.
11 For an example, see Anthony Grafton, 'How Guillaume Budé read his Homer', in *Commerce with the Classics: Ancient Books and Renaissance Readers* (Ann Arbor: University of Michigan Press, 1997), pp. 135–84.
12 See Mary Carruthers, *The Book of Memory: A Study of Memory in Medieval Culture* (Cambridge: Cambridge University Press, 1990).
13 See, for example, Johann Heinrich Alsted, *Orator* (Herborn: G. Corvinus, 1616), pp. 302–3; also Justin Stagl, *A History of Curiosity: The Theory of Travel 1550–1800* (Chur: Harwood Academic Publishers, 1995), pp. 50, 121.
14 On the development of notes of various kinds, see Anthony Grafton, *The Footnote: A Curious History* (Cambridge, Mass.: Harvard University Press, 1997).
15 For an example of this practice see Kepler's reading of Plutarch's *On the Face in the Orb of the Moon*, as discussed in Anthony Grafton, *Commerce with the Classics*, p. 210. On this method of constructing early indexes, see Francis J. Witty, 'Early indexing techniques: a study of several book indexes of the fourteenth, fifteenth and early sixteenth centuries', *The Library Quarterly*, 35 (1965): 141–8.
16 Nicolas de Livre, *Discours du tremblement de terre* (Paris: Duval, 1575).
17 See Francis Goyet, 'A propos de "ces pastissages de lieux communs" (le rôle des notes de lecture dans la genèse des *Essais*)', *Bulletin de la Société des Amis de Montaigne*, 5–6 (1986): 11–26 and 7–8 (1987):

9–30. For a detailed presentation of some of his reading notes, see M. A. Screech, *Montaigne's Annotated Copy of Lucretius. A Transcription and Study of the Manuscript, Notes and Pen-marks* (Geneva: Droz, 1998). See Montaigne, *Essais*, III, 12 (Villey ed., p. 1055).

18 Theodor Zwinger, *Theatrum Humanae Vitae* (Basel: Henricpetri, 1604), pp. 70, 120, 280, 2738 and 2755 respectively, and indexes at back.

19 For Bacon's favourable assessment of commonplaces, see Francis Bacon, *The Advancement of Learning*, ed. Arthur Johnston (Oxford: Clarendon Press, 1974), Book II, ch. 14, pp. 129–30.

20 For a similar point see Paolo Rossi, *Francis Bacon, from Magic to Science* (Chicago: University of Chicago Press, 1968), pp. 207–14.

21 On this understudied text, see Graham Rees, 'An unpublished manuscript by Francis Bacon: *Sylva Sylvarum* drafts and other working notes', *Annals of Science*, 38 (1981): pp. 377–412.

22 See Jean Bodin, *Method for the Easy Comprehension of History*, trans. Beatrice Reynolds (New York: Octagon Books, 1966), ch. 3. For more on Bodin's *Theatrum*, see the references in 'Further reading'.

23 Jean Bodin, *Universae naturae theatrum* (Frankfurt: Wechel, 1597), p. 212.

24 These annotations are reproduced in full in the appendix to my 1990 Princeton Ph.D. thesis: 'Restaging Jean Bodin: the *Universae naturae theatrum* in historical context', pp. 554–609, notably p. 577, annotations to p. 284. Cf. Bodin, *Theatrum*, pp. 279, 284.

25 For an example of the censoring of an index, in which references to a Protestant historian are blacked out of the index to Sebastian Münster's *Cosmographia*, see Frank Hieronymus, *1488 Petri–Schwabe 1988: eine traditionsreiche Basler Offizin im Spiegel ihrer frühen Drucke*, 2 vols. (Basel: Schwabe and Co., 1997), #209, p. 718. I am grateful to Anthony Grafton for recommending this wonderfully rich bibliographical source.

26 See the references to works by Mary and Richard Rouse in 'Further reading'.

27 See Konrad Pellicanus, *Index bibliorum* (1537); Johannes Camers, *Prima/secunda pars Pliniani indicis* (1525); or Julius Palamede, *Index in Avicennae libros nuper venetiis editos* (1558) – I am grateful to Nancy Siraisi for this reference.

28 For examples, see Hieronymus, I, #14 and II, #525.

29 See Hieronymus, II, #378, illustration p. 1087.

30 Hans H. Wellisch, 'Early multilingual and multiscript indexes in herbals', *The Indexer*, 11 (1978): 81–102, at 83–4.

31 Conrad Gesner, *Pandectae* (Zurich: Froschauer, 1548), book I, titulus 13 (de grammatica), part 2 (de indicibus), ff. 19r–20v; for an English translation and discussion of this passage see Hans H. Wellisch, 'How to make an index – 16th century style: Conrad Gessner on indexes and catalogs', *International Classification*, 8 (1981): 10–15; I am grateful to Sachiko Kusukawa for this reference.

32 In Hartmann Schedel, *Liber Chronicarum* (1493), as cited in Witty, 'Early indexing techniques', p. 143.

33 See Hieronymus, II, #342, p. 993; and Theodor Zwinger, *Theatrum vitae humanae* (Basel: Oporinus, 1565), 'Typographus ad lectorem'.

34 See, for example, an edition from the series of works 'ad usum delphini', designed for the instruction of the dauphin (son of Louis XIV), which consistently included multiple and copious indexes. A list of 'omissa' is appended to the index in Pliny, *Naturalis historiae libri 37*, ed. Jean Hardouin (Paris: Franciscus Muguet, 1685), tome 5, 1011ff. I am grateful to Nicholas Dew for the suggestion to consult this pedagogically oriented and very successful series of editions with indexes.

35 See *Das Rechnungsbuch der Froben und Episcopius 1557–1564*, ed. Rudolf Wackernagel (Basel: Benno Schwabe, 1881), e.g. p. 58 – corrector Bartolomaeus Varolle was paid for indexing a *Speculum iuris*; or p. 41 – the theologian Ulrich Coccius was paid for indexing Eusebius, *Ecclesiasticae historiae autores*.

36 Hieronymus, II, #374, p. 1071: Natale Conti, author of a 1556 Latin translation of Athenaeus' *Deipnosophistae*, accompanied the manuscript to be printed with a letter asking that the index he supplied be expanded and a list of authorities added.

37 See, for example, Camers, *Prima pars Plyniani indicis* and Erasmus, 'De duo indicibus', in editions of the *Adages* starting with the (posthumous) edition of 1541 (Basel: Gryphius).

38 A copy of a 1526 edition of the *Adages* bears annotations in the hands of Erasmus and of his 'famulus' Nicolaus Cannius revising the index for a subsequent edition; see Luigi Michelini Tocci, *In officina Erasmi: l'apparato autografo di Erasmo per l'edizione 1528 degli Adagia* . . . (Rome: Edizioni di Storia e Letteratura, 1989), p. 41 and figures 6 and 7.

39 Anthony Grafton, *Joseph Scaliger: A Study in the History of Classical Scholarship*, vol. II (Oxford: Clarendon Press, 1993), pp. 504–6.

40 Laurent Pinon, 'Les livres de zoologie de la Renaissance, objets de mémoire et instruments d'observation (1460–1605)', doctoral dissertation, Centre d'Etudes Supérieures sur la Renaissance, Tours (2000). I am most grateful to the author for sharing this information with me.

41 See Hieronymus, II, #433–4, p. 1262, for a comparison of a folio edition of Sallust, 1564, and an octavo edition of 1571, both published by Petri in Basel.

42 I am indebted here to discussions with Howard Hotson; see his *Johann Heinrich Alsted: between Renaissance, Reformation and Universal Reform* (Oxford: Oxford University Press, 2000).

43 On this theme see Edward Grant, 'Aristotelianism and the longevity of the medieval world view', *History of Science*, 16 (1978): 93–106.

44 John Locke, 'Nouvelle méthode de dresser des recueils', *Bibliothèque universelle et historique de l'année 1686*, vol. II, 2nd edn (Amsterdam: Wolfgang, Waesberge, Boom and van Someren, 1687), pp. 315–39. Ann Moss notes a new emphasis in Locke's practice of commonplacing, in that he took less interest in the authorities cited than in the content of the excerpts; see Ann Moss, *Printed Commonplace-Books and the Structuring of Renaissance Thought* (Oxford: Clarendon Press, 1996), p. 279.

Further reading

Ann Blair, 'Humanist methods in natural philosophy: the commonplace book', *Journal of the History of Ideas*, 53 (1992): 541–51; and *The Theater of Nature: Jean Bodin and Renaissance Science* (Princeton, NJ: Princeton University Press, 1997)

Jean-Marc Chatelain, 'Humanisme et la culture de la note', *Revue de la Bibliothèque Nationale de France*, 2 (June 1999): 26–36, among other articles of interest in this special issue on annotation

Lloyd W. Daly, *Contributions to a History of Alphabetization in Antiquity and the Middle Ages* (Brussels: Latomus, 1967)

Ann Moss, *Printed Commonplace-Books and the Structuring of Renaissance Thought* (Oxford: Clarendon Press, 1996)

Mary and Richard Rouse, 'La naissance des index', in H.-J. Martin, R. Chartier and F.-P. Viret (eds.), *Histoire de l'édition française* (Paris: Promodis, 1982–), I, pp. 77–85

'Concordances et index', in H.-J. Martin and J. Vezin (eds.), *Mise en page et mise en texte du livre manuscrit* (Paris: Editions du Cercle de la Libraire, Promodis, 1990), pp. 219–28

Margaret M. Smith, 'Printed foliation: forerunner to printed page-numbers?', *Gutenberg-Jahrbuch*, 63 (1988): 54–70.

Hans H. Wellisch, *Indexing from A to Z* (Bronx, NY: H. W. Wilson Company, 1991)

5 Illustrating nature

When studying books on nature published in the first 150 years of movable-type printing, it is tempting to use their illustrations as evidence for observational or empirical concerns of the author, or at least of the work they are assumed to illustrate. Whilst studies on the history of the book have highlighted interpretive problems surrounding early modern printed texts, images accompanying those texts have not been so closely scrutinised until relatively recently. Illustrations do not appear to have received the same caution and care which historians of science now exercise when quoting sentences and texts from works of the past. Yet, as part of a printed book which is a result of the efforts of authors, correctors, proof-readers, paper-makers, designers, wood-carvers, pressmen, printers and bankers, images deserve as much scrutiny and care as texts when used as historical evidence. This chapter aims to outline the processes by which illustrations came to be included in some early modern publications and the range of attitudes towards visual illustration amongst authors of those books.

Techniques of replicating illustrations

Before the advent of the movable-type printing press, two ways of replicating images became available in Europe. Originating as stamps for printing patterns on cloth, wood-blocks were widely used for replicating images on paper from the late fourteenth century. These early prints were mainly single-sheet pictures of holy images, printed in convents and monasteries and sold as mementoes and talismans to pilgrims. Pictures in early books printed with these wood-blocks retained their devotional origin, as in the example of the illustrations in the *Bibla pauperum*.[1]

Another means of replicating images used incised metal plates and was widespread in Europe by the 1430s. This was an intaglio process (the incised and therefore sunken parts of the plate forming the lines to be printed) which required the skill of a metal smith. These prints often retained the non-religious, courtly, decorative themes of a goldsmith's artefacts. Incised metal plates were also used to produce pattern masters for playing cards. Engraved prints

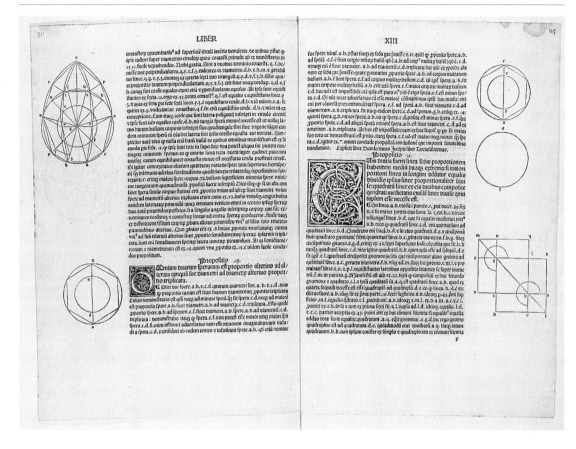

Figure 5.1 Diagrams produced with bent metal on plaster, Euclid, *Elements* (Venice: E. Ratdolt, 1482), fols. 114r–115v. (Reproduced by permission of the Syndics of Cambridge University Library)

thus tend to suggest a wealthier and more urbane audience than that of the early wood-block prints.

For technical and financial reasons, however, it was the wood-block that first became the standard medium for replicating images in books printed with movable type. Since both movable types and wood-blocks print in relief (that is, parts that were not cut away and thus left raised formed the lines and letters to be printed), they could readily be fitted into the same forme. From around the 1470s, texts and images were printed simultaneously in this way.[2] In comparison, it was more cumbersome and expensive to use copperplate images in a printed book, since copper plates were fairly expensive, and the movable-type text and the picture had to be printed separately.[3] Hence copper plates tended to be used for stand-alone images, maps or title-pages in books.

It seems that in the early years of the movable-type printing press, cutting away all but the finest of lines from the wood-block was a skill that was hard to come by. For a Latin edition of Euclid's *Elements* (Venice, 1470, *editio princeps*), the printer Erhard Ratdolt improvised by mounting bent metal on plaster to produce diagrams, an invention of which he was justifiably proud (Figure 5.1).[4]

By the middle of the sixteenth century, however, the art of woodblock cutting had developed dramatically, as can be seen from Jan van Calcar's splendid illustration to Andreas Vesalius' *Seven Books on the Fabric of the Human body* (Figure 5.2).[5] In the meantime, Albrecht Dürer had demonstrated how copper-engraving could produce a wider range of hue and finer lines than woodcuts, but Dürer was a rare case where the artist was also the engraver.[6]

In printed book productions, the person who cut the wood-block rarely designed the original picture. Wood-cutting belonged to the skills of cabinet-makers and sculptors, and was employed for producing pictures in books, which were originally the province of illuminators. Engraving used the skills and tools of the metalsmith. Neither the carpenter nor the metalsmith necessarily possessed the skill to compose or design an elaborate picture. The portrait of the three artists involved in the production of Leonhard Fuchs' *History of Plants* (Basle: M. Isingren, 1542) presents a division of labour in producing illustrations that was typical for the period (Figure 5.3): a draughtsman draws the picture, another traces it on to the woodblock and then the carpenter or the sculptor cuts the wood-block. Of these, it was the sculptor who could expect to be paid most; at least three to five times more per picture than a draughtsman.[7] Hence the sculptor's assuming centre stage in this portrait.

Despite claims by authors for the illustrations in their books as having been drawn from life (*ad vivum*) or being 'live' pictures (*vivae eicones*), it was not necessarily the case that artists always drew from life.[8] Conrad Gesner reports how his artists worked in the field in the summer, but used dried flowers dipped in water to brighten up the colours during the winter months. Some of Fuchs' pictures represented plants which would never have been found in nature – in some pictures, various stages of the plant or a variety of flowers were incorporated into one bush (Figure 5.4). Jan van Calcar inserted his illustration of the viscera on a fragment of classical sculpture known as the Belvedere Torso.[9] Calcar also seems to have adjusted proportions of skeletons and enhanced certain features in order to fit with renaissance theories of proportion (Figure 5.2).[10] Artists frequently copied each other, or used older manuscripts. Thus Hans Weiditz's illustrations in Otto Brunfels' *Live Images of Plants* (Strasburg: J. Schott, 1530) were copied in Eucharius Rösslin's *Kreuterbuch* (Frankfurt: C. Egenolf, 1533) and in Leonhard Fuchs' *History of Plants*; Dürer's rhinoceros (which itself was a copy of a broadsheet) was copied in Gesner's *Histories of Animals* (Zurich: C. Froschauer, 1551) (Figure 5.5).[11] This work by Gesner also contained images copied from manuscripts of Olaus Magnus as well as from the works of Pierre Belon and Guillaume Rondelet. Gesner also accepted pictures sent to him from his friends on trust.

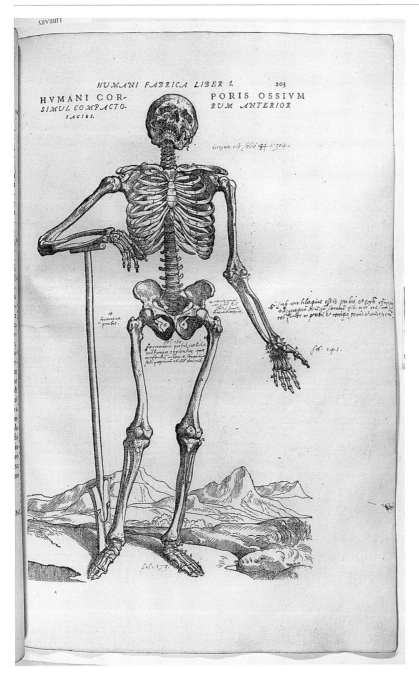

Figure 5.2 A. Vesalius, *Seven Books on the Fabric of the Human Body* (Basle: J. Oporinus, 1543), p.203. The proportion of the skeleton was adjusted to fit classical canons of ideal human proportion. (Reproduced by permission of the Syndics of Cambridge University Library)

Figure 5.3 A typical division of labour for producing illustrations for printed books, L. Fuchs, *History of Plants* (Basle: M. Isengrin, 1542), fol. 897r. (Reproduced by permission of the Syndics of Cambridge University Library)

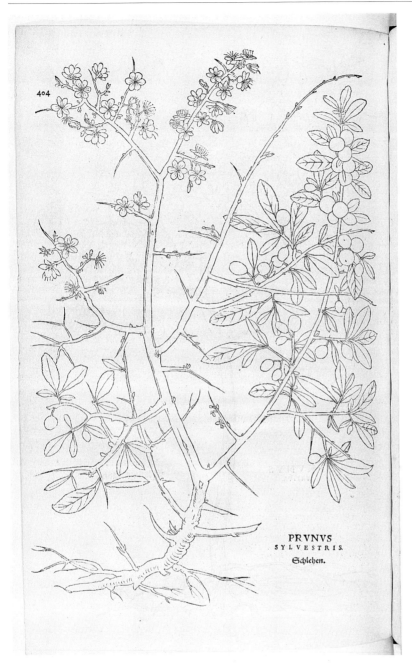

Figure 5.4 'Prunus Sylvestris', showing in one bush different stages of the development of the plant, L. Fuchs, *History of Plants* (Basle: M. Isengrin, 1542), fol. 404. (Reproduced by permission of the Syndics of Cambridge University Library)

Figure 5.5 A copy of Dürer's image of the rhinoceros, C. Gesner, *Histories of Animals* (Zurich: C. Froschauer, 1551), p. 953. (Reproduced by permission of the Syndics of Cambridge University Library)

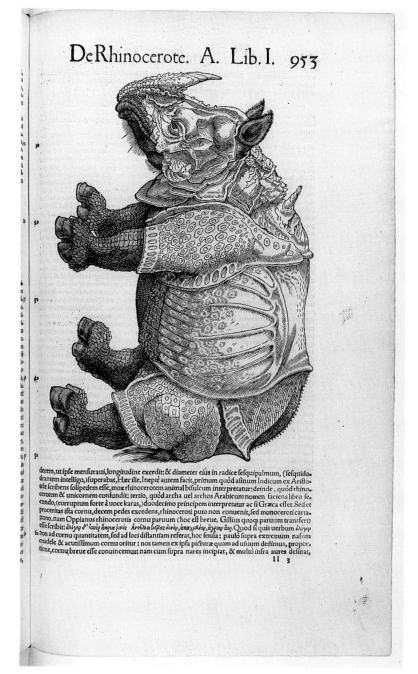

In fact, it is important to note that authors did not necessarily have control over what the artists drew or how they were produced. Gesner, again, complained that pictures were not drawn to a uniform scale, and that a comparison of the relative sizes of animals and insects was therefore impossible from his book. In the case of Brunfels' *Live Images of Plants*, the artist and the author were clearly working separately. The artist Hans Weiditz sketched in detail (including blemishes and withering leaves) the plants he found around Strasburg, at the behest of the printer Johannes Schott. Otto Brunfels, the Carthusian-turned-Lutheran humanist who had worked frequently for Schott, began compiling descriptions and medicinal uses of plants from ancient and scholastic authorities before Weiditz had started his sketches. Thus, the final product, the *Live Images of Plants*, contains several plants described by Brunfels but not illustrated by Weiditz, and plants illustrated by Weiditz on which Brunfels had little to say. Several illustrations were re-used for different plants (Figure 5.6).[12] This case serves as a cautionary reminder that it would be highly misleading to assume that an author could and did conceive, plan, write and produce the final product of a printed book in exactly the way he intended it to be. In the case of the *Live Images of Plants*, it is more like a project of a skilled artist and a learned humanist brought together by an enterprising printer. It would thus be misleading to attribute the naturalistic qualities of Weiditz's illustrations to Brunfel's knowledge and practice of botany.[13] Some authors, however, did draw or design the illustrations for their work. Pierre Belon drew the pictures of fish that were included in his *Histoire naturelle des estranges poissons marins* (1551). Fuchs and Gesner are also known to have drawn pictures of plants, but they were in the end subject to the vagaries of printers.[14] In the early modern period, relatively few authors seem to have had control over the illustrations that went into their books.

From the point of view of the printers, the main problem with illustrations was that they were expensive. The cost of producing illustrations could form up to three-quarters of the capital investment of producing a book, and therefore wood-blocks and plates were regarded as valuable assets by printers. Printers were at liberty to re-use material in other publications, loan them to other printers or sell them off. They often bought up older wood-blocks simply to stop their rivals printing similar works.[15] In the case of the *New Kreütter Buch* (Strasburg: W. Rihel, 1539) by Hieronymus Tragus, both the author and the printer excuse themselves from including illustrations, since that would have put the books out of the reach of the poor students.[16]

Printed books with illustrations were relatively pricey, but they could be made even more expensive by using more expensive material. Vellum or coloured paper could be used instead of ordi-

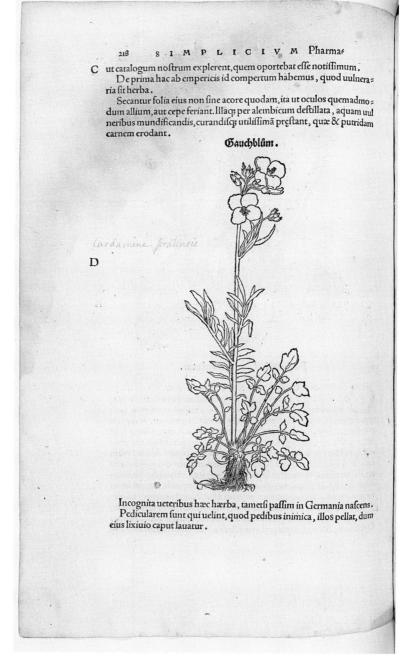

Figure 5.6 'Gauchblum' or a plant without a classical counterpart, O. Brunfels, *Live images of plants* (Strasburg: J. Schott, 1530), p. 218. (Reproduced by permission of the Syndics of Cambridge University Library)

nary paper; gold ink instead of the normal black ink; illustrations could be coloured and illuminated lavishly, retaining in many ways the appearance and value of the manuscript book.[17] Such de luxe copies were usually created as dedication copies to patrons. The advent of the movable-type printing press, therefore, did not necessarily lead to the diffusion of completely identical copies of a given title. Thus, illustrations in them may not necessarily look identical either.

From the early stages of the movable-type printing press, efforts were made to replicate colours. Stencilling techniques were used by Ratdolt to print three colours in his edition of Sacrobosco's *De sphaera* (1485), but many printers were happy to hand-colour individual copies on request. In the case of Brunfels' *Live Pictures of Plants* and Gesner's *Histories of Animals*, it is known that colouring was done in the printer's workshop following a coloured 'archetype', but colouring invariably pushed up the price of a book.[18]

As the printing press began to develop new forms of lettering, the physical appearance of a printed book began to diverge from that of its medieval ancestors, and illustrations too began to appear in new places. For instance, title-pages of incunables often carried the opening words of the book (the *incipit*) alone, as was the case in manuscripts, but during the sixteenth century, more decorative title-pages with architectural façades and classical columns appeared in humanist books.[19] These were followed in the baroque period by the highly allegorical pictures intended to summarise the entire content of a book (Figure 5.7).[20]

The technical and financial constraints that operated in producing a printed book thus also governed the quality and quantity of illustrations present in such books. In other words, there were many reasons for the presence, or absence, of illustrations in a printed book – the intentions of the author rarely being the decisive one.

Word and image

As I have argued, the presence (or absence) of illustrations with certain qualities (such as naturalistic representation) does not necessarily imply the presence (or absence) of certain kinds of qualities in the studies of nature described in the accompanying texts. It is not surprising, then, that there existed a multitude of ways in which the image and the text of a printed book were related to each other. Pictures could serve as mnemonics for the text by diagrammatically accentuating certain features or attributes described in the text, just as medieval herbals were illustrated in manuscripts.[21] An image could also 'illustrate' a piece of text in a general way: a picture of a church, a river and houses could be used for both Nuremberg and

Figure 5.7 C. Scheiner, *Rosa ursina* (Bracciani, 1630), title-page. This is an emblematic title-page, based on Bonaventurian ideas. The divine light of Jesus at the top of the page illuminates sacred authority to the right and the light of reason at the left. The light of reason draws a crisp image of the sunspots, the subject-matter of the book. At a lower and therefore inferior position is the sun, which illuminates profane authority to the right and light of the senses to the left. Using a telescope, however, the light of senses on its own can only depict a fuzzy image of the sunspots. See W. B. Ashworth Jr, 'Light of Reason, Light of Nature: Catholic and Protestant metaphors of scientific knowledge', *Science in Context*, 3 (1989): 94–7. (Reproduced by permission of the Syndics of Cambridge University Library)

Augsburg, because the image portrays a generic 'city'. This is much the way that early printed herbals were illustrated. Exotic simple medicines such as aloe and anacardus were illustrated with a woodcut of jars (Figures 5.8 and 5.9).[22]

Leonhard Fuchs criticised such repeated use of the same woodcut as a scourge of the miserly printer (Figure 5.10). In the preface to the *History of Plants* (1542), Fuchs was unusual for his time in insisting that there ought to be a different picture for each plant. A one-to-one correspondence between an illustration and an object is a rarely achieved standard in early modern printed books.[23] Fuchs placed great importance on the 'accidents' (non-essential features) of plants – the shapes and colours of the leaves, the numbers and colours of the petals and so on – in establishing the identity of plants described by classical authors such as Galen, Dioscorides and Pliny the Elder. Each feature, the greenness or the roundness of a leaf could be shared across different kinds of plants, but an aggregate of these accidental features could be used to distinguish plants. A picture represented these accidental features of plants; thus each picture of a plant had to look different and no re-use of pictures could be allowed.[24] Fuchs could thus establish the identity of a classical plant by matching up classical descriptions and illustrations of contemporary plants.

Many people, as we shall see below, objected to the way Fuchs used his illustrations, and his was by no means the dominant way in which text and image became related in printed texts. Although he insisted that pictures of animals were necessary for his *Histories of Animals* because many of them were difficult to come across in life, Gesner understood visual illustrations hieroglyphically. A picture of an elephant, for instance, served as a juncture for a multitude of meanings associated with the elephant: the elephant had various culinary and medicinal uses (for rheumatism and sciatica); its name could be used to describe diseases (such as leprosy), and mountains and towns in Egypt; and was used with various metaphorical and proverbial meanings (e.g. 'the Indian elephant cares not of the biting of a Gnat').[25] Moreover, cosmic powers of plants could be sought through the shapes and physical attributes of plants in Giambattista della Porta's *Phytognomica* (Naples: H. Salvianus, 1588).[26]

There are several illustrated printed books in the early modern period where texts and images seem simply to coexist alongside each other without any rigorous link or coherence. For instance, in the commentary on Mondino's *Anatomy* by the surgeon and art connoisseur, Berengario da Carpi (1521), anatomical images were based on contemporary artworks of Michelangelo and Giorgio Ghisi, without much concern to ensure that the illustration depicted exactly what was explained in the text.[27]

On the other hand, there were books in which a closer tie between the text and the image was sought. For instance, in anatomical books of the sixteenth century, several devices were developed to link the image and text. Alphabet captions were distributed all over the illustrations of Vesalius' *Seven Books on the Fabric of the Human Body*, and were used in the text and margins to refer the reader to the visual representation of the particular part of the body under discussion in the text (Figure 5.11).[28] There is no doubt that in Vesalius' case, the artist and the author had worked closely together – a conscientious reader would have had to refer

Figures 5.8 and 5.9 Pictures for the plants aloe and anacardus, showing the same jars, T. Dorsten, *Botanicon* (Frankfurt: C. Egenolff, 1540), fol. 25r (aloe) and fol. 23v (anacardus). (Reproduced by permission of the Syndics of Cambridge University Library)

SIMPLICIB. MEDIC. 25

C A monum inflammatione, quam peripneumoniam dicunt, unde ph- C
thysis prouenire solet. In eclegmate auellanæ nucis magnitudine
additis melle & lacte: iocinorosis, tussientibus, & maioris intestini,
quod Colon dicunt, inflationibus opitulantur.
    Gummi ex Amygdalo, in uino utiliter bibitur contra sanguinis
excreationes.
    Idem summæ cutis impetigines illitum ex Aceto rapit.
    Oleum de Amygdalis amaris suffocationi & præcipitationi ma
tricis subuenit, si circa umbilicum inungatur.
    Idem temporibus illitū, capitis doloribus succurrit, & quietem
adfert.
    Calidum auribus idem instillatum, sonitum & tinnitum aurium
pellit, auditum restituit.
    Idem mixtum cum melle, dulci radice, oleo rosaceo, & cera, inde
unguentum si fiat, & illinatur circum oculos, expurgat & claros
eos reddit.
    Conducit quoq; aduersus calculum, dolores renum, ac strangu
riam, si lumbis & umbilico illinatur. Amygdalæ amaræ comestę ad
idem conducunt, Quæ etiam Lumbricos necant & pellunt, si quis
sæpius eis utatur. Emplastrum inde factum si umbilico imponatur
idem præstat.

D B ### ALOE. D

NOMENCLATVRAE.

Ἀλόη.
Aloe Succotrinum.
Epaticum.
Gaballinum.
Ein bitter safft in der
Apoteck.

ANNOTATIO IN ALOEN.

LOE & herba est & eiusdem herbæ succus, quę, ut in
quit Galenus, est similis scillę, folijs pinguibus, crassis,
in latitudinem se modicum fundentibus, rotundis &
retrouersum repandis , Cuius descriptionem etiam
apud Dioscor. uide. Nascitur plurima in India, unde & liquor ad
nos affertur. Nascitur & in Arabia, Asiaq;, & maritimis quibusdam
locis, glutinandis uulneribus accomodata, sed ad exprimendum ex
ea liquorem inutilis, Duplex genus, ut idem Diosco. scribit, liquoris
eius est, alterum Arenosum, & ueluti purissimæ Aloes sedimentū,
alterum iocineris modo coactum. Eligito pinguem Aloem, synce-
ram, & sine calculis & arenis, splendentem, ruffo colore, friabilem,
                                           E

DE HERBIS, CAETERISQVE

A Eadem ad iij. uel iiij. uncias pota mane & uesperi, omnibus interioribus uulneribus, & rupturis medetur.

ANACARDVS.

NOMENCLATVRAE.

Anacardus.

𝔈lephanten lauß.

ANNOTATIO IN ANACARDVM.

NACARDVS apud Diosco. & alios Græcos autores non inuenitur. Officinę perperam appellant Pediculum Elephantis, cũ Anacardus, ut inquit Serapion, est fructus arboris, similis cordi auis, cuius color est parum rubens, & inter ipsum est res similis sanguini, & est id quod administratur in eo.

B Ruellius Anacardium appellat, hoc modo de eo scribens: Anacardium recentiori Græciæ, nam uetus non neminit huius, arbos est Indis familiaris, prouenit quoq̃ in Siciliæ montibus, qui flammas ignis eructant, fructu auiculæ corculo non dissimili, unde nomen putatur inditnm, rubente intus cruore, quo ueluti sanguine scatet.

TEMPERAMENTVM.

Anacardi fructus calefacit atq̃ desiccat in quarto ordine. Ruellius inquit quòd quidam tertiam excalfacientium & resiccantium classem huic assignant. Durat triginta annos, nihil de sua uirtute perdens, si tantum in locis ualde humidis aut siccis non seruetur.

Si quis solo Anacardo usus fuerit absq̃ additamento, quod malitiam eius corrigat, illum mori oportet, aut euadet leprosus.

VIRES AC IVVAMENTA.

Si quis debili memoria & hebeti ingenio præditus fuerit, Accipiat ij. uncias Pimpinellæ & decoquat in xij. uncijs acidi aceti, fructusq̃ huius decorticati iij. uncias addat, ac simul misceat dum acquirat conuenientem spissitudinem, & inde se in occipitio mane & uesperi inungat, tum bonam & tenacem parabit memoriam.

Accipe

Figure 5.10 Aloe, L. Fuchs, *History of Plants* (Basle: M. Isengrin, 1542), fol. 138. Previously exotic plants like aloe tended to be represented by the jars in which their extracts were stored. (Reproduced by permission of the Syndics of Cambridge University Library)

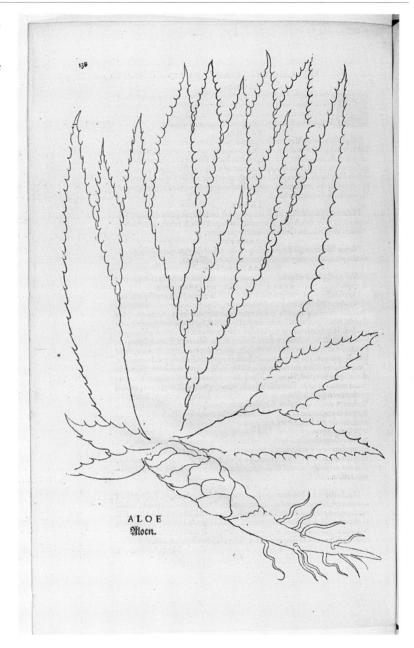

back to the picture and then back to the text again about 150 times in one section.[29] Vesalius' teacher, Jacobus Sylvius, however, complained that the captions were hardly legible in the illustrations, and in the second edition of the *Seven Books on the Fabric of the Human Body*, the background to the captions were duly cut out more clearly in order to make them more legible. Bartholmaeo Eustachio used another system of reference, adopting a coordinate system

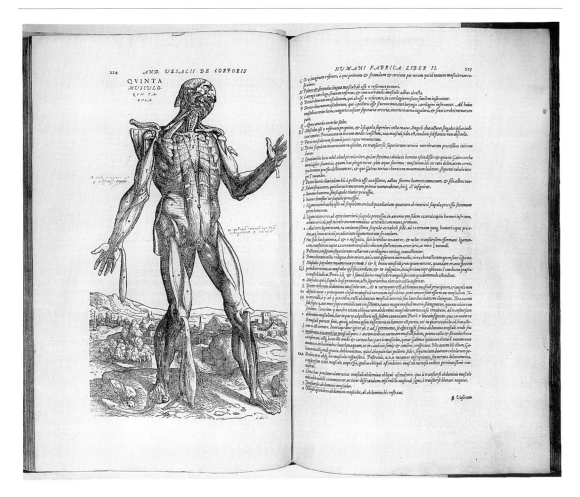

used in geographical maps (Figure 5.12). This was to enable Eustachio to refer to specific places in the illustration with greater precision.[30]

Figure 5.11 Alphabets as a reference index, A. Vesalius, *Seven Books on the Fabric of the Human Body* (Basle: J. Oporinus, 1543), pp. 224f. (Reproduced by permission of the Syndics of Cambridge University Library)

Uses of pictures

If there were several ways in which the text and the image were related in the early-modern printed book, there were also a variety of positions in the period over how useful pictures could be. Many people in the period felt that pictures could not be helpful in acquiring knowledge about plants. The physician Sebastianus Montuus, one of Fuchs' enemies, argued that a pictorial knowledge of plants formed from gathering together accidental features could not lead to knowledge about the medicinal knowledge of the substance of the plant.[31] Janus Cornarius, another physician, also objected to Fuchs' use of pictures on the basis that plants could change their appearances whilst retaining their medicinal efficacy. A picture,

Figure 5.12 Co-ordinates as a reference system, showing a variety of kidneys, B. Eustachio, *Opuscula anatomica* (Venice, 1564), fol. 1v. (Reproduced by permission of the Syndics of Cambridge University Library)

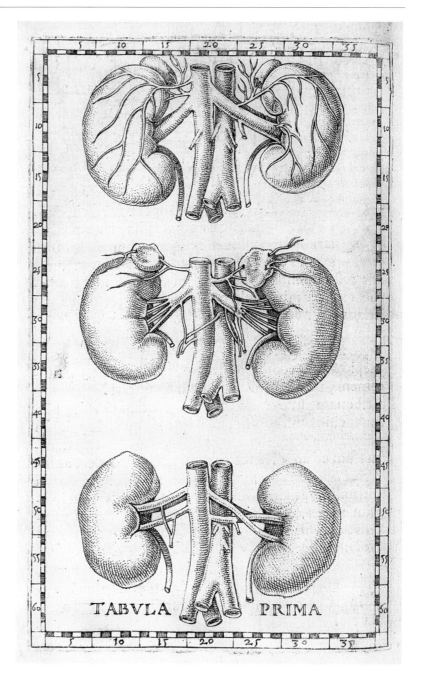

according to Cornarius, depicted a particular plant at a given place and a given time, and one would hardly ever come across in the fields a plant at a stage and place similar to the one depicted. Moreover, in order to be able to recognise a picture of a plant, one had to know what the plant looked like. Cornarius was perhaps extreme in his denial of the use of pictures in his commentary of Dioscorides and in his insistence on portraying verbal pictures of plants culled from ancient authorities.[32]

Pier Andrea Mattioli, whose commentary of Dioscorides was the most popular in the early modern period, certainly thought that Cornarius' objection was perverse and based on a misreading of a passage in Galen which seemed to disparage the use of pictures for simple medicines. Although Mattioli himself had scorned illustrations in his 1550 edition as of little use because of their inability to capture the ever-changing appearance of a plant across seasons, by 1558 he was employing two artists in Prague and five engravers in Vienna to produce full-page illustrations for his Dioscorides.[33] But the flooding of the market with Mattioli's picture-filled editions of Dioscorides' *Materia medica* did not settle the issue of illustrations. Pietro Antonio Michiel criticised Mattioli's work and his use of dried specimens, insisting that the whole course of the development of a plant had to be described in words and illustrations, through the laborious toil of raising plants in the garden.[34] Andrea Cesalpino, one-time director of the botanical garden at Pisa, offered the *herbarium* (an album of dried and pressed plants) as a visual aid to his book, *Sixteen Books on Plants* (Florence: G. Marescott, 1583), which had no illustrations owing to the cost.

There were various reasons why people questioned the usefulness of illustrations in the period. For Cornarius, the study of plants meant that of the classical authors and he saw his task as interpreting and translating the words of the ancients. Since the ancients did not approve of the use of pictures, he argued that pictures were not necessary. Cesalpino, who was following the Aristotelian philosophy of nature, felt on the other hand that pictures were of limited use to the way he wanted to study plants. Cesalpino was concerned to establish the common features of a group of plants and the differentiating features between these groups. No picture was capable of showing at once the common as well as distinct features of plant-groups.

An assumption underlying objections against the use of pictures was that pictures of natural objects were portrayals of singular objects, representing all their accidental qualities, but not their substantial forms or essences. This can clearly be seen in the term 'counterfeit', frequently employed in vernacular broadsides and texts to convey the sense of an exact portrayal of an object.[35] Fantastic creatures, such as a half-human and half-animal beast or a monstrous monkfish, were printed in single broadsheets as portents of terrible things to come; but they were also eventually included in the natural historical works of Gesner, Rondelet and Belon. Just as the unicorn and the mandrake had a long line of scholarly authorities attesting to their existence, reports of monstrous or preternatural beings, when coming from creditable eye-witnesses, could be found a place in the scholarly literature of nature.[36]

Problems of representation were also discussed in anatomy in terms of the variability of the human body.[37] In order to gain a proper knowledge of what a standard human body is like, Alessandro Benedetti argued that the cadavers for dissection had to be as standard as possible, while Berengario da Carpi urged frequent practice in dissection to gain knowledge of individual variations in human bodies. In the *Seven Books on the Fabric of the Human Body*, Vesalius, following Galen, hoped to deal with the 'canonical' human body, a perfect male body, devoid of individual variations. Hence, the illustrations in the *Fabric of the Human Body* show traces of careful enhancement and readjustments (Figure 5.2). Instead of coming up with a picture of the 'canonical' body, Eustachio decided to show as many different examples of kidneys as possible (Figure 5.12).[38]

Although he used diagrams and illustrations in his lectures, Sylvius doubted they were of much use for discovery. He believed that anatomy had to be practised by one's own hands as well as eyes, not by merely looking at the body superficially, which is how pictures express the body.[39] This is an objection Sylvius repeated later against his student Vesalius.[40] Vesalius seems to have tried to counter this criticism too, when he printed the *Epitome*, a picture book designed to be cut out and stuck on a card-board, one on top of another, in order to represent the layers of the human body.[41]

The period between 1450 and 1600 may thus be characterised as a time when there was no established consensus as to what illustrations represented and how they might be used in gaining knowledge about nature, and indeed a time when people experimented with different ways to represent nature on two-dimensional paper and devised rules of representation.

Replication and privileges

Since illustrations were expensive to produce, printers often took out privileges to guard against unlawful copying; yet privileges usually ran for only a specified period between five and ten years, applied only to areas over which the issuer of privileges had jurisdiction, and was rarely enforceable. Such privileges could cover the pictures as well as the text.[42] It seems, however, that privileges were considered worth securing, since Walter Ryff and the Strasburg printer Beck went as far as forging imperial recommendation in 1540.[43] A famous case was the lawsuit brought by the printer Johannes Schott against the Frankfurt printer Christian Egenolff for copying the pictures in Brunfels' *Live Images of Plants* in Rösslin's *Kreuterbuch*. Egenolff gave a spirited defence of his case: only a portion of the pictures in the *Kreuterbuch* was derived from Schott's book; others were derived from an earlier edition of the *Kreuterbuch* published by himself; therefore, his *Kreuterbuch* was not a plagiarism of Brunfels' work. Moreover, if one were to depict plants responsibly, from nature, one cannot avoid a certain similarity of appearance. Surely, simply because Dürer had skilfully painted Adam and Eve does not mean that nobody else is allowed to paint them ever again? It seems, in the end, that the court found in favour of Schott, who was in possession of Egenolff's woodcuts soon afterwards.[44] This, however, does not seem to have set Egenolff back financially. He continued to pirate illustrations from other herbals, much to the chagrin of men like Fuchs, and died a very rich man.

Conclusion

There were many different reasons why illustrations of nature came to be included in early-modern printed books. Whether, or what kind of illustration was included in a printed book frequently depended on the printers and the artists. Authors of the printed text held a wide range of views on the usefulness and function of those illustrations: there were those who reluctantly accepted the presence of illustrations next to their texts, while some enthusiastically experimented with various ways to establish a visual convention of representing nature. There were still others who simply did not accept the value of visual representation when learning about nature. And they frequently quarrelled over pictures. That is, in the first 150 years of the movable-type printing press, there was little consensus as to what illustrations could represent, and how they related to the text that they accompanied. This implies that the task of a historian of science who wishes to use visual illustration from this period as evidence for contemporary scientific practice should include the study of the history of the book. What those images were meant to represent, why they came to be included in a printed book, whether and how the images are related to the text are issues that have to be established for each case, precisely because there was little consensus. A study of illustrations of nature based on the history of the book can be expanded to include all non-textual representation of the period – for instance, it will be interesting to ask how representations of flora, fauna and the human body may be related to representations of planetary movement, machine and mathematical objects.[45] Illustrations may then become an important primary source for shedding light on the practice of the study of nature in the period.

Notes

1. In this section, I follow the excellent introduction to the subject of printing images in D. Landau and P. Parshall, *The Renaissance Print: 1470–1550* (New Haven, 1994). For a study of the 'blockbooks' (books printed with wood-blocks, but not with movable type), see also N. F. Palmer, 'Junius's blockbooks: copies of the *Biblia pauperum* and *Cantica canticorum* in the Bodleian Library and their place in the history of printing', *Renaissance Studies*, 9–2 (1995): 137–65.
2. Until then, illustrations were often stamped into a printed book by hand.
3. The Antwerp printer Plantin, for instance, used specialist workshops for the printing of copperplate illustrations; L. Voet, *The Golden Compasses: A History and Evaluation of the Printing and Publishing Activities of the Officiana Plantinana at Antwerp* (Amsterdam, 1969 and 1972), vol. II, p. 226.
4. A. H. Mayor, *Prints and People: A Social History of Printed Pictures* (New York, 1971), page before illustration 74 (no pagination). For Ratdolt's works, see G. R. Redgrave, *Ratdolt and His Work in Venice* (London, 1899).
5. These lines were achieved by softening the blocks by dipping them in hot linseed oil before cutting them; K. B. Roberts and J. D. W. Tomlinson, *The Fabric of the Body: European Traditions of Anatomical Illustrations* (Oxford, 1992), p. 137.
6. Dürer's prowess at engraving was identified by Erasmus with the power to replicate images; A. Hayum, 'Dürer's portrait of Erasmus and the *ars typographorum*', *Renaissance Quarterly*, 38 (1985): 650–87.
7. On top of that, the carpenter or sculptor frequently provided the wood-block, for which he was paid separately.
8. See for instance, Hieronymus Bock, *Verae atque ad vivum expressae imagines omnium herbarum, fructicum, et arborum . . .* (Strasburg, 1553); Otto Brunfels, *Herbarum vivae eicones ad naturae imitationem, summa cum diligentia et artificio effigiatae* (Strasburg, 1530); Leonhart Fuchs, *De historia stirpium commentarii insignes, maximis impensis et vigiliis elaborati, adjectis earundem vivis plusquam quingentis imaginibus* (Basle, 1542). This issue is also tackled in James S. Ackerman, 'Early Renaissance "naturalism" and scientific illustration', in A. Ellenius (ed.), *The Natural Sciences and the Arts* (Uppsala, 1985), pp. 1–17, also reprinted with updated notes, in idem, *Distance Points: Essays in Theory and Renaissance Art and Architecture* (Cambridge, Mass. and London, 1991).
9. G. Harcourt, 'Andreas Vesalius and the anatomy of antique sculpture', *Representations*, 17 (1987): 28–61.

10 According to J. B. de C. M. Saunders and C. D. O'Malley, *The Anatomical Drawings of Andreas Vesalius* (New York, 1982), p. 84
11 For the enduring influence of Dürer's Rhinoceros, see T. H. Clarke, *The Rhinoceros from Dürer to Stubbs 1515–1799* (London, 1986). For the extent of copying in zoological illustrations in this period, see W. B. Ashworth Jr, 'The persistent beast: recurring images in early zoological illustrations', in Ellenius, *The Natural Sciences and the Arts*, pp. 46–66.
12 Landau and Parshall, *The Renaissance Print*, pp. 247–52. Cf. A. Arber, *Herbals: Their Origin and Evolution. A Chapter in the History of Botany* (orig. 1912, Cambridge, 1990), p. 55, for Brunfels' idea of 'herbae nudae', plants without classical counterparts.
13 Cf. F. D. Hoeniger, 'How plants and animals were studied in the mid-sixteenth century', in J. W. Shirley and F. D. Hoeniger (eds.), *Science and the Arts in the Renaissance* (Washington, 1985), pp. 130f.
14 Fuchs left behind 900 pictures of flowers and, despite the efforts of his heirs to have them published, they remain unpublished to this day, at the National Library of Austria (Vienna); see Kurt Ganzinger, 'Ein Kraüterbuchmanuskript des Leonhart Fuchs in der Wiener Nationalbibliothek', *Sudhoffs Archiv*, 43 (1959): 213–24. Gesner's drawings of plants, which numbered about 1,500, were partially printed by Joachim Camerarius Jr, and C. J. Trew; see Arber, *Herbals*, pp. 111–13.
15 For the logistics and economics of printing illustrations at the Plantin press, see Voet, *The Golden Compasses*, vol. II, pp. 194–243.
16 K. M. Reeds, 'Renaissance humanism and botany', *Annals of Science*, 33 (1976): 530f.
17 A vellum copy of Vesalius' *Fabrica* exists in the British Library (C.18.e.4); see further R. C. Alston and B. S. Hill, *Books Printed on Vellum in the Collections of the British Library* (London, 1996). A spectacular example of a de luxe copy is Ratdolt's dedication copy to the Senate of Venice of Euclid's *Elements*, whose preface was printed in gold ink and illuminated lavishly; see V. Carter, L. Hellinga et al., 'Printing with gold in the fifteenth century', *The British Library Journal*, 9 (1983): 1–13. A copy of Federico Commandino's commentary on Euclid's *Elements*, the *Elementorum Libri XV* (Pesaro, 1572) in Cambridge University Library (F.157.a.2.1) is printed on blue paper, which was normally used for printed Hebraic books; see B. S. Hill, *Carta Azzurra: Hebrew Printing on Blue Paper* (London, 1995). I thank Dr Liba Taub for drawing my attention to this pamphlet.
18 A. Arber, 'The colouring of sixteenth-century herbals', in idem, *Herbals*, pp. 315–18.
19 For the revival of Vitruvian theories of architecture in this period, see Vaughan Hart and Peter Hicks (eds.), *Paper Palaces: The Rise of the Renaissance Architectural Treatise* (New Haven, 1998).
20 For examples, see W. B. Ashworth Jr, 'Divine reflections and profane refractions: images of a scientific impasse in seventeenth-century Italy', in Irving Lavin (ed.), *Gianlorenzo Bernini: New Aspects of His Art and Thought* (University Park, PA, 1985), pp. 179–206, and 'Light of reason, light of nature: Catholic and Protestant metaphors of scientific knowledge', *Science in Context*, 3 (1989): 89–107.

21 For medieval illustrations, see P. M. Jones, *Medieval Medicine in Illuminated Manuscripts* (London, 1998).
22 For the replication of illustrations of containers of medicine, see for instance Theodore Dorsten, *Botanicon* (Frankfurt, 1540), 25r (aloe), 23v (anacardus) and 96r (cubebe) use the same woodcut of jars. Compare also 64r, 49r, 41r and 226v.
23 For a study of repeated use of illustrations in English books, see R. S. Luborsky, 'Connections and disconnection between images and texts: the case of secular Tudor book illustration', *Word and Image*, 3 (1987): 74–85.
24 For a further analysis, see S. Kusukawa, 'Leonhart Fuchs on the importance of pictures', *Journal of the History of Ideas*, 58(3) (1997): pp. 403–27.
25 W. B. Ashworth Jr, 'Natural history and the emblematic world view', in D. C. Lindberg and R. S. Westman (eds.), *Reappraisals of the Scientific Revolution* (Cambridge, 1990), pp. 303–32.
26 For astrological botany, see further Arber, *Herbals*, pp. 247–63.
27 A. Hyatt Mayor, *Artists and Anatomists* (New York, 1984), pp. 90–3. For a social historical reading of Berengario's pictures, see K. Park, 'The criminal and the saintly body: autopsy and dissection in Renaissance Italy', *Renaissance Quarterly*, 47(1)(1994): 1–33.
28 This was not the first time that such indexing was used, however; see the cases of Magnus Hundt (1501) and Johannes Dryander (1537) in Roberts and Tomlinson, *Fabric of the Body*, p. 85.
29 As pointed out in N. G. Siraisi, 'Vesalius and human diversity in De humani corporis fabrica', *Journal of the Warburg and Courtauld Institutes*, 57 (1994): 64.
30 'De usu tabularum', *Opuscula anatomica* (Venice, 1564), 1^r.
31 See for instance his *Dialexeon medicinalium libri duo* (Lyons, 1537).
32 Cornarius' commentary, *Pedacii Dioscoridae . . . de materia medica libri V* (Basle, 1557), is therefore completely devoid of illustrations.
33 R. Palmer, 'Medical botany in Northern Italy in the Renaissance', *Journal of the Royal Society of Medicine*, 78 (1985): 153.
34 Ibid.
35 P. Parshall, '*Imago contrafacta*: images and facts in the Northern Renaissance', *Art History*, 16 (1994): 554–79.
36 For the role of the supernatural and praeternatural in the medieval and early modern period, see L. Daston and K. Park, *Wonders and the Order of Nature, 1150–1750* (New York, 1998).
37 In this section I have followed Siraisi, 'Vesalius and human diversity'.
38 Figure 5.12 represents three different specimens of a kidney: the figure at the top shows the case of a right kidney larger than the left; in the middle figure the left kidney larger and placed higher than the right, and in the bottom image, the right kidney is placed higher than the left side (note that the left and right are reversed in the illustration). Eustachio, *Opuscula Anatomica* (Venice, 1564), 1^v.
39 Jacobus Sylvius, *Ordo et ordinis ratio in legendis Hippocratis et Galeni libris* (Paris, 1539), p. 13.
40 Jacobus Sylvius, *Vaesani cuiusdam calumniarum in Hippocratis Galenique rem anatomicam Depulsio*, as reproduced in Renato Henerus,

Adversus Jacobi Sylvii Depulsionem anatomicarum calumnias, pro Andrea Vesalio Apologia (Venice, 1555), p. 134. For Sylvius, see C. E. Kellett, 'Sylvius and the reform of anatomy', *Medical History*, 5 (1961): 101–16.

41 For other layered books, see S. G. Lindberg, and W. S. T. Mitchell, 'Mobiles in books, volvelles, inserts, pyramids, divinations and children's games', *The Private Library*, 3rd Ser. 2–2 (1979).
42 Sometimes privileges were sought for the printed gores of the globe which were to be cut out and pasted onto a sphere.
43 R. Hirsch, *Printing, Selling and Reading 1450–1550* (Wiesbaden, 1967), p. 85, and K. Schottenloher, 'Die Druckprivilegien des 16 Jahrhunderts', *Gutenberg-Jahrbuch* (1933): 94–110.
44 H. Grotefend, *Christian Egenolff. Der erste ständige Buchdrucker zu Frankfurt a. M. und seine Vorläufer* (Frankfurt a. M., 1881), pp. 16f.
45 See, however, the direction taken in Martin Kemp, 'Temples of the body and temples of the Cosmos: vision and visualisation in the Vesalian and Copernican revolutions', in Brian S. Baigrie (ed.), *Picturing Knowledge: Historical and Philosophical Problems Concerning the Use of Art in Science* (Toronto 1996), pp. 40–85.

Further reading

A. Arber, *Herbals: Their Origin and Evolution. A Chapter in the History of Botany* (orig. 1912, Cambridge, 1990)

W. B. Ashworth Jr, 'Natural history and the emblematic world view', in D. C. Lindberg and R. S. Westman (eds.), *Reappraisals of the Scientific Revolution* (Cambridge, 1990), pp. 303–32

Wilifrid Blunt and William T. Stearn, *The Art of Botanical Illustration* (orig. London, 1950; Woodbridge, 1995)

William M. Ivins Jr, *Prints and Visual Communication* (London, 1953)

P. M. Jones, *Medieval Medicine in Illuminated Manuscripts* (London, 1998)

M. Kemp, '"The mark of truth": looking and learning in some anatomical illustrations from the Renaissance and eighteenth century', in W. F. Bynum and R. Porter (eds.), *Medicine and the Five Senses* (Cambridge, 1993), pp. 85–121

F. Koreny, *Albrecht Dürer and the Animal and Plant Studies* (Boston, 1988)

S. Kusukawa, 'Leonhart Fuchs on the importance of pictures', *Journal of the History of Ideas*, 58(1997): 403–27

D. Landau, and P. Parshall, *The Renaissance Print: 1470–1550* (New Haven, 1994)

R. Palmer, 'Medical botany in Northern Italy in the Renaissance', *Journal of the Royal Society of Medicine*, 78 (1985): 149–57

K. M. Reeds, 'Renaissance humanism and botany', *Annals of Science*, 33 (1976): 519–42

K. B. Roberts, and J. D. W. Tomlinson, *The Fabric of the Body: European Traditions of Anatomical Illustrations* (Oxford, 1992)

N. G. Siraisi, 'Vesalius and human diversity in *De humani corporis fabrica*', *Journal of the Warburg and Courtauld Institutes*, 57 (1994): 60–88

ADAM MOSLEY

6 Astronomical books and courtly communication

> After the birth of printing books became widespread. Hence everyone throughout Europe devoted himself to the study of literature... Every year, especially since 1563, the number of writings published in every field is greater than all those produced in the past thousand years. Through them there has today been created a new theology and a new jurisprudence; the Paracelsians have created medicine anew and the Copernicans have created astronomy anew.
>
> Johannes Kepler (1571–1630), *De stella nova* (1606), pp. 186–188.[1]

Printing and the transformation of astronomy

Written by one of its chief architects, the above verdict on the role of printing in the development of the new astronomy of the seventeenth century merits serious consideration. Addressing the impact on human affairs of great planetary conjunctions (one such having occurred in 1563), Kepler showed himself to be both a sophisticated student of history and a keen observer of change in his own time. As more recent scholars have done, he asserted that the presses were implicated not only in the transformation of astronomy, but also in the religious schisms, educational reforms, and developing philological expertise of the sixteenth century. But subtleties to his account, including the fact that these latter phenomena were interrelated in complex ways, and omissions, such as their significance in turn to the practice of astronomy, diminish its value as an outright endorsement of the thesis that the invention of printing caused both Reformation and 'Astronomical Revolution'. In fact, since they predated Gutenberg's invention, humanist literary concerns were only contingently dependent upon printed text; while claims about the distinct character of print and scribal cultures prove difficult to sustain, except in respect of the mixed blessing of the presses' fecundity. Kepler's appropriation of printing as an ally of Copernicanism, perhaps polemical, occurred at a time when few authors openly supported heliocentrism, and much of the vast output of astronomical and astrological works showed little or no concern with cosmological issues. Indeed, the transformation of astronomy was itself more thoroughgoing than the elaboration and adoption of any one new theory or practice, involving a renegotiation of the relationship between different forms of knowledge of the heavens (mathematics, philosophy, theology) which accompanied and licensed such adoption.

If it now seems problematic to reduce one protracted and involved phenomenon, the development of a new astronomy, to another of like complexity, the triumph of the book, it is nevertheless clear that the interests of historians of the book and historians of early-modern astronomy frequently converge. Books not only expressed the outcome of astronomical investigations, they were also important to their prosecution. Moreover, in addition to their scholarly and commercial worth, astronomical books were frequently appreciated as tokens representative of heavenly knowledge, whose production and display rewarded and stimulated princely patronage. This characteristic of multivalency was one which books shared with other objects issued by astronomers, among them letters, manuscript texts and instruments. Indeed, many books were related to such objects in one or more ways. A work like Peter Apian's *Astronomicum Caesareum* (1540), for example, constituted an eminently collectable volume of paper calculating instruments, while globes and other devices were frequently issued with manuals explaining their construction and use.[2] Thus, study of books provides one point of entry into the broader culture of early-modern astronomy; a culture in which the circulation of objects helped define communities of practitioners and, in this way too, contributed to the future course of astronomical and cosmological thought.

Tycho Brahe, publisher and 'Prince of Astronomers'

A rewarding figure on whom to focus such a study is the Danish astronomer and nobleman Tycho Brahe (1546–1601). Tycho's mastery of a printing press rapidly became part of his mythology; it excited the envy and admiration of John Flamsteed (1646–1719), England's first Astronomer Royal, and has since contributed to his characterisation as an exemplary figure of the print/knowledge revolution.[3] This mastery, such as it was, was a consequence of the court patronage which enabled Tycho to pursue his interests in astronomy and alchemy. Granted the fiefdom of Hven, a small island in the Danish Sound, he constructed there, with the generous support of King Frederick II, an observatory and dwelling-place which he called Uraniborg. Between 1576 and 1597, when he quit the site, Tycho was the *de facto* royal astronomer of Denmark; he subsequently became Imperial Mathematician at the Prague court of Emperor Rudolf II.[4]

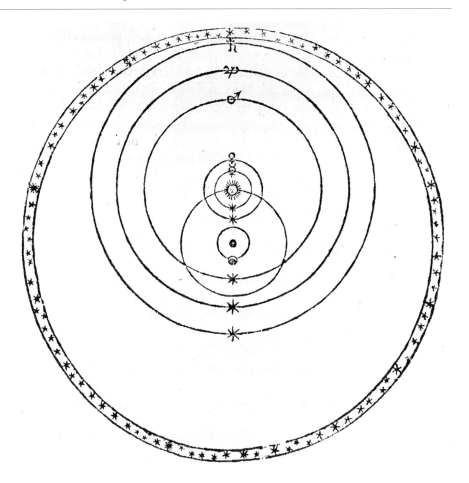

Figure 6.1 Diagram of the Tychonic world-system from T. Brahe, *De recentioribus phaenomenis* (Uraniborg, 1588): the planets orbit the Sun, which in turn orbits the stationary Earth. In variant schemes advanced by others, a daily rotation was attributed to the Earth. Dissemination of Tycho's idea was assisted by Rantzov, who requested additional copies of his book, excerpted this diagram, and forwarded it to other astronomers. (By courtesy of the Whipple Museum of the History of Science, Cambridge)

Tycho's first publication, the *De nova stella* (1573), contained astronomical and astrological analyses of the supernova of 1572, and discussed a forthcoming lunar eclipse; it was produced at the shop of a Copenhagen printer, Laurentius Benedicht. Subsequent texts were printed, or at least begun, at Tycho's own press on Hven. The *De mundi aetherei recentioribus phaenomenis* (1588) treated the comet of 1577; it also contained Tycho's conception of the cosmos (Figure 6.1), a scheme which he claimed combined the advantages of both the Ptolemaic and Copernican systems of the world. The *Epistolae astronomicae* (1596) was an edition of the letters that Tycho had exchanged with Landgrave Wilhelm IV of Hesse-Kassel (1532–92), a German prince with a strong interest in astronomy, and with Christoph Rothmann, his mathematician and astronomer. The *Astronomiae instauratae mechanica* (1598), completed at Hamburg, contained detailed descriptions of the observatories and instruments established on Hven. A fifth work, the *Astronomiae instauratae progymnasmata* (1602), finished and issued posthumously in Prague, discussed the motions of the sun, moon and stars, and returned to the topic of the 1572 supernova in more detail.

At first sight, the most curious of these publications is the letter-book of 1596. Why should Tycho have chosen to publish a selection of his correspondence? Historians have identified a range of reasons: the text was produced in response to the request of close friends; as a lasting memorial to Tycho's own astronomical work, or to Wilhelm's; as an attempt to expand his network of scholarly contacts; as a means of augmenting his reputation and undermining that of his rivals. Although, as we shall see, each of these claims is substantially correct, their presentation reinforces the impression that the *Epistolae astronomicae* is an idiosyncratic and peculiar work.

In fact, correspondence was an important part of the practice of early-modern astronomy. Tycho exchanged letters about his studies with more than eighty individuals, and the 1596 work, whose fuller title is the *Epistolarum astronomicarum liber primus*, was only the first of a projected series of letter-books. The correspondence with the Kassel astronomers dated from 1586, when Heinrich Rantzov (1526–99), governor of the Duchy of Holstein, forwarded to Tycho the Landgrave's request to see his observations of a recent comet. It was of particular significance because the astronomical programmes pursued at Hven and Kassel had much in common. Both aimed at an observational reform of astronomy, a project whose desirability had been expressed in the mid-fifteenth century by Johannes Regiomontanus (1436–76), one of the first to apply humanist concern for the recovery and emendation of ancient texts to the sources of technical astronomy. To him, and to succeeding generations of scholars, comparison of astronomical treatises and tables with one another, and with the heavens themselves, revealed many discrepancies in the places assigned to the stars and the parameters employed in modelling the celestial motions.[5] The *De revolutionibus* (1543) of Nicholas Copernicus (1473–1543) contained new values for some parameters; these and certain other modifications were incorporated into the widely distributed *Tabulae Prutenicae* (1551), elaborated by the Wittenberg professor of mathematics Erasmus Reinhold (1511–53), and dedicated to Duke Albrecht of Prussia.[6] But a systematic empirical rectification had to await the degree of financial and temporal commitment exhibited by Wilhelm and Tycho, and enabled by the resources of the Hessen and Danish states. Whereas Wilhelm's objective was the production of a new star catalogue, Tycho proposed to study the motions of the planets as well.

Partly as a consequence of the adoption of Tychonic designs conveyed to Kassel by the itinerant mathematician Paul Wittich (d. 1587), the instruments at both sites produced observations of like and unprecedented accuracy. Much of the correspondence was concerned with the comparison of results obtained by the two observatories, with a view to the calibration of their instruments and observing practices. But empirical methods could not entirely displace textual exegesis as a means of emending astronomy. Establishing the value of parameters operating over a large span of time, like the rate of precession of the equinoxes, necessitated the use of data contained in earlier sources, which in turn required that the astronomers distinguish between the observational, scribal and printing errors in such works as the *Syntaxis* of the Alexandrian astronomer Claudius Ptolemy (fl. second century AD). To Wilhelm, Tycho wrote that, 'the art of Printing was invented scarcely 150 years ago; and in the long interval prior to that many things, especially the numerical figures, could have been falsely assigned through being rewritten so many times. For error is easily admitted even in the art of printing itself, especially in this matter, unless a careful corrector is employed' (VI, pp. 71–2).[7] And in a letter to Rothmann, he declared that he preferentially used a 1512 edition of the *Syntaxis*, as it expressed numerical values in words, not figures (VI, p. 95).

A slight disagreement about the stellar positions determined at the two observatories developed into a debate about physical phenomena affecting the interpretation of data, particularly atmospheric refraction. These exchanges then fed into a cosmological dispute, in which Tycho endeavoured to persuade Rothmann to surrender his allegiance to the Copernican world-system.[8] Drawing not only on astronomical texts, but also published works in geometry, optics, natural philosophy and theology, these epistolary discussions illustrate astronomy's reconstitution as a moral discipline combining mathematics and physics, showing the importance to this process of the interpretation of new observational data in the light of existing written sources. Thus, by reproducing the Hven–Kassel correspondence in the *Epistolae astronomicae*, Tycho published a text which charted the path from the construction and use of observatory instruments to the advancement of claims about the nature of the cosmos.

Manuscript and print

Concern with observational data and its physical implications was central not only to the *Epistolae astronomicae*, but also to the Tychonic texts of 1573, 1588 and 1602. Furthermore, in both the *De recentioribus phaenomenis* and the *Progymnasmata*, Tycho's desire to establish the superiority of his instruments and observing techniques led him to cite, as evidence of their accuracy, the slightness of the discrepancy between his stellar positions and those obtained at Kassel. As for the *Mechanica*, this book was prefigured in the correspondence by a description of the Uraniborg instruments sent to Wilhelm in 1591. Because Tycho had been planning to publish such a work since 1584, many of the engravings were already to hand; four which later appeared in the *Mechanica* were printed in the 1588 text, several were distributed to correspondents as gifts (Figure 6.2), and those showing Hven and Uraniborg were reproduced in the *Epistolae astronomicae*. The *Progymnasmata* was another work of long gestation; largely completed by 1591, it was again illustrated with the woodcuts used elsewhere. And since this book reproduced in full the astronomical section of the *De nova stella*, it is evident that there was a considerable degree of overlap across the entire corpus of Tychonic publications.

In other respects, too, the *Epistolae astronomicae* is similar to Tycho's other works. Sections of the 1588 and 1602 texts that are devoted to discussion of the observations of other astronomers simulate, in more partisan fashion, the critical dialogue developed in the Hven–Kassel correspondence through the exchange and evaluation of data. Not all of the material that Tycho analysed had previously been published; indeed, some of the data included in the *Progymnasmata* were reproduced from letters, including ones not originally addressed to the astronomer.

It was, in any case, not unusual for letters to be found in printed works. Modelling themselves on the likes of Cicero, Pliny and St Jerome, early-modern scholars consciously composed epistles both as didactic texts and means of self-promotion. Humanists supplemented the classical works used in the teaching of epistolography, a branch of rhetoric, with writing-manuals and books of exemplars, and colluded in the circulation of other letters in both manuscript and print.[9] Letter-collections also served as instructive texts in disciplines such as law and medicine. But epistles could also be reproduced or excerpted within larger works, or published as pamphlets in their own right, and in this fashion many which touched on astronomical matters came to be printed. The *Narratio prima* of the mathematician Georg Joachim Rheticus (1514–74), a description of the Copernican world-system published in 1540, 1541 and as an appendix to Kepler's *Mysterium cosmographicum* (1596), and the latter's *Dissertatio cum sidereo nuncio* (1610), a response to the *Sidereus nuncius* (1610) of Galileo Galilei (1564–1642), constitute two of the most famous examples; while in his *De natura caeli* (1597), Cort Aslakssön (1564–1624), for three years an assistant at Uraniborg, quoted from several items of Tycho's correspondence, not all contained in the *Epistolae astronomicae*. Acknowledging the conventional nature of his own publication, Tycho wrote that, 'many renowned for the learning, both among the Ancients and the Moderns, have not unprofitably used this way of writing and conveying their findings to posterity' (VI, p. 22).

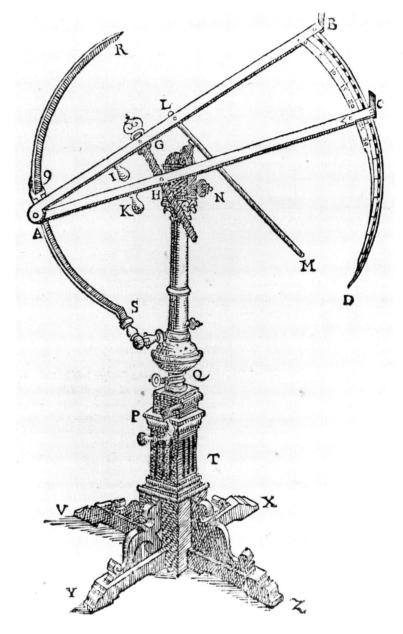

Figure 6.2 Woodcut of one of Tycho's sextants. This engraving, which appeared in both his *De recentioribus phaenomenis* (1588) and his *Mechanica* (1598), was among those sent as gifts to correspondents such as Landgrave Wilhelm IV and Rantzov, and included as an appendix to some of the presented copies of the *Epistolae astronomicae* (1596). (By courtesy of the Whipple Museum of the History of Science, Cambridge)

The epistolary format was also customary for many of the introductory and supplementary elements of a printed book: dedications, prefaces and addresses to the reader. Such devices were among the components of the *Epistolae astronomicae* extraneous to the original correspondence. Tycho opened the book with six poems eulogising his astronomical endeavours, a dedicatory letter to Moritz, Wilhelm's son and heir, and two introductory prefaces. He interrupted the (roughly) chronological sequence of letters with a handful of explanatory notes, and completed the work, except for the errata, with a verse of his own commemorating Wilhelm. While these additions were highly conventional, they played a correspondingly important role in shaping perceptions of the text. Poetry, for example, was widely cultivated as a medium suited to displays of decorous erudition, and its use as an expressive idiom was common across disciplines otherwise divided by their respective technical vocabularies. Employed in the *De nova stella*, which contained Tycho's 234-line elegy to Urania, the muse of astronomy, in the *Progymnasmata*, which included two commendatory poems contributed by James VI of Scotland, or in Kepler's *Mysterium cosmographicum* and *Astronomia nova* (1609), it not only embellished the text but situated it within the broader enterprise of polite scholarship. Moreover, like other exhibitions of skill addressed to individuals (including letters), commendatory verse served to glorify both subject and composer. While the poetic praise of Wilhelm in the *Epistolae astronomicae* helped to establish the text as a memorial to the Landgrave, the delicacy of this gesture could only enhance Tycho's own reputation. However, the Dane also contrived to speak of Wilhelm in terms echoed in the opening verses, which praised his own life and work; prompting his rival, the Imperial Mathematician Ursus (1551–1600), or Nicolai Reymers Baer, to complain with some justification that Tycho dared to boast of being the 'Prince of Astronomers'.[10]

In the first of the introductory prefaces, Tycho solicited contributions to further books of letters, and stated that he had been exhorted to publish his correspondence by his friends: in particular by Rantzov, the Wittenberg polymath Caspar Peucer (1525–1602), and the Imperial physician Thaddaeus Hayek (d. 1600). The variance between this justification of the text and that provided in the dedication exposes the twofold nature of the book's intended audience. Such equivocation was common to dedicated works of any technical content: as a means of participating in the culture of gift-exchange which characterised and regulated relations between patrons and clients, early-modern books were often addressed to individuals of rank with little or no interest in their subject matter. With flattery as their chief purpose, and visibility their greatest asset, printed dedications might engage with the body of a text only as much as was necessary to sketch a plausible reason for the choice of dedicatee. In the *Epistolae astronomicae*, the commemoration of Wilhelm constituted this explanation, even though the edition had been planned several years before the Landgrave's demise. Yet if there was an element of misdirection in this claim, such may have been present in the preface as well: the claim that a work had reluctantly been issued in response to eager requests was a long-standing way of feigning authorial modesty,[11] and one in support of which friends might be induced to compose their petitions after the fact. An artifice of this sort had in fact been employed when justifying the publication of the *De nova stella*.[12]

The act of dedication was descended from the medieval custom of marking publication of a work by presentation to a higher authority, and in the early-modern period, the two practices frequently occurred in tandem. On the occasion of a work's presentation, some indication of its donor's expectations of reward might be appropriate, and the dedicatory letter became a way of incorporating such a petition within the gift itself. But not all dedications clearly reveal their author's aspirations. If Tycho hoped for something specific from Moritz in response to the *Epistolae astronomicae*, it is not obvious what it was. Moreover, not every presentation was similarly motivated: Rothmann wrote of dedicating books to the King of Denmark, or his Chancellor, merely to provide a pretext for visiting Uraniborg (VI, pp. 118–19); whereas in 1598, having left Denmark in search of a new patron, Tycho dedicated the *Mechanica* to Emperor Rudolf II with the express hope, soon fulfilled, of obtaining a post in Prague.

Tycho's gift to Rudolf could be appreciated on a number of levels over and above mere recognition of the Emperor's status. Lavishly illustrated and coloured by hand, the book was undeniably aesthetically pleasing; it was presented alongside two manuscript works, the Tychonic star catalogue (III, pp. 331–77) and a pamphlet on the motions of the sun and moon (V, pp. 165–89), and all three books, beautifully bound, found their way into Rudolf's *Kunstkammer*. But like the globes and other scientific instruments in princely collections, such works were also valued because possession of objects that embodied knowledge was treated as a visible manifestation of the erudition and mastery attributed to the Renaissance ruler,[13] and astronomical books and devices were particularly suited to serving as allegories of universal dominion, or of providential harmony on a celestial scale. Moreover, in the *Mechanica* dedication, Tycho made it clear that the books he presented to Rudolf collectively represented a much greater gift, that of the Tychonic astronomical project in all its parts; a symbolic function made somewhat easier by the fact that the instruments of Uraniborg had themselves been furnished, according to the text, with decorations and emblems that signified the moral value of astronomical study and the scope of the Tychonic programme of reform. In its capacity to reflect glory on the recipient, this present was comparable with Galileo's later naming of the four satellites of Jupiter after the Medici brothers; the gift that, publicised through his *Sidereus Nuncius*, won him the post of Mathematician and Philosopher to the Grand Duke of Tuscany. From Tycho's own testimony, which reveals that by the time of the presentation of his texts, the decision to appoint him to a position at the Imperial Court had already been taken, it would appear that the *Mechanica* actually played a similar role to Galileo's work in making manifest a gift whose acceptability had already been established via behind-the-scenes negotiation.[14]

Production and distribution

Although Tycho gave several reasons for installing a press on the island of Hven (VII, p. 81), it is likely that a determining factor was the inconvenience which would have attended any attempt to correct proofs produced elsewhere whilst continuing his astronomical work. His situation appears to contrast favourably with that of Kepler, who encountered considerable difficulties in seeing through the presses his many publications, particularly the *Tabulae Rudolphinae* (1627), the final legacy of the Tychonic observational project. Yet Kepler's problems were exacerbated by lack of funds, the political turmoil of the Thirty Years War, and the demands of his patrons; the practical difficulties of locating type, paper and skilled labour were ones which he and Tycho shared. Printing was not as straightforward a process as the sheer quantity of early-modern books, including many astronomical and astrological works, might be taken to imply, even for authors sustained by court patronage.[15]

Tycho's operation of a press was aided by his friends and correspondents. Thus, he obtained paper and type from Rantzov, a keen author, financier and collector of books, and through the agency of Henry van den Brock (1531–98), professor of medicine and mathematics at Rostock. In 1590, having constructed a paper-mill on Hven, he asked if Rantzov could spare the services of a paper-maker; mention of his difficulties to Wilhelm prompted the Landgrave to write that, while he only employed one such craftsman, he would be happy to make enquiries in Frankfurt on Tycho's behalf. Ultimately, the arrival of a papermaker created another problem, that of obtaining the linen which constituted paper's raw material. It is clear, nevertheless, that correspondence allowed Tycho to overcome the isolation of his location, otherwise so conducive to astronomical work.

The phenomenon of early-modern privileges is little understood; ostensibly devices for preventing piracy, printing privileges were also bound up with local practices of licensing and censorship. Their origins lay in the medieval strategy of rewarding innovation by granting limited monopolies to the inventors of new devices or processes, and instrument-makers, as well as publishers, were recipients of privileges similar in nature. Occasionally, most frequently in the case of globes, a privilege was awarded jointly for the production of an instrument and its accompanying manual.[16] But the value of these documents may have derived more from the prestige and legitimacy they conferred on the texts or objects to which they applied, rather than any utility as enforceable guarantors of commercial advantage. It is noteworthy that the full text of two privileges, granted to Tycho by Rudolf II and James VI of Scotland, was prominently displayed in the *Progymnasmata*; while the title-page of Ursus' *Tractatus* (1597), an outrageous attack on Tycho and several other astronomers, emphasised the work's maverick status with the boast that it was produced 'without any privilege'. In obtaining privileges from an astonishing variety of authorities, Tycho again relied upon the assistance of his friends and correspondents.

The same constituency of scholars and 'prince-practitioners' participated in the exchange of books and papers which became Tycho's chief means of issuing his works. As his own publisher, Tycho was not obligated to participate in the commercial book-trade, although he expressed the intention of distributing his works by sale eventually (VII, p. 122), and frequently employed book-merchants as couriers for his letters. Of the many copies of the 1588 and 1596 texts printed, only a few were distributed before his death, these being sent to and via his correspondents in Germany, Bohemia and Italy; the *Mechanica* was issued in similar fashion, but sent to a greater number of dukes and princes. Several copies were bound in a cover that emphasised Tycho's nobility (Figure 6.3, a, b), and contained hand-coloured engravings; most carried autograph dedications.

It was not uncommon for authors to be involved in the circulation of their writings. Even when publication was undertaken as a commercial enterprise, the author might be paid in copies of the text (if commissioned by the printer), or contracted to purchase a certain number. At a cost of more than a fifth of his then annual salary, Kepler bought 200 of his *Mysterium cosmographicum*, gave 50 to his mentor Michael Mästlin (1550–1631), who had seen the work through the press, and sent complimentary copies to several scholars, Tycho and Ursus among them. Patrons might also become involved in the dissemination of books. Copies of Galileo's *Sidereus nuncius* were distributed, along with telescopes, through Medici diplomatic channels, while Kepler, whose *Astronomia nova* was not only dedicated to Rudolf II, but also financed by him, was at first forbidden to sell or present copies without his consent.

Works were distributed in manuscript as well as print: Tycho's star catalogue accompanied the *Mechanica* copies presented to potential patrons, and Rothmann's treatise on the comet of 1585 was sent to Hven during the correspondence, but not published until 1618.[17] Similarly, in the wider acquisition and circulation of texts mediated by this community of corresponding scholars, manuscripts and annotated texts featured as well as pristine printed copies. Sending an armillary sphere to Tycho in 1575, one correspondent wrote that 'I wanted to add to this instrument Johannes Regiomontanus' exposition of its construction and use; which, in case you lacked the book, I had copied from his writings edited at Nuremberg in 1543, with additions by Johannes Schöner...' (VII, p. 19). Among the texts most keenly sought by Tycho were individual copies of *De revolutionibus* which contained the annotations of both Erasmus Reinhold and Paul Wittich.[18]

Debate and disputation

Perennially implicated in the attribution of brilliance and erudition, oral debates and literary disputations had particular significance in the scholarly and courtly cultures of early-modern Europe. Princes, cast as the arbiters of truth, played an ambiguous role. Kepler learnt from Mästlin in 1597 that a dedication to a duke could inhibit criticism, and in Prague, with the support of the Emperor, Tycho managed to have Ursus' *Tractatus* condemned to the pyre.[19] Yet when invoked by an author matching an opponent's dedication, as Ursus did in addressing the *Tractatus* to Moritz, or Kepler in the case of the *Tertius interveniens* (1610), princes typically declined to deliver a verdict, at least in so far as epistemological matters were concerned.[20]

Tycho's quarrel with Ursus centred on the claim that the latter had appropriated a faulty version of the Tychonic world-system, publishing it as his own in the *Fundamentum astronomicum* (1588). Largely in virtue of the contents of the original correspondence, publication of the *Epistolae astronomicae* intensified the dispute,[21] although the charge of plagiarism contained in the letters was compounded by Tycho's failure to conform to etiquette and conceal his rival's identity; excused as an editorial oversight, it was viewed by Ursus as a calculated insult. However, the suspicion that Tycho's decision to publish the correspondence may have affected the content and composition of his letters seems borne out by the last of those sent to Rothmann. Prepared in 1594, this epistle contained a lengthy attack on 'a certain Scotsman, addicted to Aristotelian philosophy' (VI, p. 317); that is, the physician John Craig (d. 1620), with whom Tycho was engaged in a public, if not yet published, epistolary dispute concerning his 1588 text and the true nature of comets. For all Rothmann's prior interest in this subject, there can be little doubt that the topic was introduced to the letter so as to be swiftly printed.

One important Tychonic dispute was 'settled' in the *Epistolae astronomicae* not through the correspondence, but within one of the editorial addresses to the reader. Upon visiting Hven in 1590,

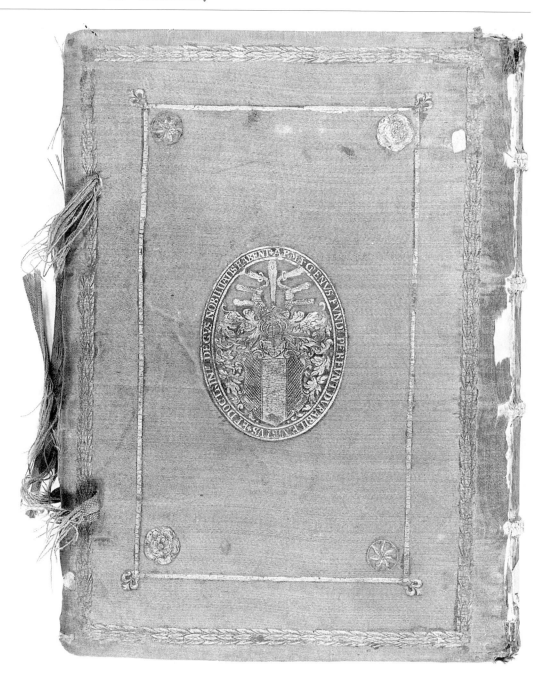

Figure 6.3 Back and front of the binding of a presentation copy of the *Epistolae astronomicae* (1596) in the Herzog August Bibliothek, Wolfenbüttel [8 Astron.]. The *quarto* binding is apple-green silk, with gold tooling, gilt edges and silk ties. On the front cover, the portrait of Tycho stamped in gold is surrounded by the inscription: 'This book displays on the outside the figure

of Tycho Brahe / More beautiful shines that which is contained within.' The back cover shows Tycho's coat of arms, with the legend: 'Arms, lines and estates perish, but virtue and erudition preserve the honour of the nobleman.' Identical bindings have been found on gift-copies of the *Mechanica* (1598). (By courtesy of the Herzog August Bibliothek, Wolfenbüttel)

Tycho reported, Rothmann had finally been persuaded of the absurdity of Copernicanism. The arguments Tycho presented as conclusive in settling their arguments were subsequently cited by other authors, and came to be employed by the Italian cleric Franciscus Ingoli (1578–1649) in debate with Galileo Galilei, another famous astronomical disputant.[22] Such contests may seem dominated by the forceful personalities involved, but did in fact make important contributions to the reconstitution of the discipline of astronomy. When, for example, Tycho delegated to Kepler the chore of responding to the *Tractatus*, Ursus' reply to the *Epistolae astronomicae*, he produced an important rejoinder to another outgrowth of sixteenth-century humanism, the radical scepticism about empirical routes to knowledge of the heavens that had been elaborated by the Frenchman Petrus Ramus (1515–72), and which both Tycho and Rothmann had also striven to refute.[23]

It is perhaps fitting to conclude by reflecting on the status of this *Apologia* of Kepler's. Although viewed by its own author as a work of philology, not astronomy, it was intended as a contribution to a public quarrel mediated by letters, printed books and letters in printed books, and one in which the theft of a manuscript diagram was a contested claim. Instruments were also involved, in the form both of the observing apparatus that Tycho claimed legitimated his new cosmology, and of models constructed to display the rival world-systems. It was a work that engaged closely with earlier sources in order to construct an elegant defence of the status of astronomical hypotheses via a history of the discipline of studying the heavens. And, despite being written for publication, it was not printed until the nineteenth century. Thus, like the other texts discussed, it supports the view that writing and reading, not always of printed matter, were practices as central to the pursuit of early-modern astronomy as those of observing, calculating and theorising. At the same time, however, the circumstances of the *Apologia*'s composition illustrate the extent to which one form of astronomical work, a book or manuscript, might result from, and draw upon, the production and reception not just of other examples of the same medium and genre, but also those of a whole range of astronomical objects; objects whose circulation amongst astronomers and their patrons, accomplished for a variety of reasons, was of the greatest significance for the development of the art.

Notes

I would like to thank Nicholas Jardine for comments on an earlier draft of this essay, Miguel Granada for access to unpublished transcriptions of Rothmann's manuscripts, and Nicholas Pickwood for bringing the copy of *Epistolae astronomicae* at Wolfenbüttel to my attention.

1 Translated in N. Jardine, *The Birth of History and Philosophy of Science: Kepler's* A Defence of Tycho against Ursus *with Essays on its Provenance and Significance* (Cambridge, 1984; corrected ed. 1988), pp. 277–8.
2 See O. Gingerich, 'Apianus' Astronomicum Caesareum and its Leipzig facsimile', *Journal for the History of Astronomy*, 2 (1971): 168–77; and on globe-manuals, the many examples in P. van der Krogt, *Globi Neerlandici. The Production of Globes in the Low Countries* (Utrecht, 1993).
3 See A. Johns, *The Nature of the Book: Print and Knowledge in the Making* (Chicago, 1998), pp. 6–28.
4 For biographical details see V. Thoren, *The Lord of Uraniborg: A Biography of Tycho Brahe* (Cambridge, 1990); J. Dreyer, *Tycho Brahe: A Picture of Scientific Life and Work in the Sixteenth Century* (Edinburgh, 1890).
5 On Regiomontanus, see N. Swerdlow, 'Astronomy in the Renaissance', in C. Walker (ed.), *Astronomy Before the Telescope* (London, 1996), pp. 188–95.
6 See O. Gingerich, 'The role of Erasmus Reinhold and the Prutenic tables in the dissemination of Copernican theory', *Studia Copernicana*, 6 (1973): 43–63, 123–5.
7 References are to J. Dreyer (ed.) *Tychonis Brahe Dani opera omnia* (Copenhagen, 1913–1929), 15 vols.
8 See A. Blair, 'Tycho Brahe's critique of Copernicus and the Copernican system', *Journal for the History of Ideas*, 51 (1990): 355–77; M. Granada, *El debate cosmológico en 1588. Bruno, Brahe, Rothmann, Ursus, Röslin* (Naples, 1996), pp. 45–76; B. Moran, 'Christoph Rothmann, the Copernican theory, and institutional and technical influences on the criticism of Aristotelian cosmology', *Sixteenth Century Journal*, 13 (1982): 85–108.
9 A good introduction is J. Rice Henderson, 'Defining the genre of the letter: Juan Luis Vives *De conscribendis epistolis*', *Renaissance and Reformation*, 7 (1983): 89–105; see also L. Jardine's *Erasmus, Man of Letters: The Construction of Charisma in Print* (Princeton, 1993), pp. 148–74.
10 N. Jardine, *The Birth of History and Philosophy of Science*, p. 35.
11 See T. Janson, *Latin Prose Prefaces: Studies in Literary Conventions* (Stockholm, 1964).
12 Thoren, *The Lord of Uraniborg*, pp. 63–5.
13 See P. Findlen, 'Cabinets, collecting and natural philosophy', in E. Fučikova et al. (eds.), *Rudolf II and Prague: The Imperial Court and Residential City as the Cultural and Spiritual Heart of Central Europe* (Prague and London, 1997), pp. 209–19.
14 Thoren, *The Lord of Uraniborg*, pp. 410–13.
15 On Kepler's difficulties see C. D. Hellmann's translation of M. Caspar's *Kepler* (London and New York, 1959). Hellmann's own *The Comet of 1577: Its Place in the History of Astronomy* (New York, 1944; reprinted 1971), gives a good indication of the magnitude of sixteenth-century astronomical/astrological publications.
16 See P. van der Krogt, *Globi Neerlandici*; the text of a privilege awarded

by Rudolf II to Landgrave Wilhelm's instrument-maker Jost Bürgi for a surveying device and accompanying manual is reproduced in L. von Mackensen (ed.), *Die erste Sternwarte Europas mit ihren Instrumenten und Uhren. 400 Jahre Jost Bürgi in Kassel*, 3rd edn. (Munich, 1988), pp. 33–4.
17. In W. Snel's *Descriptio cometae* (Leiden, 1618), pp. 69–156.
18. See O. Gingerich and R. Westman (1978), 'The Wittich connection', *Transactions of the American Philosophical Society*, 78 (1988), part 7.
19. See E. Aiton, 'Johannes Kepler and the "Mysterium Cosmographicum"', *Sudhoffs Archiv*, 61 (1977), p. 179; Thoren, *The Lord of Uraniborg*, pp. 453–5.
20. On this see M. Biagioli, *Galileo, Courtier: The Practice of Science in the Culture of Absolutism* (Chicago, 1993), pp. 159–209, especially pp. 208–9. The *Tertius interveniens* is discussed in Caspar's *Kepler*, pp. 181–5.
21. See also O. Gingerich and J. Voelkel, 'Tycho Brahe's Copernican campaign', *Journal for the History of Astronomy*, 29 (1998), especially pp. 23–8.
22. For Ingoli's 'De situ et quiete Terre' and Galileo's response see A. Favaro (ed.), *Le Opere di Galileo Galilei* (Florence, 1890–1907), 21 vols., vol. IV, pp. 403–12, and vol. VI, pp. 509–61.
23. See N. Jardine, *The Birth of History and Philosophy of Science*. Rothmann's attack on Ramist scepticism appears in his unpublished *Observationum stellarum fixarum liber primus*, Murchardsche Bibliothek, Kassel, 2° Ms. Astr. 5. nr. 7.

Further reading

J. Bennett and D. Bertoloni Meli, *Sphaera Mundi: Astronomy Books in the Whipple Museum 1478–1600* (Cambridge, 1994)

N. Z. Davis, 'Beyond the market: books as gifts in sixteenth-century France', *Transactions of the Royal Historical Society*, 5th Series, 33 (1983): 69–88

G. Galilei, *Sidereus Nuncius or The Sidereal Messenger*, translated with introduction, conclusion and notes by A. van Helden (Chicago, 1989)

A. Grafton, 'Humanism and science in Rudolphine Prague', in his *Defenders of the Text: The Traditions of Scholarship in an Age of Science, 1450–1800* (Cambridge, Mass., 1991)

N. Jardine, *The Birth of History and Philosophy of Science: Kepler's A Defence of Tycho against Ursus with Essays on its Provenance and Significance* (Cambridge, 1984; corrected 1988)

'The places of astronomy in early-modern culture', *Journal for the History of Astronomy*, 29 (1998): 49–62

S. Kettering, 'Gift-giving and patronage in early-modern France', *French History*, 2 (1988): 131–51

B. Moran, 'Wilhelm IV of Hesse-Kassel: informal communication and the aristocratic context of discovery', in T. Nickles (ed.), *Scientific Discovery: Case Studies* (Dordrecht, 1980), pp. 67–96

N. Swerdlow, 'Astronomy in the Renaissance', in C. Walker (ed.), *Astronomy Before the Telescope* (London, 1996), pp. 187–230

E. Tennant, 'The protection of invention: printing privileges in early modern Germany', in S. Schindler and G. Williams (eds.), *Knowledge, Science, and Literature in Early Modern Germany* (Chapel Hill and London, 1996), pp. 7–48

V. Thoren, *The Lord of Uraniborg: A Biography of Tycho Brahe* (Cambridge, 1990)

R. Westman, 'The astronomer's role in the sixteenth century: a preliminary study', *History of Science*, 18 (1980): 105–47

7 Reading for the philosophers' stone

In 1584 John Dee, the astrologer and mathematician, and Edward Kelley, the alchemist, alleged forger and necromancer, arrived at the court of Rudolf II in Prague. They brought with them a red powder, which Kelley used to demonstrate the transmutation of base metals into gold. He gave some of the powder to one 'Count Sellich', a privy councillor to Rudolf II, who gave it to the prior of an Augustinian convent, who gave it to Wenzel Seyler. Seyler was an alchemist at the court of Leopold I in the 1670s, and, like Kelley, he used the powder to demonstrate alchemical transmutation. Robert Boyle, the gentleman philosopher and a 'founder of modern science', recorded these details along with other eye-witness accounts of Seyler's transmutations as evidence for the existence of the philosophers' stone.[1]

According to Daniel Georg Morhof, the German historian who met Boyle in 1670, Boyle had pursued the history of the red powder further than his extant notes record. Boyle had interviewed one of Kelley's relatives, and was told that Kelley, who had been working as a scrivener in London, fled from a warrant for arrest for forgery to an inn in Wales where the innkeeper sold him an ancient book and an ivory box full of red powder for one pound sterling.[2] In the early 1650s Elias Ashmole, the astrologer and antiquary, recorded a different story in which Kelley and Dee together had found the powder in the ruined tomb of a bishop in Glastonbury.[3] Ashmole later recorded that Kelley and Dee had received a black powder from an Italian while they were in Prague, noting that he had heard this account from William Backhouse, his alchemical master, who had heard the story from someone who had been told it by Kelley himself.[4] In neither version did Ashmole mention the ancient book. Two years later, in 1654, Sir Thomas Browne wrote to Ashmole with another account. Browne described some powder and a book which were found together 'in some old Place', and noted that Arthur Dee, John Dee's son, had told him this.[5]

In John Dee's notes we find yet another version of the story. In March 1583 Edward Kelley and one Mr John Husey came to see Dee with a scroll, a book and some red powder that they had unearthed at Northwick Hill, near Blockley in the Cotswolds. They had found these items 'by the direction and leading of some such a spiritual Creature', who then told them to return to Dee for an exposition of the meaning of the scroll, which was written in strange characters. The scroll contained the names of ten places where the Danes had buried treasure in England, the powder was a form of the philosophers' stone, and the book was written by the tenth-century archbishop of Canterbury, St Dunstan, and contained the secrets of the philosophers' stone.[6] Over the next three or four years Dee would receive twenty-eight volumes of angelic dictates and numerous other magical items.

While the versions of this story differ about the location and the presence of the book and the scroll, they all focus on the powder. In accounts of transmutations, known as transmutation histories, the philosophers' stone, usually in a powdered form, played a central role. The working of the powder proved the existence of the stone. Ransacking tombs for the red powder was one approach to verifying the existence of the stone; but in order to learn how to make it and how to use it, the alchemist needed to unlock the texts which contained these secrets. Study, along with prayer, were essential activities for anyone who wished to conduct successful alchemical experiments (see Figure 7.1).

Alchemy was the art of combining a series of ingredients at the correct times and temperatures. The result was the philosophers' stone, a substance which could transmute base metals into gold, and prolong life. Alchemical secrets were conventionally passed from master to pupil orally; they were also written down in a range of genres, in verse and in prose, in allegories and emblems. Some alchemists, such as George Ripley, and later Robert Boyle and George Starkey (alias Eirenaeus Philalethes) devoted their writings to ingeniously encrypted expositions of alchemical procedures. An alternative view, adopted in many of the writings of Paracelsus and his followers, focused instead on the philosophy of alchemy and used hermetic and gnostic concepts to explain alchemy as analogous to creation, and disease as a corruption of the body to which man had been susceptible since the Fall. In this context the pursuit of the philosophers' stone was a pursuit of the secrets of creation, and a quest to return to the Edenic state of purity and health.[7]

By the second half of the sixteenth century, numerous alchemical manuscripts were in circulation in England both in Latin and the vernacular. Some of these were ancient, some medieval, others more recent; some were genuine, other spurious. Towards the end of the century alchemical books began to be translated into the vernacular, but most of the corpus would have had to be read either in Latin books or manuscripts or in translation in manuscripts. A surge of publications began in the 1650s which would climax with the efforts of the publisher William Cooper in the 1680s.

Figure 7.1 An emblem from Heinrich Khunrath's *Amphitheatrum sapientiae aeternae* (Hanover, 1609) in which the alchemist's combined pursuits of manual labour and meditative prayer and study are represented. (Reproduced by permission of the Syndics of Cambridge University Library)

Alchemical texts became more accessible, but print did not triumph.[8] A combination of the reluctance of English printers before the 1650s to publish alchemical texts, and the secretive nature of alchemy resulted in an enduring culture in which manuscripts were sought after and read alongside printed books. What an alchemist could read depended on when and where he lived, and how educated and wealthy he was. It might have been a printed book or a tattered manuscript. He might have bought it or borrowed it from a friend. Whatever form the text took, it might have borne the name of an ancient alchemist, a pseudonym, or no name at all. It might have recorded when and where it was written, or it might have been an unanchored fragment. It might have described the magical powers of the stone, or how to transmute copper into gold. An alchemist needed to know how to decipher the obscure tropes or mystical ventures of a text, and also how to assess its authority.

This essay will explore some of the ways in which alchemical texts were read in early-modern England by charting the history of the book which Kelley brought to Dee – the Book of Dunstan. Or rather, since Dee is the last known person to have seen this book, we will trace its legacy. This will lead us through the readings and writings of Elias Ashmole and Isaac Newton, and to a corpus of manuscripts which constitute an English tradition which I call hermetic alchemy. These texts describe the alchemist's quest as the pursuit of a substance which, through revelation, frees man from the corruption of the Fall, restoring his health, and enabling him to commune with angels. For some people this knowledge was sought by reading alchemical texts, written or printed; for others an ancient alchemical manuscript, such as the Book of Dunstan, might itself act as a vehicle for divine inspiration.

Elias Ashmole and 'a Booke which E.G.A.I made much use of'

Elias Ashmole's *Theatrum chemicum Britannicum* (1652) contains the seminal works of a tradition of English alchemical poetry (a planned volume of prose was not completed), and marks the beginning of a period in which substantial numbers of alchemical texts would be printed in English. Ashmole's antiquarian pursuits, in addition, resulted in the preservation of thousands of pages of alchemical manuscripts. In the prolegomena to *Theatrum chemicum Britannicum*, Ashmole signalled the importance of the Book of Dunstan in his account of the types of the philosophers' stone. Conventionally, there were three species of the stone. The mineral stone could be used to transmute base metals into gold, usually through a process called projection; the vegetable stone had the power to improve plants and animals and to make them grow; the animal stone was often called the elixir of life and healed man's body. Ashmole described a variation of this scheme in which the animal stone was re-named the angelic stone and a magical stone was added. The angelic stone contained the secret of eternal life and enabled man to see and to converse with angels. It could not be seen, but was identifiable by its sweet aroma and flavour, and was accordingly called the food of angels. The magical or prospective stone provided the power to see the past and the future, near and far, indeed the whole world all at once, and to understand the language of creatures. Ashmole attributed this four-part scheme to 'S. Dunstans Worke De Occulta Philosophia'.[9]

An alchemical treatise attributed to Dunstan had circulated since at least the fifteenth century. This treatise is devoted to the vegetable stone, and is related to the writings of George Ripley, the fifteenth-century alchemist whose numerous writings circulated widely in England and abroad. While the treatise mentions the elixir of life, its emphasis is on a series of procedures for producing the vegetable stone. It is referred to variously as Liber Dunstani and 'A treatis of the most excelent Mr Dunstan Bishop of Canterburie, a true philosopher, concerninge the philosophers stone'; it is never called 'De Occulta Philosophia', and does not mention anything resembling the angelic or magical stones.

When Ashmole cited Dunstan's 'De Occulta Philosophia', he parenthetically noted that this was 'a Booke which E.G.A.I. made much use of'. This alludes to a work by Edwardus Generosus Anglicus Innominatus, 'The Epitome of the Treasure of Health' (see Figure 7.2).[10] This treatise is dated 1562, two decades before Edward Kelley found the Book of Dunstan. Either the book which Edwardus read and the book which Kelley unearthed were one and the same, or they were copies of the same text, or they contained the same information in a different form. One possibility is that the dating of Edwardus' manuscript is false, and that it was written around the time that Kelley found the Book of Dunstan. Another possibility is that, for instance, Kelley fabricated the Book of Dunstan, and wrote the Epitome in order to authenticate and promote Dunstan's secrets.[11] This is conjecture. Even if this, or a similar scenario, could be documented, it would not change how Ashmole and other early-modern readers judged these texts.

Ashmole was not looking for the Book of Dunstan. In the two accounts of how Dee and Kelley found the powder, described above, Ashmole does not mention a book, let alone the Book of Dunstan. In Dee's diary for 1583, which Ashmole owned during the 1650s, there is no mention of the book or scroll. Ashmole did not have access to Dee's angelic conversations, in which the powder is accompanied by the scroll and the book, until he acquired one of Dee's angelic notebooks in 1672 (see Figure 7.3).[12] Even after this date, Ashmole left no signs of having associated the Book of Dunstan possessed by Dee and Kelley with the work described in the Epitome, and it remains unclear why Ashmole referred to Dunstan's work as 'De Occulta Philosophia'. For Ashmole, that the Epitome was dated 1562, two decades before the Book of Dunstan was unearthed, was not a problem. Nor, it seems, was relying on the Epitome for an account of Dunstan's ideas.

The Epitome falls roughly into three sections. The first section discusses the alchemical regimen, that is the sequence of processes of heating and cooling materials, and draws on authors such as Ripley and his contemporary, Thomas Norton. The final section

Reading for the philosophers' stone 137

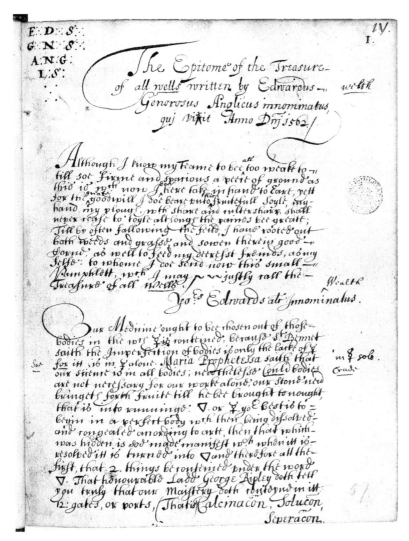

Figure 7.2 Ashmole's copy of 'The Epitome of the Treasure of Health' (Ashm. 1419, fol. 57). Note that this copy has a singular variation in its title, 'The Epitome of the Treasure of all wells', meaning Wales, which Ashmole twice corrected to read 'Wealth'. (Reproduced by permission of the Bodleian Library, Oxford)

discusses the creation of the world and of Adam, the nature of the soul, and the analogies between these and alchemical processes in hermetic terms. The middle section is devoted to the types of the stone, and contains the exposition of Dunstan's four-part scheme as noted by Ashmole. After an account of how to make what he describes as a particularly invaluable stone, Edwardus notes that he has heard about another sort. This is Dunstan's angelic stone. Edwardus then describes Dunstan's definitions of the magical, vegetable and mineral stones, and notes that the angelic stone was possessed by Hermes and the magical stone by Moses. Then he returns to his own experiments, and describes the properties of the golden and silver stones; these, he implies, were related to Dunstan's magical stone. They were derived from one substance but had

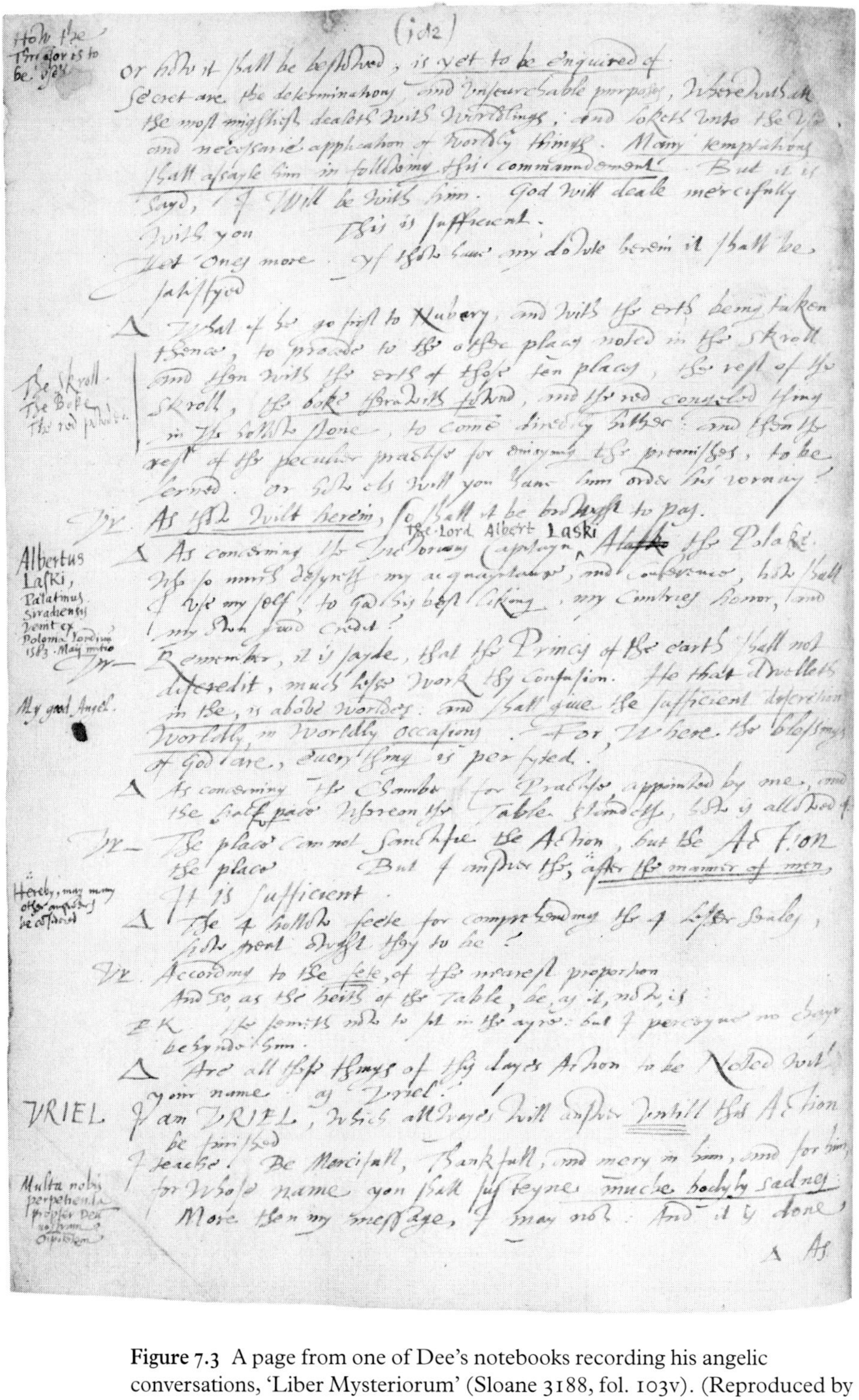

Figure 7.3 A page from one of Dee's notebooks recording his angelic conversations, 'Liber Mysteriorum' (Sloane 3188, fol. 103v). (Reproduced by permission of the British Library)

opposite properties, the former hot, the latter cold. While such an account of the different types of the stone is atypical, in what follows this treatise becomes highly unusual.

Edwardus recounted how he made the silver stone, which produced a most remarkable light. He took it to bed with him as a reading-light, and to illuminate under the covers so that he might catch fleas. But he soon began to feel chronically cold, and the fleas began to die. Eventually he realised that the stone was the cause of this: it benumbed the whole body, congealed all of the vital fluids, and eventually would have killed him. Luckily, he had remembered something which St Dunstan had written, and following his recipe, prepared a remedy. This was the golden stone, in the form of a red oil, and it had its own perils. He had put it in a tower with lots of windows in order to dry the luting on the glass before continuing with the experiment, and went to dine with a neighbour. But the location of the sun, and the conjunction of the planets, were such that the tower 'was to all likelyhood of a light bright burneing red flame of [fire] as not only itt seemed ^ soe ^ to all the neighbours, dwellers & country thereabouts, whereby they came running and wondring to the tower by the way crying [fire], [fire], [fire] as if they had binn madde'. Edwardus ran home, insisted that he go into the tower alone, and discovered that the apparent fire was the red glow of the oil. It had become so hot that the leading in the windows of the tower had melted. He removed the container from the window, burning his hands, and had to explain to the local justice how he had quenched the (apparent) fire so quickly, and without water.[13]

Ashmole did not believe everything he read. He warned the reader of his translation of Arthur Dee's *Fasciculus chemicus* (1650) that not everything that was written about the philosophers' stone was true: no one publicly professed its existence and no one had seen it.[14] He divided the authors of alchemical texts into four types. The first type 'lay down the whole Mystery *faithfully* and *plainly*; giving you a *Clew*, as well as shewing you a *Labyrinth*'. The second type wrote to demonstrate their expertise, not to instruct. The third type 'out of *Ignorance* or *Mistake*, have delivered blind and unbottomed *Fictions*'. And lastly, the 'worst sort of all, are those, who through *Envy* have scattered abroad their unfaithful *recipes* and false *glosses*'.[15] The first two sorts were to be trusted, and the difference between them and the second two sorts was experience. Two years later, in the prolegomena to *Theatrum chemicum Britannicum*, Ashmole likewise warned the reader that many philosophers had written 'what their *Fancies*, not their *Hands* had wrought'; that is they had written before they had seen transmutation. These works had been of great harm to students who were 'at first not able to distinguish, who have written upon their undeceivable *Experience*, who not'. It is unclear whether Ashmole had himself seen projection; he thought he knew enough to distinguish between true and false texts, but had failed to set himself 'effectually upon the Manual practice'.[16] Instead of writing an alchemical treatise of his own, he furthered the course of the art by publishing the texts of others.

If Ashmole could not verify alchemical experiments, how did he distinguish between true and false treatises? His editorial practices reveal three criteria for judging the veracity of a text. Firstly, Ashmole assessed the processes which it described. For instance, he included extracts from a treatise by John Gower, a contemporary of Chaucer's, in *Theatrum chemicum Britannicum*, and in the annotations described him as 'placed in the *Register* of our *Hermetique Philosophers*', noting 'In this litle *Fragment* it appeares he fully understood the Secret, for he gives you a faithfull account of the *Properties* of the *Minerall*, *Vegitable*, and *Animall Stones*, and affirmes the *Art* to be true.'[17] Even though Ashmole had not seen the stone, he knew in theory how to make it. The second and third criteria by which Ashmole evaluated a text were external. He included histories of how numerous authors, including Kelley and Dee, had acquired the secret in his annotations to *Theatrum chemicum Britannicum*. Ashmole also stressed the importance of the history of the text. In his annotations to George Ripley's 'Compound of Alchemy', for instance, Ashmole noted that it had been printed in 1591, and explained that he had relied on a manuscript that was written 'about the time Ripley lived' because 'in these Streames of *Learning* the more clearest and without the least of *Mixture* is to be found nearest the *Spring-heed*'.[18] Throughout his collection of manuscripts he underlined words and phrases to indicate textual variants. In his copy of *Theatrum chemicum Britannicum*, next to the text of Norton's 'Ordinal of Alchemy', Ashmole noted 'what words are scored under; I have thereby noted the severall readings, as they are in other Coppies; but they are not all to be trusted; for they are not soe true as ye Print'.[19] This attention to bibliographical details is evident throughout the papers and books collected by Ashmole. Where possible he evaluated an alchemical tract according to the accuracy of its contents, the history of the text, and the circumstances of its composition.

Ashmole applied these bibliographical standards to the Epitome. In 1652 he compared his copy of the text with one ostensibly in the hand of Edwardus and noted variant words and phrases. He had an authentic text, but, it seems, did not know the identity of its author. Except for his citation of the account of Dunstan's types of the philosophers' stone, we do not know how he read this manuscript, except that he noted variants between two different versions; perhaps Ashmole's silence about the powers of Edwardus' golden and silver stones is evidence of his scepticism. Dunstan, in contrast to Edwardus, was a renowned divine, legendary prophet and a notable alchemist, but Ashmole lacked the text written by him, let alone a manuscript dating from his lifetime.[20]

An explanation for why Ashmole endorsed the ideas attributed to Dunstan and recounted by Edwardus does not lie in bibliography, but in the other texts which Ashmole read. The hundreds of alchemical books and manuscripts which he acquired, and how he read them, merit further study. It is possible to situate the ideas attributed to Dunstan within a tradition of hermetic alchemy current in England from the 1570s, and to ascertain Ashmole's exposure to texts containing such ideas. As already noted, the Dunstanian account of the philosophers' stone emphasised the prolongation of life, associated this with the ability to commune with angels, and added a magical stone by which men could increase their natural powers. These ideas were not new, but the combination and emphasis were. In the 1590s, for instance, Simon Forman, the astrologer physician, described a four-part stone which echoed the account in the Epitome, though he did not cite Dunstan. Forman's manuscripts document a tradition of alchemy which combined gnostic, hermetic and Paracelsian ideas with more conventional notions and stressed the powers of the philosophers' stone to free the body of the corruption that had ensued since the Fall of Adam and Eve.[21]

Further evidence of a tradition of hermetic alchemy is found in a number of treatises which are of unknown authorship and uncertain date, though most seem to have been written in the late sixteenth or early seventeenth centuries. For instance, a treatise entitled 'The Apocalypse or Revelacon of the Secret Spirit' focuses on the powers of the philosophers' stone to prolong life, and, like the Epitome, describes an invisible substance. In this case the secrets of nature are described as having 'neyther culler nor forme'.[22] Although this treatise was printed in Latin with an Italian commentary in London in 1566, this, and the English manuscript versions of it, are extremely scarce.[23] Similarly, a treatise entitled 'The Ph[ilosoph]ers Physicke' which focuses solely on the medicinal virtues of the stone, seems to have survived in a single manuscript.[24] Other treatises which we might consider as part of this tradition of medical and spiritual alchemy had a relatively larger circulation. For instance, a treatise entitled 'Manna', or spiritual food, circulated in manuscript and was printed in 1680.[25] This describes the healing virtues and magical uses of the stone. Likewise, 'The Way to Bliss', a lengthy treatise which meditates on the state of man in hermetic terms and advocates the use of alchemy to prolong life and transmute base metals into gold, circulated in manuscript until it was published by Ashmole in 1658.[26] Although Ashmole did not acquire Forman's manuscripts until 1677, he copied Manna and the Epitome into the same notebook in the early 1650s. Along with the Epitome, he referred to Manna in the prolegomena to *Theatrum chemicum Britannicum*, noting that the 'incomparable author' stated that the mineral stone could be used for more than making gold.[27] The Epitome and Manna were associated with each other, and with Dunstan's treatise on the vegetable stone, in a number of manuscript compilations.[28]

Ashmole's prolegomena drew on and endorsed a tradition of hermetic alchemy. This tradition, however, is not explicit in the collection of poems which comprise the bulk of *Theatrum chemicum Britannicum*. In his annotations Ashmole provided the link between these alchemical texts and hermetic ideas about the prolongation of life and natural magic. For instance, he glossed a mention of the red stone in Norton's 'Ordinal of Alchemy' with a long description of the workings of physic. He drew on Paracelsian treatises such as Richard Bostocke's *The Difference Betwene the Aunceient Phisicke . . . and the Latter Phisicke* (1585) and Petrus Severinus' *Idea medicinae philosophicae* (1570).[29] Ashmole framed seminal English alchemical poems with an exposition of alchemical ideas conventionally attributed to Paracelsus and his followers; Ashmole, however, implied that this tradition derived from St Dunstan. For Ashmole, the Dunstanian scheme was part of the tradition of hermetic, revealed alchemy; a tradition rooted in tenth-century England. This history of a philosophers' stone which could be invisible and aromatic or magical was not confirmed by a red powder or an old book; it was documented in the Epitome and corroborated by a number of anonymous Elizabethan and Jacobean manuscripts.

Isaac Newton and 'The Epitome of the Treasure of Health'

Isaac Newton was not interested in Dee and Kelley. He did, however, possess one of the four other copies of the Epitome now extant. Newton's accumulation of alchemical works and composition of alchemical notes from the late 1660s through 1696 (and perhaps beyond) have been rehearsed by several scholars. Until recently Newton has been the locus of discussions about the place of alchemy in early-modern natural philosophy, and it is generally accepted that for Newton alchemy and physics, mediated by theology, were complementary projects in exploring the workings of the universe. While Newton's engagement with mystical, vitalistic and corpuscularian alchemical traditions has been studied, his reading of the corpus of texts from late sixteenth- and early seventeenth-century England has not been explored, except in Frances Yates' unfounded conjectures about the history of Rosicrucianism from Dee to Newton.[30] Despite, or perhaps because of, Yates' provocative conclusions, a historiographical rift remains in histories of alchemy in early-modern England. Studies of the period from the revival of alchemy in the 1570s through the 1650s emphasise Paracelsian doctrines of medical innovation and the political contexts in which such ideas thrived; studies of the second half of the seventeenth century focus on the procedures of transmutational alchemy and argue for the influences of alchemical theories of matter on physics and chemistry.

Newton began studying alchemy in the 1660s, and some time before April 1669 he included Dunstan's treatise on the vegetable stone in a list of works which he hoped to read.[31] Between 1668 and 1675 Newton acquired two of the anonymous treatises noted above, 'The Apocalypse or Revelacon of the Secret Spirit' and 'The Ph[ilosoph]ers Physicke', though he seems not to have paid them much attention after 1680.[32] Some time before 1675 Newton transcribed the Epitome and acquired a copy of Manna, and, like Ashmole, he noted variations between these manuscripts and others. He frequently cited these treatises in the alchemical notes which he compiled through the 1680s.

Newton left copious, if intractable, traces of how he read alchemical texts. Most famously, he 'dogeared', or turned down, the corners of pages, aiming the point at a word or passage he wished to note, and for the majority of texts he left no other traces of reading. He also occasionally noted references to other authors or interpretations in the margins; he compiled a commonplace book, the 'Index Chemicus'; he wrote treatments of specific subjects, such as the regimen, in attempts to reconcile information from different texts; and he kept records of experiments.[33] The mammoth task of exploring the complex interrelationships between these documents has not yet been undertaken. In addition, Newton's methods for reading and writing alchemical texts should be considered alongside his other scholarly pursuits. Bearing in mind these difficulties, there is some evidence for how he read the Epitome. He did not dogear or annotate his manuscript, but cited it in his alchemical notes.

The Index Chemicus consists of a list of alchemical headings followed by citations of relevant pages in texts, and occasionally a brief exposition. These notes, and Newton's changes to them during a series of at least four redraftings during the 1680s, reveal some of his reading habits.[34] In his citations of the Epitome throughout the various versions of the Index, Newton drew on its opening and concluding sections, which focus on the regimen and the hermetic exposition of alchemy, and he neglected the middle section which discusses the types of the stone and includes the exposition of Dunstan's ideas and Edwardus' own alchemical adventures. There is one exception. In what seems to be the earliest version of the Index, under the heading for the mineral, vegetable and animal stones, Newton cited five items (see Figure 7.4).[35] The first is to the section of the Epitome describing Dunstan's four-part philosophers' stone. The third, fourth and fifth are to pages in *Theatrum chemicum Britannicum* in which the poems describe the three sorts of the stone.[36] The second reference notes 'El Ashm', is then obliterated by a drop of liquid which has dissolved the ink, and continues 'f. Theatr en ~~Dunstan~~.<E.Gen.>'. This appears to be a reference to Ashmole's account of Edwardus' account of Dunstan. Perhaps Newton replaced Dunstan with Edwardus because he thought Ashmole based his account on the Epitome, or perhaps he attributed the scheme to Edwardus. Either way Newton, unlike Ashmole, allowed Edwardus to supersede Dunstan.

When we consider this brief list of words in relation to the thousands of pages of alchemical material that Newton read, and the hundreds of pages that he wrote, his eclecticism comes to the fore. He read a number of texts which constitute an English tradition of hermetic alchemy, some of which he cited, others which he apparently dismissed. There is no evidence that he associated these treatises as part of an alchemical tradition. But then Newton's alchemical notes focus on procedures, procedures which seem not to have produced an invisible, aromatic substance. Newton's notes do not discuss the hermeneutics of alchemy. They nonetheless document the energies which he, like many alchemists, devoted to charting the true path through a labyrinth of obscure alchemical instructions. The goal was the correct procedure for making the philosophers' stone. What this pursuit meant to Newton might become clearer if we turn to his theological writings. Newton was trying to discover the true language of the Bible which had been corrupted over the centuries.[37] If for Newton reading alchemy was analogous to reading the Bible, then the encoded language of the alchemists had likewise to be rationalised. Perhaps for Newton alchemy, like theology, was evidence of revealed knowledge.

John Dee, Edward Kelley and the Book of Dunstan

Our pursuit of the Book of Dunstan has followed the path of the Epitome, and the reading of the ideas attributed to Dunstan therein. Ashmole edited texts; Newton compiled vast notebooks. Ashmole read alchemy nationalistically and historically, and Newton read it eclectically and rationally. Ashmole identified a tradition of hermetic alchemy; Newton did not make such a distinction. Despite their different approaches, for Ashmole and Newton alike, alchemical texts were receptacles of knowledge, albeit veiled in tropes and obsolete language.

Neither Ashmole nor Newton read the Book of Dunstan; John Dee did, or at least he tried to. When Kelley brought the scroll, the

Figure 7.4 The entry for 'lapis' in Newton's 'Index Chemicus' (Keynes 30/2, fols. 2v–3r). (Reproduced by kind permission of the Provost and Fellows of King's College, Cambridge)

book and the powder to Dee in March 1583, he did not give Dee the last two items. A month later Dee noted that the spirits advised Kelley 'to communicate to me [Dee] the boke, and the powder, and so all the rest of the roll, which was there fownd: saying, true friendes use not to hide any thing eche from other'.[38] Three years later, in April 1586, an angel instructed Dee to present the twenty-eight volumes of dictates and other objects which the angels had given him, as well as the book and powder in Kelley's possession. Here Dee describes these treasures as 'assigned by God not just to the two of us, and our children, but also to other servants of God', but 'entrusted' to Kelley's care. Dee and Kelley were then instructed to commit these items to the furnace, which they did, only for them to be miraculously restored a fortnight later.[39] The powder and the book were apparently returned to Kelley, as in April 1587 he told Dee that several little spiritual creatures had appeared to him, and one called Ben had told him that he, the spirit, had provided Kelley with the powder and the Book of Dunstan. These would be rendered powerless if Dee and Kelley did not agree to swap wives, as the angels had previously instructed them to do.[40] At some point over the next two years Kelley did give the book and the powder to Dee, and in February 1589 Dee returned these and other items to Kelley, so that he might pass them on to Lord Rosenburg, one of Rudolf's councillors.[41] Kelley and Dee parted in 1589, and Kelley stayed in Prague and continued practising alchemy. One practitioner there noted that he learned the secrets of Dunstan from Rudolf, who had them from Kelley.[42] Dee returned to England, and continued pursuing the secrets of Dunstan. In 1607, through the skryer Bartholomew Hickman, the angel Raphael told Dee that at last he would reveal 'the secret knowledge and understanding of the philosophers' stone, of the Book of Dunstans'.[43] Dee died within two years and whether or not he succeeded in reading the Book of Dunstan remains uncertain. Sir Thomas Browne was told by

Figure 7.5 The title-page of Dee's *Monas Hieroglyphica* (Antwerp, 1564), at the centre of which is his Monas. (Reproduced by permission of the Bodleian Library, Oxford)

Arthur Dee that the book contained 'nothing but Hieroglyphics, which Booke his Father bestowed much tyme upon', and added 'but I could not heare, that he could make it out'.[44] Hieroglyphics probably did not indicate a strange or antiquated script, but some sort of diagram, emblem or symbol, such as Dee's Monas, which represented the structure and genesis of the universe (see Figure 7.5). These images conveyed knowledge through divine language and inspiration.

The Book of Dunstan was found by the direction of an angel, and could not be read without angelic inspiration: it was a vehicle to angelic wisdom. Its totemic status was perhaps reinforced when, as

Dee noted in his diary, in December 1587 Kelley reported to him that he had spilled a lamp and some wine, damaging several books. These included '40 leaves in quarto, entitled "extractiones Dunstani": which he had himself extracted and noted out of Dunstan his book. And the very book of Dunstan was but cast on the bed hard by, from the table.'[45] The Book of Dunstan was miraculously preserved, much as it had been miraculously restored the year before, while extracts from its text were not. The power of the alchemical book is similarly illustrated in stories about the red powder which specify that it was found with or lodged in books. These books were not vehicles of alchemical secrets to be decoded by the adept; they were testaments to the necessity of alchemical revelation. More than half a century after Kelley brought the scroll, powder and book to Dee, Ashmole expounded the tenets of an inspired alchemy while promoting a didactic alchemical literature. For Ashmole, reading alchemical texts was a matter of mastering the language of the alchemists, not invoking spirits; at the same time, the ends of alchemy were the achievement of the angelic and magical stones. Ashmole's publication of *Theatrum chemicum Britannicum* signifies not only the presence of alchemical books in the rise of didactic literature, but their demystification.

Coda

Whatever the power of the book as an object, a tradition of hermetic alchemy could have thrived; the notion of an invisible stone, however, contained an inherent paradox. From Dee to Newton the existence of the philosophers' stone was supported by eye-witness testimony. As Sir Thomas Browne reported to Ashmole, Arthur Dee had 'ocularly undeceavably & frequently beheld' the transmutation of tin into gold.[46] A stone that could not be seen could not be proved to exist, or not to exist. An anonymous letter dated 1660 described an angelic stone with powers akin to those of the angelic and magical stones attributed to Dunstan. This stone was the way to the Tree of Life and the Garden of Eden; it was 'an oraculous method or Instrument for the seeing opening & discovering of spirits in generall, & of such invisible Powers & Substances, as the Eye can no way see'. Those who wished to achieve this state, however, had to overcome the one thing that the stone lacked: 'The palpability, security, & irrefragable certainty conveyed to the minde by it, that this rich, glorious, & beautifull order it serves of all things that they see, is what it is & what it is discovered to be; and can be noe other.'[47] The historian of alchemy faces these same obstacles. We can see and read the alchemical books and manuscripts which have survived, but any history of missing texts, invisible substances, and the ethereal spirits which governed them is by definition uncertain.

Notes

I am indebted to audiences of an early version of this paper at seminars in Cambridge and Oxford for incisive questions, to the editors of this volume for detailed comments on successive versions of this essay, and to Scott Mandelbrote, Margaret Pelling, Joad Raymond and Charles Webster for saving me from many mistakes. Unfortunately D. Harkness' *John Dee's Conversations With Angels* (Cambridge, 1999) appeared when this was in the press.

1. M. Hunter, 'Alchemy, magic and moralism in the thought of Robert Boyle', *British Journal for the History of Science*, 23 (1990): 387–410, esp. 401–6; L. Principe, *The Aspiring Adept: Robert Boyle and His Alchemical Quest* (Princeton, 1998), pp. 95–7.
2. Principe, *Aspiring Adept*, p. 196.
3. Elias Ashmole (ed.), *Theatrum chemicum Britannicum* (London, 1652), p. 481 (hereafter *TCB*).
4. C. H. Josten (ed.), *Elias Ashmole 1617–1692: His Autobiographical and Historical Notes, his Correspondence, and Other Contemporary Sources Relating to his Life and Work* (Oxford, 1966), vol. II, pp. 603–5.
5. Josten, *Ashmole*, vol. II, p. 662.
6. London, British Library, Sloane MS 3188, ff. 61, 63; 101v–3v.
7. For the history of the elixir in England see M. Pereira, 'Mater medecinarum: English physicians and the alchemical elixir in the fifteenth century', in R. French, J. Arrizabalaga, A. Cunningham and L. García-Ballester (eds.), *Medicine from the Black Death to the French Disease* (Aldershot, 1998), pp. 26–52.
8. For the circulation of literary texts in manuscript see D. Carlson, *English Humanist Books: Writers and Patrons, Manuscript and Print* (Toronto, 1993); H. Love, *Scribal Publication in Seventeenth-Century England* (Oxford and New York, 1993); A. Marotti, *Manuscript, Print, and the English Renaissance Lyric* (London, 1995); and H. R. Woudhuysen, *Sir Philip Sidney and the Circulation of Manuscripts* (Oxford, 1996).
9. Ashmole, *TCB*, sigs. A4v-B1v.
10. Oxford, Bodleian Library, Ashmole MS 1419, Item IV, ff. 57–82v. Other copies are Cambridge, King's College, Keynes MS 22, Glasgow, Glasgow University Library, Ferguson MS 199, Item IV, pp. 19–70; Sloane 2502, ff. 54–69v, 70–81v (two copies).
11. For the suggestion that Edwardus was Edward Dyer see W. Black, *A Descriptive, Analytical and Critical Catalogue of the Manuscripts Bequeathed Unto the University of Oxford by Elias Ashmole* (Oxford, 1845), col. 1144, and W. D. Macray, *Index to the Catalogue of the Manuscripts of Elias Ashmole* (Oxford, 1866), p. 51.
12. Sloane 3188, ff. 61, 63, 101v–103v; Josten, *Ashmole*, vol. III, pp. 1264–6.
13. Ashm. 1491, Item IV, f. 21.
14. Arthur Dee, *Fasciculus Chemicus: Or Chymical Collections*, trans. James Hasolle [Elias Ashmole] (London, 1650), sig. **8v.
15. Dee, *Fasciculus Chemicus*, sigs. **[1]v–**2.
16. Ashmole, *TCB*, sig. B2v.
17. Ashmole, *TCB*, p. 484.

18 Ashmole, *TCB*, p. 455.
19 Josten, *Ashmole*, vol. II, p. 600.
20 See for instance A. Clark (ed.), *'Brief Lives'* . . . *by John Aubrey* (Oxford, 1898), vol. I, pp. 242–3; R. Holinshed, *Chronicles* (London, 1587), vol. II, pp. 165–6.
21 Ashm. 1494, p. 623; L. Kassell, '"The food of angels": Simon Forman's alchemical medicine', in A. Grafton and W. R. Newman (eds.), *The Occult Sciences in the Renaissance, Archimedes*, 3 (2000).
22 Keynes 67, f. 7v.
23 Giovanni Battista Agnello, *Espositione . . . sopra un libro* (London, 1566); Ashm. 1490, ff. 15–17.
24 Keynes 67, ff. 64–99v.
25 Keynes 33; Ferguson 9, Item II, ff. 14–24; Ferguson 199, Item V, pp. 72–8 (partial copy); Sloane 2194, Item 9, ff. 77–84v; Sloane 2222, ff. 128v–136; Sloane 2585, ff. 90–105; Ashm. 1419, Item III, ff. 45–56; John Frederick Houpreght (ed.), *Aurifontina Chymica: Or a Collection of Fourteen Small Treatises Concerning the First Matter of Philosophers* (London, 1680).
26 Elias Ashmole (ed.), *The Way to Bliss* (London, 1658).
27 Ashmole, *TCB*, sig. [A4v]. For this identification see Principe, *Aspiring Adept*, p. 199.
28 Ashm. 1419; Ferguson 199; Ferguson 9.
29 Ashmole, *TCB*, pp. 448–9.
30 F. Yates, *The Rosicrucian Enlightenment* (Routledge, 1972).
31 K. Figala, J. Harrison and U. Petzold, '*De Scriptoribus Chemicus*: sources for the establishment of Isaac Newton's (al)chemical library', in P. D. Harman and A. Shapiro (eds.), *The Investigation of Difficult Things: Essays on Newton and the History of the Exact Sciences* (Cambridge, 1992), pp. 135–79.
32 Keynes 67.
33 A catalogue which includes most of Newton's alchemical papers was drawn up by Sotheby's in 1936, and is reproduced as Appendix A in Dobbs, *The Foundations of Newton's Alchemy*. A full catalogue of Newton's theological, alchemical and Mint papers is being prepared by Robert Iliffe.
34 Keynes 30/1–5.
35 Keynes 30/2, f. [2v] (lapis).
36 Ashmole, *TCB*, pp. 370, 389, 411.
37 S. Mandelbrote, '"A duty of the greatest moment": Isaac Newton and the writing of biblical criticism', *British Journal for the History of Science*, 26 (1993): 281–302, esp. 292–8.
38 Sloane 3188, f. 90v.
39 Ashm. 1790, ff. 9v–11; E. Fenton (ed.), *The Diaries of John Dee* (Charlbury, 1998), pp. 188–91.
40 Sloane 3188, f. 63; Fenton, *Diaries*, p. 217.
41 Fenton, *Diaries*, p. 238.
42 R. J. W. Evans, *Rudolf II and His World* (Oxford, 1973), p. 226.
43 Meric Casaubon (ed.), *A True & Faithful Relation of What Passed for Many Years Between Dr. John Dee and Some Spirits* (London, 1659), pt. 3, p. *34; Fenton, *Diaries*, p. 296.
44 Josten, *Ashmole*, vol. II, p. 662.

45 Fenton, *Diaries*, pp. 231–2.
46 Josten, *Ashmole*, vol. II, p. 755.
47 Sloane 648, f. 100. Principe identifies this as a hand often found in manuscripts circulating within the Hartlib circle: *Aspiring Adept*, p. 199.

Further reading

N. Clulee, *John Dee's Natural Philosophy: Between Science and Religion* (London, 1988)

A. Debus, *The English Paracelsians* (London, 1965)

A. Debus and M. T. Walton (eds.), *Reading the Book of Nature: The Other Side of the Scientific Revolution* (Kirksville, Mo., 1998)

B. J. T. Dobbs, *The Foundations of Newton's Alchemy, or 'The Hunting of the Greene Lyon'* (Cambridge, 1975)

The Janus Faces of Genius: The Role of Alchemy in Newton's Thought (Cambridge, 1991)

D. Harkness, *John Dee's Conversations with Angels: Cabala, Alchemy and the End of Nature* (Cambridge, 1999).

M. Hunter, 'Alchemy, magic and moralism in the thought of Robert Boyle', *British Journal for the History of Science*, 23 (1990): 387–410

L. Kassell, '"The food of angels": Simon Forman's alchemical medicine', in A. Grafton and W. R. Newman (eds.), *The Occult Sciences in the Renaissance, Archimedes*, 3 (2000)

H. Love, *Scribal Publication in Seventeenth-Century England* (Oxford and New York, 1993)

W. R. Newman, *Gehennical Fire: The Lives of George Starkey, an American Alchemist in the Scientific Revolution* (Cambridge, Mass., 1994)

L. Principe, *The Aspiring Adept: Robert Boyle and His Alchemical Quest* (Princeton, 1998)

P. H. Smith, *The Business of Alchemy: Science and Culture in the Holy Roman Empire* (Princeton, 1994)

C. Webster, 'Alchemical and Paracelsian medicine', in Charles Webster (ed.), *Health, Medicine and Mortality in the Sixteenth Century* (Cambridge, 1979), pp. 301–34

R. S. Westfall, 'Isaac Newton's Index Chemicus', *Ambix*, 22 (1975): 174–85

F. Yates, *Giordano Bruno and the Hermetic Tradition* (Chicago, 1964)

For general and bibliographical information about alchemy see:

L. Abraham, *A Dictionary of Alchemical Imagery* (Cambridge, 1998)

G. Roberts, *The Mirror of Alchemy: Alchemical Ideas and Images in Manuscripts and Books, from Antiquity to the Seventeenth Century* (London, 1994)

L. Thorndike, *A History of Magic and Experimental Science*, 8 vols. (New York, 1923–41)

The Alchemy Website and Virtual Library: <www.levity.com/alchemy>

8 Writing and talking of exotic animals

In 1651 a book on Mexican flora and fauna was published in Rome.¹ An arch surrounded by half-naked native Americans frames the title-page (Figure 8.1) and supports the coat of arms of Spain. Two putti hold a curtain containing the title and partially obscuring a map of New Spain; readers are invited to move the curtain aside and embark on a 1,000-page journey through images and descriptions of animals and plants of New Spain. The book's title-page is dense and exciting; its patron none less than Philip IV, King of Spain; its size impressive. And yet the *Rerum medicarum Novae Hispaniae thesaurus* (hereafter *Thesaurus*) was hardly the fresh account of a recent voyage. The journey in which it originated had taken place nearly eighty years earlier; producing the book took so long that by the time it was finally completed, most of the authors had died and the illustrious Accademia dei Lincei ('Academy of the Lynxes') to which they belonged had long since ceased its activities.

To scholars of early-modern natural investigations, the *Thesaurus* represents at once the Academy's project to follow their most famous member, Galileo, in carrying out innovative inquiries, and a disappointing outcome, a belated work which, save for short sections, was still deeply entangled in Renaissance erudition. The point has been driven home by unfavourable comparison to the *Historia naturalis Brasiliae* (1648; hereafter *Historia*), the account of the natural investigations carried out by Georg Markgraf and supported by the Dutch Governor of Brazil, Maurits von Nassau.² However, the complex process by which the *Thesaurus* was produced raises important questions about the transmission of natural knowledge and the publication of works of natural history in the early-modern period. And from this point of view, the *Thesaurus* and the *Historia* have more in common than we might expect.

The European encounter with the new world generated a vast body of travellers', missionaries' and naturalists' reports. These have recently become the focus of historians interested in the media which made possible both the encounter between Europeans and indigenous peoples, and its recounting to those who had never set foot in the new world. On the one hand, scholars have dissected the

Figure 8.1 Title-page of *Rerum medicarum Novae Hispaniae thesaurus seu plantarum animalium mineralium Mexicanorum historia ex Francisci Hernandez* (Romae, Ex Typographeio Vitalis Mascardi, 1651). (By permission of the Syndics of Cambridge University Library)

rhetorical strategies of the reports and analysed the construction of the reporters' reliability. On the other, they have examined the production and circulation of pictures of the new specimens and their impact on the discipline of *materia medica* and on collecting. Here I combine this type of investigation into forms of cultural transmission with recent work in the field of book history. This complements the more traditional analysis of the material aspects of book production with an interest in the interplay in different social settings between print and other modes of communication, especially orality.³ How did specimens to see, manuscripts to copy, woodblocks to cut, books to print and words to hear interact along the route from New Spain back to Spain and Baroque Rome? While breaking down boundaries between the oral, written, iconographical and printed communication which resulted in the *Thesaurus*, I assess the quantum of reliability which those involved in its production attributed to each medium. I argue that in the process of adjudicating testimonies, crucial in early-modern natural investigations, this type of evaluation complemented that of the social status of witnesses.

First I explore the long and complex editorial history of the *Thesaurus*. It began with the notes and images of Mexican flora and fauna which the Spanish physician Francisco Hernandez took during his journey to New Spain in the 1570s. While mapping the various forms in which these notes circulated in the seventeenth century, I focus in particular on the transformations they underwent as members of the Academy of the Lincei contributed to the production of the *Thesaurus*. I then look at the chapters of the book which Johannes Faber (1574–1629), secretary of the Academy and professor of *materia medica* at the University of Rome, devoted to twenty Mexican animals. I use the problems he faced in identifying a civet, an animal known for its scented secretion, to highlight his attempt to sift true facts from the web of words which had been transmitted in the traditional encyclopaedia. Providing first-hand accounts was held the best way to prove a naturalist's trustworthiness. But Faber had never been to New Spain and never seen a Mexican civet, and so for first-hand reports he had to substitute all the available descriptions. I discuss his strategy of incorporating a wide range of sources, including letters and woodcuts, and the credit he allowed to each. Among other sources, Faber reported the oral accounts of Father Gregorio Bolivar, a Franciscan missionary who happened to have returned from America to Rome just as Faber was writing his chapters. Bolivar was a preacher whose sophisticated oratory sought to convert the non-Christian population of the new world. In Counter-Reformation Rome more generally, oratorical abilities were highly praised as the best tools for imposing the new devotion. In the final section I focus on how, even within the commitment to visual representations which Faber shared with his fellow Lincei, he used the voice and memory of Gregorio Bolivar to recapture the wealth and variety of Mexican nature.

Mexican nature in the press

During his five-year trip to New Spain on behalf of King Philip II, Hernandez gathered natural specimens and their images, some produced by native draughtsmen, and interviewed local people extensively to discover their medical uses.[4] He did not name the healers and interpreters to whom he talked, but was impressed by their knowledge and botanical nomenclature. Translating and recording the oral knowledge of the indigenous people was an important feature of the cultural colonisation of the new world. But in the case of Hernandez, it was accompanied by a genuine interest in the variety of their languages. The plan was to publish Hernandez's notes once he was back in Spain. He had deliberately kept text and drawings separate, so that the former could be easily corrected. He also knew that only in the preparation for printing would he be able to order his untidy notes: his presence would be essential.[5] But publishing the now massive work probably proved prohibitively expensive, and the volumes of notes including hundreds of brightly coloured images were laid to rest in the Library of the Escorial, while a copy with corrections by the same Hernandez was placed in the Colegio Imperial in Madrid. The King limited himself to asking the physician Antonio Leonardo Recchi to prepare an abridgement. Hernandez's complaints about Recchi's unsuitability were in vain. Recchi cut what he considered to be too personal annotations and replaced the Aztec images with allegedly more accurate pictures.[6] The amazing history of the dissemination of Hernandez's work had started.

When Recchi returned with his own manuscript to Naples, then under Spanish rule, the founder of the Academy of the Lincei, Federico Cesi, decided to fund the publication of an expanded version.[7] He asked members of the Academy to put flesh on Recchi's sketchy chapters. But the manuscript came without the illustrations, which remained in the possession of one of Recchi's heirs. From time to time the Lincei were allowed to have a look at and copy the pictures, but only two of them had direct access to the precious original in the Escorial. Meanwhile, European naturalists never ceased to regard Hernandez's work as an extraordinary repository of information about the new world, and other versions of Hernandez's notes were produced (Figure 8.2). An augmented version of Recchi's summary was published in New Spain; the Jesuit Juan Eusebio Nieremberg incorporated portions of Hernandez's notes into his volume on natural history; scholars continued to copy extracts of the original notes which circulated in manuscript.[8] The body of data Hernandez had collected fed the work of generations of naturalists, but mainly in dismembered form, while in 1671 a fire in the Escorial would destroy his volumes.[9] Obviously, the project of the Lincei stirred great interest and in the 1610s publication looked imminent. But it was not until 1628 that a few copies came out, and of Faber's section on animals only. In 1630 the book was finally ready. Prince Cesi's lack of capital

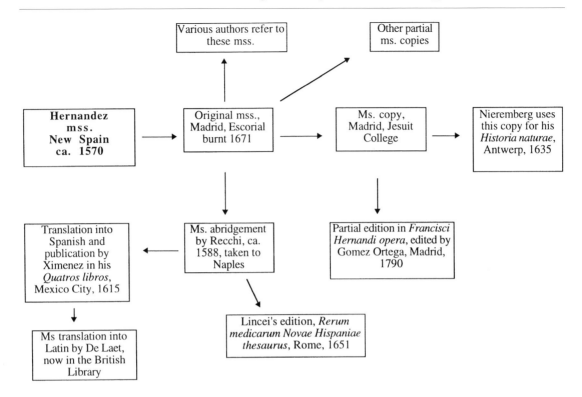

Figure 8.2 Diagram of the dissemination of Hernandez's work in the seventeenth century.

may have slowed its publication, but it was certainly his death in the same year which hindered distribution of the belatedly printed book.[10]

For almost twenty years loose printed sheets languished somewhere in Cesi's palace. In 1649, Alfonso Las Torres, the Secretary of the Spanish ambassador in Rome, bought all the copies at a bargain price. A new long chapter, indexes and dedications were added, parts which had deteriorated were replaced, more copies were produced. Finally in 1651, the *Thesaurus* started its life as a book. It comprised two parts and had a very complex structure. The first and longer part included the brief introduction which Recchi had written for his abridgement; nine chapters on plants, animals and minerals based on a selection of Recchi's abridgement and Linceo Johannes Schreck's annotations to it; another short section which he wrote on plants; Faber's 400-page commentary on animals; a brief section with more annotations on plants by the Neapolitan botanist Fabio Colonna; and twenty botanical tables by Cesi. The second part, of only ninety pages, reproduced a portion of Hernandez's notes which had been copied at the request of the Linceo Cassiano Dal Pozzo in 1626.

Production of the *Thesaurus* is fascinating because it allows us to see how its numerous participants contributed to building the multiple authorship of the volume. Hernandez's reputation and direct experience of Mexican nature certainly bestowed authority on the various versions in which his notes circulated; his name is prominent on the title-page of the *Thesaurus*. Yet his original notes had completely disappeared from the book. Recchi's role as an editor had been crucial to guaranteeing their circulation. But he never meant to offer a thorough and reliable version of the original notes and his editing weakened the link with Hernandez's experience; to the Lincei his abridgement was so sketchy that it could not possibly be published as it stood. Their editing ranged from the minimum of annotations supplied by Schreck to Faber's wealth of annotations. The former made a clear typographical distinction between Recchi's text and his brief comments; the latter followed the model of Renaissance commentaries more closely and surrounded the Mexican specimens with his own vast erudition. In addition to comparing the exotic animals to more familiar European specimens, he used the commentary to discuss the results of his research as a physician. Perspectives on Mexican nature were multiplied still further by the contribution of Colonna, who worked on some of the specimens already described by Schreck. Reading the *Thesaurus* was not a linear process: as its authors had embroidered information onto the original canvas, so were readers meant to construct their knowledge shuttling back and forth between the pages.

The *Thesaurus* is a revealing case for the study of multiple authorship and the media through which natural knowledge was communicated. But it was by no means exceptional, as a brief survey of the production of the *Historia* shows.[11] When Georg Markgraf departed from Brazil he left his notes to his patron, Johan Maurits von Nassau; soon after Markgraf died in Africa, and Maurits asked Johan de Laet, who never went to Brazil, to edit and publish the notes. They were very untidy and, more problematically, were coded, because Markgraf had been concerned about theft of his work. Though De Laet had a key, he made intrusive and inaccurate interventions; his annotations muddled things up. The *Historia* was hailed by naturalists as a marvellous source for studying Brazilian nature, but it left Francisco Piso, Markgraf's colleague in Brazil, deeply disappointed. Ten years later he published a second edition of the book, rearranging Markgraf's notes; it was heavily criticised by naturalists, who accused Piso of plagiarism.

The storage and circulation of accounts of exotic specimens was a key feature of early-modern natural history, a body of knowledge based on the co-operation of scholars, collectors and travellers but dependent on extremely unpredictable factors, including the sudden meanness of a patron, the frailty of papers and professional jealousy. The publication histories of both the *Thesaurus* and the *Historia* show that material constraints deeply affected the circula-

Figure 8.3 Woodcut of the Mexican civet, *Thesaurus*, p. 538. (By permission of the Syndics of Cambridge University Library)

tion of information, while editorial discretion which blurred distinctions between authors and editors was part of the game. Print could make knowledge more accessible and authoritative, but financial obstacles and other difficulties in printing could add yet another link to an already long chain between readers and the original source. Contemporaries were well aware of these problems as they struggled to give order to a growing mountain of information, remove mistakes from the ancient tradition of natural history and build their own credibility.

Animals or chimeras?

Faber's section on animals in the *Thesaurus* shows clearly what it meant to edit a work on Mexican nature without ever having crossed the Ocean. An illustration of the specimen accompanied by its Aztec name and a brief description open each chapter, but in Faber's survey of all the sources of the traditional zoological encyclopaedia, the vividness of the Mexican animal fades away (Figure 8.3). He aimed to establish to what European species the Mexican specimen could be assimilated. In the case of the civet naturalists disagreed even about whether it was feline or a kind of wolf.[12] Was it by any chance what the ancients used to call 'panther'? Or was it what they had called 'hyena'? In what relation did the civet stand to the musk, another animal which can produce a scented substance?

Faber was overwhelmed by the task of disentangling the mass of annotations and comments which had proliferated around the texts of Aristotelian zoology and of clearing up mistakes in their transmission. We can recognise a double process in Faber's work. On the one hand, he interrogated the authority of ancient as well as recent writers, and when they proved to be wrong, he would dismiss them with the epithet 'fabulosus' or 'hallucinatus' (p. 586). On the other, he was well aware that, hidden behind thousands of words and long-standing controversies, living animals could acquire the uncertain status of poets' creations, or, worse, of those unreal mental entities of Scholastic philosophy (p. 525). For Faber and his contemporaries poetry could provide natural knowledge on a par with other genres, and he referred to Ovid and Virgil as reliable sources of encyclopaedic knowledge.[13] But he acknowledged that the inconsistency of the information on animals was such that readers might wonder whether, for example, a lynx really existed or was rather like a sphinx or a creature out of Thomas More's *Utopia* (p. 525).

Faber himself was sometimes baffled by the nature of certain specimens, the existence of which, if not fabulous, appeared at least dubious. Dealing, for example, with an *amphisbena*, a supposedly double-headed snake, he had come to the conclusion that it did not exist: naturalists had been confused by a simple snake which could slither backwards and forwards. He had just sent his piece on the *amphisbena* to the printer, when he received an image of precisely such a double-headed snake from his friend Cassiano Dal Pozzo. Faber found this image compelling, recanted his view with the argument that nothing is impossible to God, and decided to add the woodcut to his commentary (Figure 8.4). Renaissance naturalists celebrated the variety of a nature which to them was much greater than it would be to their successors; especially at the margins of the known world nature could produce the most surprising creatures.[14] As Nieremberg put it, truth may be so amazing as to appear implausible, but it need not meet criteria of verisimilitude.[15] In this framework, to be an open-minded scholar meant being willing to accept what might have appeared, on the basis of limited experience, as unlikely as a double-headed snake. Yet Faber acknowledged that caution was needed: belief in wonders could be the result of some disgraceful diabolical intervention. And a place such as New Spain, where the Christian religion had not yet replaced superstition, was ideal territory for Satan. Banned from Europe by the victory of true religion, the devil had migrated there and instilled in the indigenous populations all sorts of wrong beliefs (p. 704). Naturalists dealing with American nature should beware.

Figure 8.4 Woodcut of the *amphisbena* as a double-headed snake, *Thesaurus*, p. 797. (By permission of the Syndics of Cambridge University Library)

Scholars have recently engaged with Shapin's argument that in deciding questions of credit and competence the social status of witnesses in early-modern natural history and natural philosophy was of paramount importance.[16] While Faber never forgot to report this status, I want to focus here on the various media through which testimonies were supplied to him. Arguing that these media affected the way he evaluated his sources, I now discuss how their specificity could be maintained and reproduced on a printed page. Faber's commentary is dotted with warnings that witnesses who testify to what they have heard are much less reliable than eye-witnesses: ears are never as trustworthy as eyes. He combined this stock maxim of jurists and historians with strong statements against ancient authorities and in support of direct experience. It had taken the image of the *amphisbena* to convince him of its existence, and in the case of the civet, he was happy to incorporate a woodcut after a European civet which his fellow Linceo Colonna had sent him. A close-up of the organs which produced the scented substance was also enclosed (Figure 8.5). Not only was Colonna a renowned botanist, he also skilfully engraved the plates of his own books.[17]

Figure 8.5 Woodcuts of the European civet and its genital organs, *Thesaurus*, p. 580. (By permission of the Syndics of Cambridge University Library)

Faber could rely on him. And yet Colonna himself warned Faber not to trust pictures: quoting a famous passage from Pliny, Colonna stressed that, especially when colours were important, pictures were problematic (p. 581). Faber knew this well: in the transmission of Hernandez's notes, the images had suffered most. And colours were among those details important for the identification of specimens which images, once they were reproduced on a printed page, could not convey. Pictures, however faithful, were not always self-evident. Written descriptions of the size and especially of the colours of the specimen were necessary for the reader to figure out what the specimen or the original drawing looked like. In the case of the European civet, Faber incorporated into his text the two long letters which Colonna had sent him with his woodcuts. Other friends and colleagues had sent him letters with details on animals which Faber reproduced in his text in their entirety. Not just the contents but the form of these letters and their informal register had to be preserved so as to add credit and authority. Such letters were the main media through which collectors and naturalists communicated and were the best surrogates for the informal conversations on which they based their co-operation when they could meet in their cabinets. In this period the power of orality as a rhetorical tool was fully exploited in the Galilean dialogues, but the monological style of the commentary still retained a strong grip on writers of natural history.[18] A compromise was to incorporate letters, or even the exchange of conversation, and capitalise on their rhetorical impact, without the dramatic change in the function of the author which a fully dialogical form would imply. This was precisely what Faber did with Father Bolivar's account.

The voice and memory of Gregorio

Faber opened his chapter on the civet with his translation into Latin of the lengthy description given him by the Spaniard Gregorio

Bolivar in the course of a meeting at his house. In weighing up dozens of accounts and descriptions of animals, Faber never failed to turn to Gregorio with absolute confidence in his ability to settle once and for all those controversial questions which illustrious and authoritative naturalists had confused. Who was this man?

Bolivar was a Franciscan, who, following the tradition of his Order, had lived for twenty-five years in the most remote areas of South America converting the indigenous population to Catholicism.[19] For this work he had learnt local languages, which proved useful in acquiring knowledge about Mexican flora and fauna. This he was happy to share with Faber when, in 1625, he came to Rome to campaign on behalf of the American Indians in front of the recently established Congregation of Propaganda Fide. To Faber, Gregorio's fervent religiosity and moral commitment – he called the preacher a 'buccinator Dei', or 'trumpeter of God' (p. 557) – bestowed on him an indisputable charisma and a matchless trustworthiness. This showed in other qualities of the friar which Faber described as follows:

I am full of admiration for the diligence of Gregorio's investigations of the animals and for his extremely persistent memory of all the animals he has ever seen. I swear to you, my reader, that what I derived up to now from his mouth and notes, and the most rare things which I will be saying, he described with a good recollection and without the aid of any book. And that he brought them together with the aid of a sort of compendium only (indeed during our long conversations he tells me many more things).[20]

Gregorio was reliable because he had seen, and because, without the filter of other people's books, he could remember. In Faber's view, his exceptional memory marked him as trustworthy; his brief notes played the relatively marginal role of triggering his recollections. Gregorio's memory re-established the contact with New Spain which Hernandez had had but was lost in the complicated history of his manuscripts. To Faber's ears, nothing could match the liveliness of Gregorio's voice:

I will add to my notes those wonders and many other things of the new world which he described and to which we used to listen, hanging anxiously on his words.[21]

Episodes of Gregorio's adventurous life, for example a dangerous encounter with a tiger, reported in the first person, crept into Faber's text (p. 508).

Faber was aware that some of his readers would find Gregorio's reports suspicious because, being oral, they shared the dubious status of hearsay. To meet this concern, Faber recalled that the annotations which Gregorio carried with him might be published soon (p. 506). He seemed to concede that the transformation of

Gregorio's account into a written and, possibly, a printed text, would enhance his credibility. And yet Faber's culture as a courtier provided him with the resources to justify his great appreciation of Gregorio's voice as the most reliable medium of communication. Against the long-standing tradition which considered hearing a dubious source of knowledge, Faber could play Guazzo's influential *Della civile conversazione*:

> It would be wrong to believe that we can acquire knowledge more in the solitude of reading than in conversation with learned men. It is proven that it is easier and better to learn by means of ears than of eyes. Nobody would ever have weak eyes or thin fingers by leafing through books, if only one could constantly enjoy the presence of their authors and receive through the ears their living voice which would impress the mind very strongly.[22]

While the function of ears was fully rehabilitated by the liveliness of the conversation, that of eyes was associated not with first-hand testimony, but with the health-threatening art of reading. The appeal of orality as an antidote to bookish and stale knowledge had led Galileo to write dialogues. As a writer of a natural historical commentary, Faber was less keen on experimenting with innovative genres, but by introducing Gregorio as a powerful counterweight to reams written on animals he pursued an analogous battle against the supposed authority of print. Faber's praise of Gregorio's memory contrasts with Buffon's comment in the chapter of his *Histoire naturelle* devoted to the civet, where he mentioned Faber only to reject his work. It was precisely because Faber had relied so heavily on Gregorio's memory that his account, in Buffon's view, had to be dismissed. In the late eighteenth century memory was no longer the mark of unquestionable experience, while Gregorio represented a kind of knowledge as worthless as it was obsolete.[23]

Gregorio was probably not a refined courtier. He stayed at the luxurious papal court for only a few months and soon returned to America, where he lost his life in a final reckless expedition to the edges of the known world. His conversations with Faber are very unlikely to have followed Guazzo's etiquette. Gregorio's oratorical skills were grounded in a different practice. To him orality was a crucial tool of proselytisation: it was inextricably linked to the art of preaching which he had cultivated deeply in his missionary career. The main media through which to indoctrinate illiterate people were sermons and images, and Counter-Reformation Catholicism particularly valued the oratorical skills of its champions (see Figure 8.6). Memory was an essential tool both for preachers and for the newly converted, who were asked to learn by heart what they could not read.[24] Faber was a courtier but he was also a devout Catholic who had been intensely involved in converting Protestant travellers to Rome: he could appreciate Gregorio's oratorical skills.

Writing and talking of exotic animals 163

Figure 8.6 D. Valades, *Rhetorica christiana ad concionandi et orandi usum accomodata utriusque facultatis exemplis suo loco insertis, quae quidem, ex Indorum maxime deprompta sunt historiis...* (Perusiae, 1579), p. 111. This image aptly conveys the co-ordination of images and words used by Franciscans preaching to the illiterate indigenous populations. (By permission of the Syndics of Cambridge University Library)

Conclusions

Before leaving Rome, Gregorio destroyed his work, possibly out of annoyance with political intrigues. His notes shared the same fate as Hernandez's: they were consumed by flames. The frailty of their material support was partly compensated by their dissemination in the writing of other people. And in this sense the editorial history of the *Thesaurus* as well as of the *Historia* are wonderful examples of the vital connections between printing and natural history. Yet this should not lead us to celebrate the contribution of print to the emergence of a stable, modern natural knowledge.[25] Questions recently raised by book historians urge us to produce a more satisfactory account of the role of print in the transmission of knowledge.[26] With regard to authorship, the control of early-modern naturalists over the circulation and publication of their work was often very limited: abridgements, translations and other manipulations of their material by editors could affect both form and content. The relations between print and other media of communication, such as pictures or speech, were in the early-modern period more complex than has been assumed. In the case of images, while producing and circulating them was one of the major activities of collectors and naturalists, they were well aware of the flaws of print, for example the impossibility of reproducing colours. In the case of orality, different traditions evaluated the reliability of spoken words in contrasting ways: they were either the medium for hearsay or the most lively and exciting way of learning and transmitting knowledge. But print was not the final stage of a process where orality was gradually effaced. In the case of the *Thesaurus*, we cannot avoid recognising the power of speech which lies at its core. The wealth of languages of the American Indians who first talked to Hernandez, Gregorio's enchanting story-telling and the appreciation of oratory by the Counter-Reformation courtier Faber surface on every page of this most praised product of Baroque printing.

Notes

For constructive comments on earlier versions of this paper, I am indebted to Nick Jardine, Jim Secord, Shelley Costa, Eugenia Roldán Vera, Aileen Fyfe, Marina Frasca-Spada and Nick Hopwood.

1 *Rerum medicarum Novae Hispaniae thesaurus seu plantarum animalium mineralium mexicanorum historia ex Francisci Hernandez . . . relationibus in ipsa Mexicana Urbe conscriptis a Nardo Antonio Recchio . . . collecta et in ordinem digesta . . .* (Romae, Ex Typographeio Vitalis Mascardi, 1651).

2 See G. Olmi, *L'inventario del mondo. Catalogazione della natura e luoghi del sapere nella prima età moderna* (Bologna, 1992), p. 249.

3 See the seminal work of D. F. McKenzie, 'Speech-manuscript-print', in D. Oliphant and R. Bradford (eds.), *New Directions in Textual Studies* (Austin, 1990), pp. 87–109.
4 G. Somolinos D'Ardois, 'Vida y obra de Francisco Hernandez', in his *Obras completas de Francisco Hernandez* (Mexico City, 1960), vol. I, pp. 95–373; J. Bustamante Garcia, 'De la naturaleza y los naturales americanos en el siglo XVI: algunas cuestiones criticas sobre la obra de Francisco Hernandez', *Revista de Indias*, 52 (1992): 297–328; J. M. Lopez Pinero and J. Pardo Tomas, *Nuevos materiales, y noticias sobre la Historia de las plantas de Nueva España de Francisco Hernandez* (Valencia, 1994).
5 Bustamante Garcia, 'De la naturaleza', p. 308.
6 G. Somolinos D'Ardois, 'Sobre la iconografia botanica de las obras de Hernandez y su sustitucion en las ediciones europeas', *Revista de la Sociedad Mexicana de Historia Natural*, 15 (1954): 73–86.
7 Further details on the publication of the *Thesaurus* in G. Gabrieli, 'Il cosiddetto Tesoro Messicano edito dai primi Lincei', in his *Contributi alla storia della Accademia dei Lincei* (Rome, 1989); A. Alessandrini, *Cimeli lincei a Montpellier* (Rome, 1978), especially pp. 143–202.
8 F. Ximenex, *Quatros libros de la naturaleza y virtutes de las plantas, y animales que estan recevidos en el uso de medecina en la Nueva España* ... (Mexico City, 1615); J. E. Nieremberg, *Historia naturae, maximae peregrinae, libris XVI distincta* (Antwerp, 1635).
9 When publication based on the copy at the Colegio Imperial was started in the following century, it ran only to the first three volumes. See *Francisci Hernandi medici ... opera cum edita tum inedita, ad autographi fidem et integritatem expressa* ... (Madrid, 1790).
10 No attempt has been made to reconstruct Prince Cesi's expenses. For a possible comparison, see C. Fahy (ed.), *Printing a Book at Verona in 1622. The Account Book of Francesco Calzolari Junior* (Paris, 1993).
11 See P. J. P. Whitehead, 'Georg Markgraf and Brazilian zoology', in E. van der Boogaart (ed., with H. R. Hoetink and P. J. P. Whitehead), *Johan Maurits van Nassau-Siegen 1604–1679. Essays on the Occasion of the Tercentenary of His Death* (The Hague, 1979), pp. 424–71.
12 The chapter 'Zibethicum animal americanum' is at pp. 538–81 of the *Thesaurus*.
13 I. Pantin, *La Poésie du ciel en France dans la seconde moitié du seizième siècle* (Geneva, 1995).
14 J. Ceard, *La Nature et les prodiges: l'insolite au XVIe siècle, en France* (Geneva, 1977); K. Park and L. Daston, *Wonders and the Order of Nature 1150–1750* (New York, 1998).
15 Nieremberg, *Historia naturae*, p. 10.
16 S. Shapin, *A Social History of Truth: Civility and Science in Seventeenth-Century England* (Chicago, 1994); L. Daston, 'L'esperienza scientifica e le sue possibili storie', *Quaderni storici*, 32 (1997): 831–8; D. Carey, 'Compiling nature's history: travellers and travel narratives in the early Royal Society', *Annals of Science*, 54 (1997): 269–92.
17 W. Blunt, *The Art of Botanical Illustration* (London, 1951), p. 88.
18 N. Jardine, 'Demonstration, dialectic, and rhetoric in Galileo's *Dialogue*', in D. R. Kelley and R. H. Popkin (eds.), *The Shapes of*

Knowledge from the Renaissance to the Enlightenment (Dordrecht, 1991), pp. 101–21.
19 G. Gabrieli, 'Un contributo dei missionari cattolici alla prima conoscenza naturalistica del Messico nel "Tesoro Messicano" edito dalla prima Accademia dei Lincei (Roma, 1651)', in his *Contributi alla storia*, pp. 1567–76; P. Gato, 'El informe del P. Gregorio Bolivar a la Congregacion de la Propaganda Fide del 1623', *Archivo Ibero-Americano*, 50 (1990): 493–548.
20 *Thesaurus*, p. 540.
21 *Thesaurus*, p. 506.
22 S. Guazzo, *La civile conversazione*, ed. A. Quondam (Modena, 1993), p. 30.
23 G. L. L. Buffon, *Histoire naturelle générale et particulière. Quadrupèdes. Tome III* (Paris, 1787), p. 272.
24 D. Valades, *Rhetorica christiana ad concionandi et orandi usum accomodata utriusque facultatis exemplis suo loco insertis, quae quidem, ex Indorum maxime deprompta sunt historiis* (Perusiae, 1579). See also F. J. McGinness, *Right Thinking and Sacred Oratory in Counter-Reformation Rome* (Princeton, 1995); A. Prosperi, *Tribunali della coscienza. Inquisitori, confessori, missionari* (Turin, 1996), especially chapters 28 and 29.
25 E. Eisenstein, *The Printing Press as an Agent of Change: Communications and Cultural Transformations in Early-Modern Europe* (Cambridge, 1979).
26 Eisenstein's influential argument has been called into question by A. Grafton, 'The importance of being printed', *Journal of Interdisciplinary History*, 11 (1980): 265–86, and recently by A. Johns, *The Nature of the Book. Print and Knowledge in the Making* (Chicago, 1998).

Further reading

M. Campbell, *The Witness and the Other World. Exotic European Travel Writing, 1400–1600* (Ithaca and London, 1988)

M. De Certeau, *The Writing of History* (New York, 1988)

M. Fumaroli, *L'Age de l'éloquence: rhétorique et 'res literaria' de la Renaissance au seuil de l'époque classique* (Geneva, 1980)

S. Greenblatt, *Marvellous Possessions. The Wonders of the New World* (Chicago, 1991)

A. Johns, *The Nature of the Book. Print and Knowledge in the Making* (Chicago, 1998)

J. M. Lopez Pinero and J. Pardo Tomas, *Nuevos materiales y noticias sobre la Historia de las plantas de Nueva Espana de Francisco Hernandez* (Valencia, 1994)

A. Lugli, *Naturalia et Mirabilia: Il collezionismo enciclopedico nelle Wunderkammern d'Europa* (Milano, 1983)

G. Olmi, *L'inventario del mondo. Catalogazione della natura e luoghi del sapere nella prima età moderna* (Bologna, 1992)

A. Pagden, *European Encounters with the New World* (New Haven and London, 1993)

J. L. Phelan, *The Millennial Kingdom of the Franciscans in the New World. A Study of the Writings of Geronimo de Mendicta (1525–1604)* (Berkeley and Los Angeles, 1956)

A. Prosperi, *Tribunali della coscienza. Inquisitori, confessori, missionari* (Turin, 1996)

R. Ricard, *The Spiritual Conquest of Mexico. An Essay on the Apostolate and the Evangelizing Method in New Spain: 1523–1572* (Berkeley, Los Angeles, 1966; 1st edn. Paris, 1933)

C. Talbot, 'Prints and the definitive image', in G. P. Tyson and S. S. Wagonheim (eds.), *Print and Culture in the Renaissance* (Newark, 1986), pp. 189–205

II Learned and conversable reading

Book-wheel from Gaspard Grollier de Servière, *Recueil d'ouvrages curieux de mathématique et de mécanique ou Description du Cabinet de Monsieur Grollier de Servière*... (Lyon, 1719).

9 Compendious footnotes

> He can make use of language without wholly giving himself up to it, he may treat it as semi-alien or completely alien to himself, while compelling language ultimately to serve all his own intentions . . . he speaks, as it were, through language, a language that has somehow more or less materialized, become objectivized, that he merely ventriloquates. (M. M. Bakhtin)

This chapter addresses ten or so pages of a book entitled *An Essay on the Origin of Evil*, published in Cambridge in 1731 (see Figure 9.1); and in doing so it reconstructs a crucial episode in the history of the discussions on space, its properties and its idea in the light of Locke's and Newton's works. The main text in the pages in question is the English translation of an original in Latin by Archbishop William King, *De origine mali*, which appeared in 1702 as a slim *octavo*. The marginal summaries and comments, and the lengthy small-print footnotes and their annotations, are by an anonymous translator/editor. *The Origin of Evil* follows a familiar model, being inspired, even in the layout of the pages, by the commentaries traditionally in use in learned and academic settings (see Figures 9.3 and 9.4). The main text is surrounded by a complex apparatus provided by an editor or commentator to guide readers and direct (or indeed to take over) their attention: the lengthy two-column footnotes under the text, the brief marginal annotations summarising or criticising the content paragraph by paragraph, and, either in the margins or below the footnotes at the very bottom of the pages, the abbreviated references to sources and authorities. In spite of the myth of Enlightenment as the age of linear narratives, learned works in this far-from-linear format were relatively common in the eighteenth century. Recent important specimens of a similar kind were, for example, Pierre Bayle's *Dictionnaire historique et critique* (Amsterdam, 1697) and the first published work of Dr Samuel Clarke, a famous annotated Latin translation of Rohault's *Physique* (London, 1697). Like Clarke, this anonymous author chose to put forward his own point of view in the margins and at the bottom of the pages, in the form of comments and additions to someone else's text, and as the testimony of a particular reader's response to that

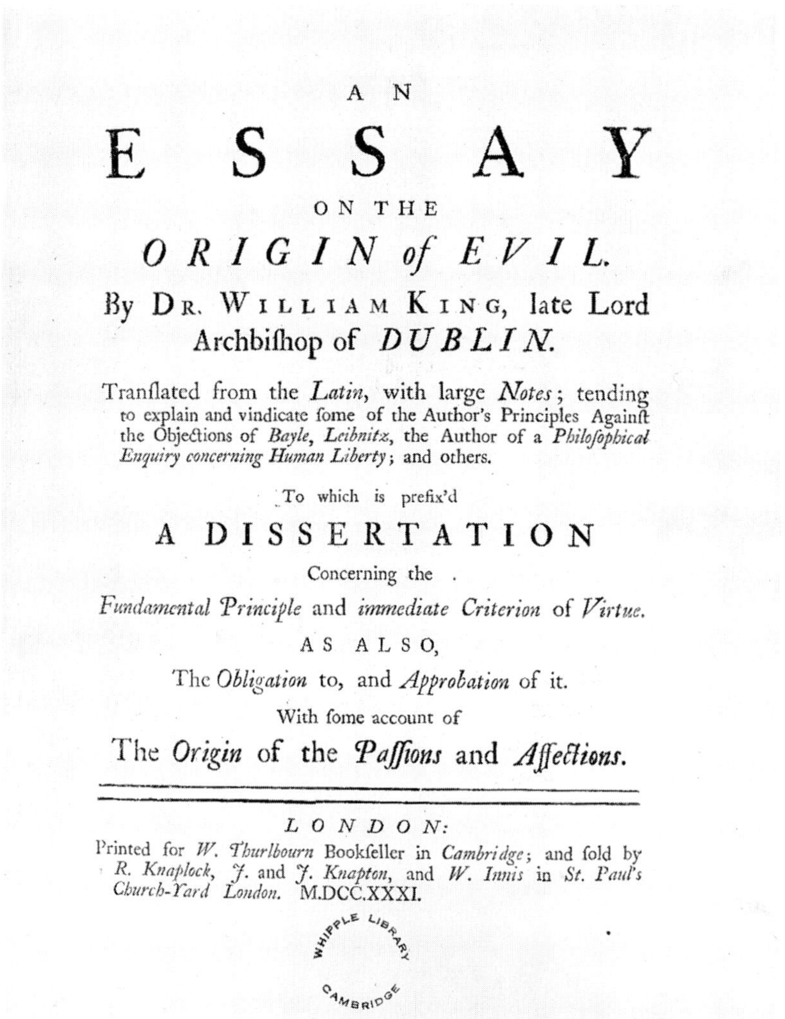
A N

E S S A Y

ON THE

ORIGIN of EVIL.

By DR. WILLIAM KING, late Lord Archbiſhop of *DUBLIN.*

Tranſlated from the *Latin*, with large *Notes*; tending to explain and vindicate ſome of the Author's Principles Againſt the Objections of *Bayle*, *Leibnitz*, the Author of a *Philoſophical Enquiry concerning Human Liberty*; and others.

To which is prefix'd

A DISSERTATION

Concerning the

Fundamental Principle and *immediate Criterion* of *Virtue.*

AS ALSO,

The *Obligation* to, and *Approbation* of it.

With ſome account of

The *Origin* of the *Paſſions* and *Affections.*

LONDON:

Printed for *W. Thurlbourn* Bookſeller in *Cambridge*; and ſold by *R. Knaplock*, *J.* and *J. Knapton*, and *W. Innis* in *St. Paul's Church-Yard London.* M.DCC.XXXI.

WHIPPLE LIBRARY CAMBRIDGE

Figure 9.1 Title-page of the first edition of *The Origin of Evil* (Cambridge, 1731). (Courtesy of the Whipple Library, Cambridge)

text. And, like Bayle, in his footnotes he made extensive use of quotations: this author is recycling other authors' texts to construct his own.

The Origin of Evil was the first book published by Edmund Law, the future editor of Locke's *Works* and Bishop of Carlisle, at the time a young fellow of Christ's College, Cambridge. Law was always very keen on footnotes and quotations: even his mature writings are crawling with them. This predilection shows a traditional notion of authority making its appearance, even among the promoters of Locke and Newton. In the words of Law's more senior friend Daniel Waterland, Master of Magdalene College, on really important matters (for instance, how to demonstrate the existence of God), if there should be any *new* Topick invented, it will probably be found much short in Value and Efficacy of the more common ones, which have been of long standing. The *commonest* Argument, in such cases, may be justly look'd upon as the *best*; because they have been proved and tried . . .[1]

Here is Law's own self-denying version, from the Preface to *The Origin of Evil*:

I chose rather to quote the very Words of the Authors, than either use worse of my own, or pretend to discover what had been often discovered before; or repeat the same things over and over again, which is endless . . . A Writer seems to me to do more good to the public by shewing the use of some of those many Volumes which we have already, than by offering new ones; tho' this be of much less Advantage to his private Character. I determin'd therefore not to say any thing myself where I could bring another conveniently to say it for me; and transcribed only so much from others as I judg'd absolutely necessary to give the Reader a short view of the Subject we were upon, and by that Sketch to induce those who have leisure, opportunity and inclination to go farther, and consult the Originals; and to afford some present Satisfaction to those who have not. (p. x)

A well-known book-lover, Law justifies the publication of his own book through his very opposition to the multiplication of books.[2] In his writings he follows to quite extreme lengths this tendency to use someone else's words and authority: for example, in his *Enquiry into the Ideas of Space, Time, Immensity and Eternity* he introduces his account of the nature of space by citing, without acknowledging them as his own, the words of 'the Author of the Notes to A.Bp. King' (p. 4). When he is writing in the first person Law always tends to present himself not as an author, but simply as a reader – a reader with pen in hand.

It is well known that footnotes, by definition modest and matter-of-fact, may be used to subversive effect; and that a tissue of quotations may be an excellent vehicle for independent views. I intend to show how this young and apparently humble author is, in fact, already a philosopher and theologian with firm views, strongly committed to a specific intellectual project. To do this I take some pages of *The Origin of Evil* as the starting point to explore different areas of significance: the text, the footnotes, their different origins and authors, the different voices with which they speak and the ways they interact with each other, their diverse receptions and readerships. In this way I map the various bookish histories which can be narrated starting from these pages and their contents. Then I focus on two of Law's footnotes, which discuss 'ideas' and space, to highlight, through a close reading of Law's prose, how that Locke-and-Newton combination that we now tend to regard as typical of eighteenth-century Cambridge and beyond, far from being a readily available and consistent doctrinal frame of reference, was then, in fact, a very complex and varied business, involving in endless disputes philosophers and divines linked or opposed to each other according to a surprisingly complicated pattern of doctrinal, political and personal relations.[3] 'Lockeanism' and 'Newtonianism' are, as it turns out, labels on the spines of later books, books published when the disputes had long subsided, leaving as their embodied trace a weighty multitude of printed pages, the pamphlets and volumes that we still find on the shelves of college libraries.

It was Law's opinion that natural philosophy, metaphysics and theology should not be 'jumbled up together'. Studying the way he put forward this view offers an occasion to reflect on an issue which is, I think, of crucial importance for historians of the sciences: namely, the artificiality of the disciplinary divisions that we, as historians, still tend to apply to the past as if they were a part of the nature of things. It has long been apparent that it was anachronistic and misleading to stick to a non-negotiable division of theology from the sciences in early modern natural-historical and natural-philosophical discussions.[4] The same applies, I submit, to another distinction which is, however, still prevalent: that between 'philosophy' in today's sense – the study of ontological, logical, epistemological, ethical questions – and 'science' – which I use here as we all commonly do in the expression 'history of science', meaning therefore, among other things, eighteenth-century natural philosophy, natural history and mathematics. This couple of footnotes by an advocate of a firm division of labour between natural philosophy, metaphysics and theology is an episode in the genealogy of this distinction. My claim is that studying this genealogy shows how important it may be to bridge the gap between today's histories of 'philosophy' and of 'science'; and that an approach combining book history and textual criticism, thus focusing on the stages of production of various layers of meaning, affords a crucial instrument to this purpose.

Layers of (controversial) meaning

When Law chose to translate it, *De origine mali* was already well-known. The author, William King, died in 1729 – the same year as Samuel Clarke, and while Law was, presumably, busy penning his translation. He had been a very prominent figure indeed. Born in 1650 at Antrim in Ireland and educated at Trinity College, Dublin, he had been a Whig and a supporter of the Orange cause. In 1691 he had published *The State of the Protestants in Ireland under the late King James's Government*, a classic defence of the principles of the Glorious Revolution, which he had prepared in 1689 while in prison for his opposition to the Jacobites. He had been Bishop of

Derry from 1691, then Archbishop of Dublin from 1703.[5] King had also played a major role in Irish intellectual life in various ways. In 1683 he had been one of the founding members, and since 1707 a vice-president, of the Dublin Philosophical Society. He had been a very active member of the Society, interested in numerous natural-philosophical and technological issues – for instance, in the phenomenon of the difference in the apparent size of the sun at the horizon and meridian, in falling weights and the force of percussion, in hydraulic engines and their possible uses to drain land, and in the composition of mineral waters. One of his contributions to the meetings, a memory 'Of the Bogs and Loughs of Ireland', had been published in the *Philosophical Transactions* in 1685, and another, 'An account of the manner of manuring Land with Sea Shells', in 1708. In November 1705 he had been elected to the Royal Society.[6] He was also a notable book collector, and his personal libraries, which he donated to the community whenever he moved from one place to another, are still renowned for their size and organisation.[7]

The problem addressed in his essay is traditional: how to reconcile evil with the existence of an infinitely good, wise and powerful God. Following the Neoplatonic and Augustinian tradition, King regards evil as the inevitable result of the limits of created beings – each creature is as perfect as possible given its own nature, and God's infinite goodness, wisdom and power display themselves in his sharing the utmost possible happiness with the largest possible number of beings. Chapter 1 gives the metaphysical frame of reference of the essay, and shows the necessity of a first cause; then, after presenting the division of the matter into three kinds – evil of privation, pain and vice – King settles down to devote a chapter to each.

The reception of King's essay was varied, and engaged different kinds of readerships. In Ireland it was very successful – a pirate edition appeared in 1702;[8] and in Trinity College, Dublin, it was apparently long read as a textbook.[9] Certain of its metaphysical and theological doctrines were, however, highly controversial, and aroused heated responses. In particular, King distinguished between the knowledge acquired directly from experience, '*per ideam*', and a less direct form of knowledge, '*per rationem*' or '*per similitudinem*' – that is, as Law was to translate them, 'by idea' and 'by analogy'. All that we can say about God is an extreme metaphorical specimen of the second kind: when we talk about divine knowledge, for example, we are in fact speaking metaphorically, as the Scriptures do when they attribute to God hands, eyes and feet – we can only talk about God as a blind man talks about colours.[10] But then, as Peter Browne, George Berkeley and many others variously pointed out, our knowledge of God is insufficient to distinguish Him from fate, chaos or natural necessity.[11]

The book also aroused great interest on the Continent, where it was involved in the controversy on Manichaeism between Pierre Bayle and (among others) Jacques Bernard, Bayle's successor in the editorship of one of the most famous review-journals of the time, *Les Nouvelles de la République des Lettres*. (At the margin of this complex theological dispute King's essay also occasioned a secondary, but bitter dispute on a doleful issue of scholarly propriety: Bayle criticised King's essay at length on the basis of Bernard's favourable review, laying himself open to universal censure for his open admission that he had not read it first-hand.)[12]

With its combination of negative theology and optimism, King's essay was to become a minor classic. Leibniz commented on it extensively in the *Théodicée*;[13] probably through Bayle and the polemics surrounding him it reached David Hume,[14] and later even Kicrkcgaard.[15] In England over a long period its reputation was comparable to Pope's 'Essay on Man': 'Pope has solved the origin of evil *per saltum*, by saying, "whatever is, is right". The prelate has done it with a pen borrowed from an angel', a historian wrote many years later.[16] There was even a widespread and long-lasting rumour, actively encouraged by Law and inevitably followed by a dispute, that Pope had borrowed from it.[17] Today it is still read and discussed by theologians.[18]

The English translation of *De origine mali* is a fat, handsome *quarto* volume. As we are informed by the title-page (see Figure 9.1), it also includes, among other things, a preface and a large apparatus of footnotes and additions written by the anonymous translator – a dedicated Lockean, as is evident from his doctrines, and an opponent of Dr Samuel Clarke's Newtonian rational theology. Let us consider the genesis of these texts. From contemporary tutorial notes and reading lists it appears that Cambridge students' work was mainly based on the practices of writing, to start with, summaries and comments, and, at a more advanced stage, of constructing personal commonplace books.[19] In the light of this it makes sense that a translation with annotations – a reader's response in the most literal sense – was not an uncommon format for the first work of a young scholar: as I mentioned, Samuel Clarke's first book, for example, was of the same sort. In a classic article Michael Hoskin has reconstructed the development of Clarke's annotations from the first edition of his *Physica* in 1697, where they are hardly more than informed comments, to the third edition of 1710, where they constitute a full-blown Newtonian subversion of Rohault's Cartesian text.[20] Thirty-five years later, the editor of King's essay was also very junior but, despite appearances, not quite so discreet. He used the space at the bottom of the pages to follow his own intellectual agenda starting with the first edition. In fact, evidently editing and translating King's essay had attracted this scholar both for the authority and fame of its author, and for the opportunities it offered of disagreement, independent assessment and polemic.

Clarke's annotated Rohault, which circulated as a reasonably sized and modest-looking *octavo*, was a highly successful textbook, in use in Cambridge from the first publication well into the mid-eighteenth century; in the 1720s it was even translated into English. Although from its very aspect – the *quarto* format, the quality of print and paper, etc. – Law's 1731 *Origin of Evil* manifestly was not intended for the student market, it is not difficult to imagine such a work being popular among students: the footnotes were a good model for them to imitate in their writings; these notes also offered their compendious expertise by citing at length the very authorities mentioned in students' reading lists, and presenting them with the additional attraction of a local polemic.[21] Indeed, with such works as *The Origin of Evil* around, one may well wonder how many students ever did 'have leisure, opportunity and inclination', as Law put it, 'to go farther and consult', say, Cudworth's massive and dense *True Intellectual System*, or even Newton's *Principia* and Locke's *Essay*, before discussing them competently and convincingly enough, or before quoting them in their own commonplace books and written assignments.

Certainly *The Origin of Evil* was a success with some of the college tutors. For example in *Quaestiones Philosophicae*, a 1730s' collection of questions with appended reading lists compiled by a fellow of Magdalene College, Thomas Johnson, with the purpose of recruiting young students to 'philosophy', it appears under a variety of general natural-philosophical, natural-historical, logical, ethical and metaphysical questions. '*Whether* the essence of bodies is to be found [*ponatur*] in extension, or in impenetrability, or in active force, and in an unknown substratum of properties? . . . *Whether* matter is infinitely divisible?' Johnson asks his student readers; and in the two reading lists following these questions refers to selected passages from classic authors such as Descartes and Rohault, Locke and Newton, and the Leibniz–Clarke correspondence, to entries in Chambers' *Cyclopaedia*, to Robert Green's and John Keill's natural philosophy textbooks – and to *The Origin of Evil*.[22]

The second edition of *The Origin of Evil* appeared, under Law's name, in 1732, with a dedication to Waterland. This time it was in a less grand *octavo* format. There are changes, cuts and additions in King's and Law's footnoted material; there are King's two best-known sermons, on predestination and on Man's fall; and there is a *post-scriptum* by Law concerning space, time and necessary existence in answer to an attack by John Clarke, who had written a defence of Dr Clarke.[23] Law prepared three more editions of *The*

Figure 9.2 Edmund Law at the peak of his career, in 1777, with his four-volume Locke, mezzotint by William Dickinson after a painting by George Romney. (Courtesy of the Fitzwilliam Museum, Cambridge)

Origin of Evil, all of them in *octavo*: one in 1739, still as 'M.A. Fellow of Christ College'; then in two volumes in 1758, as 'D.D. Master of St Peter's College, Cambridge' (since 1756); finally in 1781, as 'Lord Bishop of Carlisle' (since 1768). In the meantime he had published several other works, and had become one of the most influential theologians and metaphysicians in Cambridge (Figure 9.2).[24]

For those in the know, such as the Cambridge don Waterland, or Samuel Clarke's followers John Clarke and John Jackson, it was clear that the texts collected under the title *The Origin of Evil* constituted a strongly polemical work, to which many of them reacted accordingly, attacking or defending it in a deluge of anonymous books, essays and pamphlets.[25] The kerfuffle went beyond Cambridge:

Sometimes a man is called a *learned* man who, after a course of several years' hard study, can tell you, within a trifle, how many *degrees* of *nonentity* or *nothing* must be *annihilated*, before it comes to be *something*. See King's Origin of Evil, ch. III p. 129, with the note. That such a kind of learning as that book is filled with, and the present age is much given to admire, has done no service to the cause of *truth*, but on the contrary that it has done infinite disservice, and almost reduced us from the unity of Christian faith to the wrangling of philosophic scepticism, is the opinion of many besides ourselves, and too surely founded on fatal *experience*.

These lines, copied by an anonymous reader in a copy of *The Origin of Evil* now in the British Library, are from a book published in 1756 by George Horne, a fellow of Magdalen College, Oxford.[26] In his own time Horne was not famed for his open-mindedness; and today his name is familiar to historians of philosophy for his attack on Adam Smith, guilty of reporting in print on David Hume's godless wisdom and serenity even on the deathbed.[27] Accused of contempt for learning, he retorts that it depends on what one calls learning – there follows the above denunciation of *The Origin of Evil*. Horne's biographer, William Jones, footnoted this passage thus:

The metaphysical system alluded to above, was a book in great request at Cambridge later the years 1740 and 1750, and was extolled by some young men who studied it as a grand repository of human wisdom [. . .] Having heard so high a character of it, I once sat down to read it, with a prejudice in its favour. I afterwards shewed it to Rev. Horne; and when he had considered it, we could not but lament in secret, what he at length complained of in public, that a work so unfounded and so improfitable should have engaged the attention, and excited the admiration, of scholars, intended for the preaching of the gospel. The account here given of it has something of the caricature; but the leading principle of the book is in substance as the apologist has described it.[28]

Ideas and the nature of space

In the Preface Law wrote that he had considered eliminating from his translation all he could not fully defend – in particular Chapter 1. But since it provided the metaphysical frame of reference for King's discussion, abolishing it would have created serious problems. Therefore he did include it, but with particularly lengthy and numerous additions and criticisms. Here I intend to look at a sample of the issues discussed, focusing on the interactions between King's text and Law's patchwork of quotations at the bottom of the pages of this chapter. In this way it is possible to consider in detail the concrete ways in which *The Origin of Evil* is both a polemical work apt to give rise to strong responses, and a good compendium of 'philosophy' which contemporary Cambridge students could usefully consult.

That Law makes the best of King's text as an opportunity to expound his own version of Lockeanism is evident. In a group of footnotes devoted to uses of 'idea', for example, he disposes of King's complex and controversial account of knowledge by citing at every possible occasion Locke's theory of ideas as conclusive. His treatment of King's note A is typical. King here presents human knowledge as the possession of 'marks' or 'characters' (*notae*) which, without giving us any clues about the essences of things, help us to distinguish parts of our experience from each other. The 'marks' offered by knowledge 'by analogy' are to things like '*Algebraic Species*' to particular '*Ideas* of *Quantity*' (p. 5). But knowledge 'by idea' too, he proceeds to maintain, is no more adequate for our purposes than knowledge by analogy; also, analogy is responsible for the bulk of our knowledge (for example, it is the mechanism through which we conceive universals); and finally even definitions, far from expressing genus and difference, the principle of individuation, substance and its essence, are in fact based on the 'marks' of things (p. 6). With this note A terminates rather abruptly, and Law appends his footnote 3. And yet, this is not all King wrote here on knowledge and ideas, but only what the editor allows us to read of his note A: for Law informs us that he is omitting 'the rest of our Author's long Note, since it contains only some Arguments for *innate Ideas*; which Hypothesis is now almost universally exploded' (p. 6).[29] And he proceeds, in the rest of the footnote (a full page), to explain how King takes 'idea' in the out-of-date Platonic meaning for '*Species, Phantasm,* or *Corporeal Image,* as it were painted on the Brain'; and to offer, with the help of the *Essay* and of Isaac Watt's *Logic*, a crisp account of the right – Lockean, that is – theory of universals. Law does not here discuss King's knowledge by analogy in connection with the crucial and highly controversial question of the divine attributes: for this we must read the remarks at the end of the chapter. To complete the picture footnote 2, which preceeds King's note A, pre-empts the main point of King's discussion – our knowledge is not of substances, but of signs – by showing how Locke's approach avoids the 'absurdity of taking . . . general terms for signs of real Existences, which, we see, are signs of our Ideas only' and helps us to distinguish 'between a Collection or Combination of our own Ideas, and real Qualities, as they are found in Nature; between *Thoughts* and *Things*' (p. 4).

'If this last Distinction were sufficiently attended to', concludes Law's footnote 2, 'I believe we should not be so ready to conclude from any Ideas we may have of *infinite Space, Infinite Duration,* &c. that these must needs be real *Properties,* or *Attributes* of some

Being'. The nature and properties of space are discussed at length in these footnotes. Indeed, space is a recurrent topic in Law's writings, and this is hardly surprising: everything connected with space – its nature and idea, its infinity, extension or divisibility – was important and contentious in Newton's Cambridge. Let us consider just one footnote on space.

In his text King maintains that even though we do not have 'a *Definition* or *Idea* of space, properly so call'd' ('i.e. in his own Sense of the word Idea', clarifies Law in the margin; see Figure 9.3), we still can define it through the properties which would remain upon the annihilation of matter. Local mobility, separability and solidity are peculiar to matter, and should go, so that the residue, space, is conceived as 'something extended, immoveable, capable of *receiving* or *containing* Matter, and penetrable by it' (p. 9).[30] And here Law appends his footnote 5.

A lot of noise has been made about space, he starts. Great use has been made of it to demonstrate the divine attributes 'in a way which some style *a Priori*'; and Leibniz is right in calling it 'an Idol of some modern *English* Men' – a more than pointed allusion to Dr Clarke's celebrated and controversial Boyle Lectures published in 1705 as *A Demonstration of the Being and Attributes of God*. Nevertheless, he goes on, space is only conceivable in three ways: either as absence of body, of which – according to Locke's doctrine of privations – we do have a positive idea even though we cannot attribute anything positive to it; or as extension of body *in abstracto*, in which case it is clear that it only exists in the mind – as we know from Locke's account of abstraction; or as the imaginary substratum of the general idea of extension – the authority for this is Waterland, extensively cited in n. 16 and cross-referenced here. In all three cases space is simply an idea in the mind, to which no property can be attributed, and from which we may only infer our power to frame such ideas.

Now, consider the properties commonly attributed to space: infinity, for example. The reason why space is supposed to be absolutely infinite is – according to Gassendi, Dr Clarke and Joseph Raphson – that we cannot conceive setting bounds to it: for anything limiting it should also, in turn, occupy space. This, Law retorts, either takes for granted that space is something, or applies bounds to bare possibility. If space is, according to the first definition, mere absence of body, Clarke's and the others' argument amounts to the rather odd claim that '*beyond* this *possibility* of *Existence*' there is '*more* such *possibility* of *Existence* or more *nothing*, i.e. *non Existence*' (p. 11). If space is, according to the second definition, extension *in abstracto*, then our inability to set bounds to it means 'that we have a power of enlarging our abstract Idea *in Infinitum*', and that 'if we always find that we can add, we shall

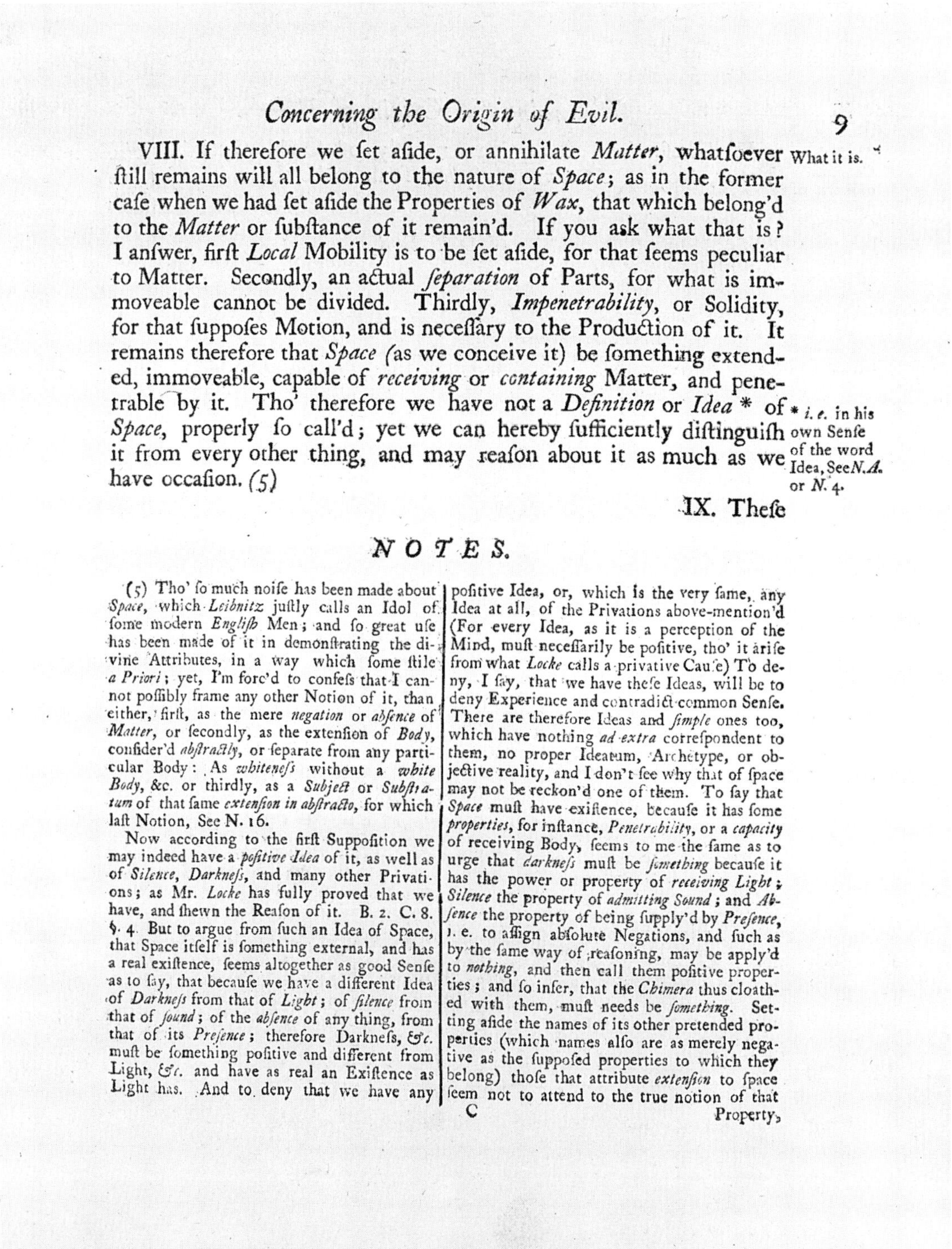

Concerning the Origin of Evil. 9

VIII. If therefore we set aside, or annihilate *Matter*, whatsoever still remains will all belong to the nature of *Space*; as in the former case when we had set aside the Properties of *Wax*, that which belong'd to the *Matter* or substance of it remain'd. If you ask what that is? I answer, first *Local* Mobility is to be set aside, for that seems peculiar to Matter. Secondly, an actual *separation* of Parts, for what is immoveable cannot be divided. Thirdly, *Impenetrability*, or Solidity, for that supposes Motion, and is necessary to the Production of it. It remains therefore that *Space* (as we conceive it) be something extended, immoveable, capable of *receiving* or *containing* Matter, and penetrable by it. Tho' therefore we have not a *Definition* or *Idea* * of *Space*, properly so call'd; yet we can hereby sufficiently distinguish it from every other thing, and may reason about it as much as we have occasion. (5)

What it is.

* *i. e.* in his own Sense of the word Idea, See *N. A.* or *N.* 4.

IX. These

NOTES.

(5) Tho' so much noise has been made about *Space*, which *Leibnitz* justly calls an Idol of some modern *English* Men; and so great use has been made of it in demonstrating the divine Attributes, in a way which some stile *a Priori*; yet, I'm forc'd to confess that I cannot possibly frame any other Notion of it, than either, first, as the mere *negation* or *absence* of *Matter*, or secondly, as the extension of *Body*, consider'd *abstractly*, or separate from any particular Body: As *whiteness* without a *white Body*, &c. or thirdly, as a *Subject* or *Substratum* of that same *extension in abstracto*, for which last Notion, See N. 16.

Now according to the first Supposition we may indeed have a *positive Idea* of it, as well as of *Silence, Darkness*, and many other Privations; as Mr. *Locke* has fully proved that we have, and shewn the Reason of it. B. 2. C. 8. §. 4. But to argue from such an Idea of Space, that Space itself is something external, and has a real existence, seems altogether as good Sense as to say, that because we have a different Idea of *Darkness* from that of *Light*; of *silence* from that of *sound*; of the *absence* of any thing, from that of its *Presence*; therefore Darkness, &c. must be something positive and different from Light, &c. and have as real an Existence as Light has. And to deny that we have any

positive Idea, or, which is the very same, any Idea at all, of the Privations above-mention'd (For every Idea, as it is a perception of the Mind, must necessarily be positive, tho' it arise from what *Locke* calls a privative Cau'e) To deny, I say, that we have these Ideas, will be to deny Experience and contradict common Sense. There are therefore Ideas and *simple* ones too, which have nothing *ad extra* correspondent to them, no proper Ideatum, Archetype, or objective reality, and I don't see why that of space may not be reckon'd one of them. To say that *Space* must have existence, because it has some *properties*, for instance, *Penetrability*, or a *capacity* of receiving Body, seems to me the same as to urge that *darkness* must be *something* because it has the power or property of *receiving Light*; *Silence* the property of *admitting Sound*; and *Absence* the property of being supply'd by *Presence*, *i. e.* to assign absolute Negations, and such as by the same way of reasoning, may be apply'd to *nothing*, and then call them positive properties; and so infer, that the *Chimera* thus cloathed with them, must needs be *something*. Setting aside the names of its other pretended properties (which names also are as merely negative as the supposed properties to which they belong) those that attribute *extension* to space seem not to attend to the true notion of that

C   Property,

Figure 9.3 A page from *The Origin of Evil*. Note Law's marginal annotation, on the left, criticising King's use of 'idea'. (Courtesy of the Whipple Library, Cambridge)

never find that we cannot add, which . . . is all the Mystery of the Matter, and all that can be understood by infinite Space'. As Waterland, duly cited at length, puts it, we can set bounds to actual extension and to the number of stars, of hairs, of men; but not to '*ideal Extension*', '*Number*, in the abstract' or '*Divisibility*', that is, not to 'the Power the Mind has of *adding* and *repeating*'. So, many arguments for the infinite divisibility of extension boil down to what Mr Locke called the 'negative' conception of infinity: 'that whenever we begin to add, we know not when to have done, because its nothing else but doing the same over and over again' (p. 12). There is also another kind of infinity, that described by Cudworth in his *True Intellectual System*: 'a positive, or *Metaphysical, Infinite*', 'which is absolutely *Perfect* in its kind, which cannot admit of *Addition*, or *Increase*' (p. 12). But it is evident that space, duration, number and quantities in general are not positively infinite – in fact, their indefinite '*Increaseableness* or *Addibility*' is the very opposite of positive infinity. The quotations are particularly numerous here: from Bentley's Boyle lectures, Sir Matthew Hale's *Primitive Origination of Mankind*, Stillingfleet's *Origines sacrae*, Samuel Colliber's *Impartial Inquiry*, etc. (p. 13).

Once again Law's target is Samuel Clarke, whose *Demonstration* also is now, finally, quoted:

The Learned Dr Clarke endeavours to evade all these Arguments about *Parts, &c.* by denying that any Number of Years, Days, and Hours; Or of Miles, Yards, or Feet, *&c.* 'can be considered as any *aliquot*, or *constituent* parts of infinite *Time* or *Space*, or be compared at all with it, or bear any kind of proportion to it, or be the foundation of any Argument in any Question concerning it' (pp. 13–14).

But this cannot be right, Law comments. The infinity attained by addition of parts is evidently inconceivable without the parts themselves; these must therefore be regarded as having a proportion to it. So – and here Law makes Leibniz speak for him – to say that infinite space has no parts means 'that infinite Space might subsist, tho' all finite spaces should be reduced to nothing' (p. 14). Of course there is 'a certain use of the term infinite among Mathematicians' which may well appear to agree with Clarke's reasoning:

When Geometricians say that one Quantity is infinitely less than another, they mean that their infinitely small Quantity is no aliquot part of, bears no proportion to, or cannot be compared with the other; but proportion is (nothing real but) purely relative, and therefore the term infinite apply'd here must be so too (p. 14).

Such, for example, is the ratio of a rectilinear angle to the angle of contact between a circle and its tangent. Thus geometers 'may have an infinite succession of increasing quantities, every one of

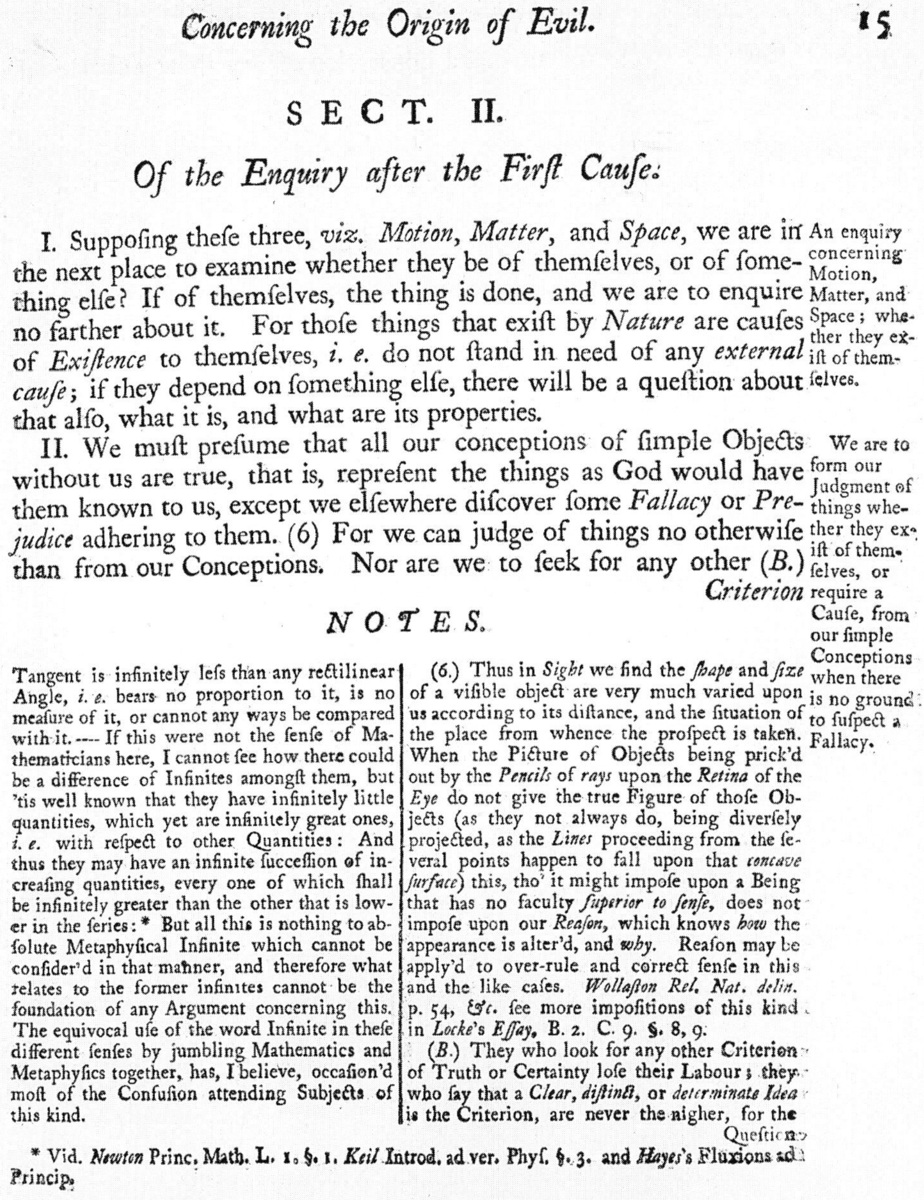

Concerning the Origin of Evil. 15

SECT. II.

Of the Enquiry after the First Cause.

I. Supposing these three, *viz. Motion, Matter,* and *Space,* we are in the next place to examine whether they be of themselves, or of something else? If of themselves, the thing is done, and we are to enquire no farther about it. For those things that exist by *Nature* are causes of *Existence* to themselves, *i. e.* do not stand in need of any *external cause*; if they depend on something else, there will be a question about that also, what it is, and what are its properties. An enquiry concerning Motion, Matter, and Space; whether they exist of themselves.

II. We must presume that all our conceptions of simple Objects without us are true, that is, represent the things as God would have them known to us, except we elsewhere discover some *Fallacy* or *Prejudice* adhering to them. (6) For we can judge of things no otherwise than from our Conceptions. Nor are we to seek for any other (B.) *Criterion* We are to form our Judgment of things whether they exist of themselves, or require a Cause, from our simple Conceptions when there is no ground to suspect a Fallacy.

NOTES.

Tangent is infinitely less than any rectilinear Angle, *i. e.* bears no proportion to it, is no measure of it, or cannot any ways be compared with it. ---- If this were not the sense of Mathematicians here, I cannot see how there could be a difference of Infinites amongst them, but 'tis well known that they have infinitely little quantities, which yet are infinitely great ones, *i. e.* with respect to other Quantities: And thus they may have an infinite succession of increasing quantities, every one of which shall be infinitely greater than the other that is lower in the series:* But all this is nothing to absolute Metaphysical Infinite which cannot be consider'd in that manner, and therefore what relates to the former infinites cannot be the foundation of any Argument concerning this. The equivocal use of the word Infinite in these different senses by jumbling Mathematics and Metaphysics together, has, I believe, occasion'd most of the Confusion attending Subjects of this kind.

(6.) Thus in *Sight* we find the *shape* and *size* of a visible object are very much varied upon us according to its distance, and the situation of the place from whence the prospect is taken. When the Picture of Objects being prick'd out by the *Pencils* of *rays* upon the *Retina* of the *Eye* do not give the true Figure of those Objects (as they not always do, being diversely projected, as the *Lines* proceeding from the several points happen to fall upon that *concave surface*) this, tho' it might impose upon a Being that has no faculty *superior to sense*, does not impose upon our *Reason*, which knows *how* the appearance is alter'd, and *why*. Reason may be apply'd to over-rule and correct sense in this and the like cases. *Wollaston Rel. Nat. delin.* p. 54, *&c.* see more impositions of this kind in *Locke's Essay*, B. 2. C. 9. §. 8, 9.

(B.) They who look for any other Criterion of Truth or Certainty lose their Labour; they who say that a *Clear, distinct*, or *determinate Idea* is the Criterion, are never the nigher, for the Question

* Vid. *Newton* Princ. Math. L. 1. §. 1. *Keil* Introd. ad ver. Phys. §. 3. and *Hayes's* Fluxions ad Princip.

Figure 9.4 A page from *The Origin of Evil*. Note at the bottom of the page the asterisked second-order footnote with the references to Newton, Keill and Hayes. (Courtesy of the Whipple Library, Cambridge)

which shall be infinitely greater than the other that is lower in the series' – Law refers, in the second-order footnote to this passage, to Newton's *Principia*, John Keill's *Introductio ad Veram Physicam* (Oxford 1701) and Charles Hayes's *Treatise of Fluxions* (London 1704) (see Figure 9.4). A lot of confusion in these matters is caused, Law concludes, by 'the equivocal use of the word Infinite in these different senses': it is the confusion which invariably arises 'by jumbling Mathematics and Metaphysics together'. The divine essence, *pace* Samuel Clarke, bears no relation whatsoever to our ideas of space and time.

Conclusion

Their Works are supposed, in great Measure, Assemblages of other Peoples; and what they take from others they do it avowedly, and in the open Sun. In effect, their Quality gives them the title to every thing that may be for their purpose, wherever they find it; and if they rob, they don't do it in any otherwise, than as the Bee does, for the publick Service. Their Occupation is not pillaging, but collecting Contributions.

Thus in his *Cyclopaedia* (1728) Ephraim Chambers, in the entry on 'Plagiary', described the work of people like himself, compilers of encyclopaedias. Law was not writing encyclopaedia entries; but he too presented himself, with deceptive modesty, as a compiler, collecting into one body what he was drawing up from various authors – his deliberate eclecticism as announced in the Preface sounds reminiscent of Chambers' description:

Some perhaps may think the frequent and long Quotations very tedious, and introduced only to stuff up. I can only answer, that I intended by the Notes, and References together, to point out a sort of *Compendium of Metaphysics*, or Speculative Divinity; but directing the Reader to a set of true Notions on the various Subjects which our Author touch'd upon; and which could not be found in any one particular Book, nor collected from several, without much Trouble and Confusion, and unnecessary Reading (p. x).

I have considered Law's strategy of juxtaposition, reference to authority and quotation in his use of King's text to present, in the footnotes, his version of Locke's theory of ideas, and strictures against Samuel Clarke on the nature of space; and to promote his own particular balance of Lockean metaphysics and Newtonian natural philosophy. By these means, Law's notes do 'hold much in a narrow room' (according to the definition of 'compendium' in Samuel Johnson's *Dictionary*): a whole cultural project is, indeed, held in the 'narrow room' between a translated text and its footnotes. This project involves the creation of a firm division between the kinds of arguments in use in 'mathematics' and in 'metaphysics'.

Law is clear in seeing Newton's natural philosophy and Locke's metaphysics as the instruments for the establishment of these all-important distinctions on a new and promising footing; and Samuel Clarke's opposite project for a rational theology provided him with the occasion for a topical way of proposing it. These distinctions, however, do themselves, at their origin, involve a specific kind of co-ordination, indeed of solidarity between the fields, which is embodied in contemporary textbooks and syllabuses; epitomised by Law's own writing, with its learned techniques of citation and reference to authorities; and expressed by the ways *The Origin of Evil* was regarded, used and responded to by its contemporary Cambridge readerships of students and dons.

Notes

I am grateful to Silvia De Renzi, Nick Jardine, Lauren Kassell, Scott Mandelbrote, Rosamond McKitterick, Victor Nuovo, Gregory Radick, Jim Secord, Richard Serjeantson, Richard Yeo, John Yolton, the members of Jim Secord's Book History Reading Group, and the members of the Cambridge Historiography Seminar, for their advice in connection with Edmund Law's footnotes and my use of them in this piece.

1 *A Dissertation Upon the Argument a Priori for Proving the Existence of a First Cause, in a Letter to Mr Law*, pp. 87–8 (published as an anonymous addition to E. Law's *An Enquiry into the Ideas of Space, Time, Immensity, and Eternity*, Cambridge 1734). On Waterland, his career and his opinions see R. T. Holtby, *Daniel Waterland 1683–1740. A Study in Eighteenth-Century Orthodoxy* (Carlisle, 1966); J. Gascoigne, *Cambridge in the Age of Enlightenment: Science, Religion and Politics from the Restoration to the French Revolution* (Cambridge, 1989); and the excellent B. W. Young, *Religion and Enlightenment in Eighteenth-Century England: Theological Debate from Locke to Burke* (Oxford, 1998).

2 On Law as a book-lover see W. Paley, *A Short Memoir of the Life of Edmund Law, D.D., extracted from Hutchinson's 'History of Cumberland'*, repr. with notes by Anonymous (London, 1800); and D. McKitterick, *Cambridge University Library. A History. The Eighteenth and Nineteenth Centuries* (Cambridge, 1986), pp. 281–7, on Law's bookish interests and activities as *Protobibliothecarius* from 1760 to 1769 and afterwards.

3 See Young, *Religion and Enlightenment*.

4 A. Cunningham, 'Getting the game right: some plain words on the identity and invention of science', *Studies in History and Philosophy of Science*, 19 (1988): 365–89; J. H. Brooke, *Science and Religion: Some Historical Perspectives* (Cambridge, 1991); J. H. Brooke and G. Cantor, *Reconstructing Nature: The Engagement of Science and Religion* (Edinburgh, 1998).

5 On King's life see, among others, *Biographia Britannica* (London, 1757), s.v., vol. IV, pp. 2839–50; and G. T. Stokes, *Some Worthies of the Irish Church*, Lectures Delivered in the Divinity School of the

University of Dublin, ed. H. J. Lawlord (London, 1900), Lectures VII–XVI, pp. 145–306.
6 K. T. Hoppen, *The Common Scientist in the Seventeenth Century. A Study of the Dublin Philosophical Society, 1683–1708* (London, 1970).
7 See R. S. Matteson, 'The early library of Archbishop William King', *The Library*, 5th Series, 30 (1975): 302–14, and id., 'Archbishop William King's library catalogue', *The Library*, 6th Series, 3 (1981): 305–19.
8 See M. Pollard, *Dublin's Trade in Books 1550–1800*, Lyell Lectures 1986/87 (Oxford, 1989), p. 69.
9 A whole century according to Stokes, *Some Worthies*, p. 305.
10 The theme of analogy is further discussed in King's most famous and controversial sermon: see A. Carpenter (ed.), *Archbishop King's Sermon on Predestination* (Dublin, 1976).
11 See D. Cupitt, 'The doctrine of analogy in the age of Locke', *Journal of Theological Studies*, N.S. 19 (1968): 186–202. E. W. Grinfield, *Vindiciæ Analogicæ* (London, 1822), an attack on Edward Copleston and Richard Whately's revival of King's doctrine of analogy in 1820–1, in Part II contains, among other things, a collection of critiques of King, together with passages from notorious sceptics in agreement with King.
12 For a splendid reconstruction of this dispute see J. P. Pittion, 'Hume's reading of Bayle: an inquiry into the source and role of the *Memoranda*', *Journal of the History of Philosophy*, 15(4) (1977): 373–86, esp. pp. 376–8.
13 *Théodicée*, Appendix 3, in *Philosophische Schriften*, ed. C. J. Gerhardt, vol. VI (Berlin, 1885), pp. 388–436.
14 Pittion, 'Hume's reading of Bayle', esp. pp. 374 ff. King is mentioned in the *Dialogues Concerning Natural Religion*, Part 10.
15 J. P. Jossua, *Pierre Bayle ou l'obsession du mal* (Paris, 1977), p. 38, fn. 44.
16 M. Noble, *A Biographical History of England* (London, 1806), vol. II, p. 103.
17 Law's *Preface* to *The Origin of Evil* (London, 1781), p. xvii; see also T. Tyers, *Historical Rhapsody on Mr. Pope*, 2nd edn. (London, 1782), pp. 79–80, and J. Nichols, *Literary Anecdotes* (London, 1814), vol. VIII, pp. 99–100.
18 See for example D. Cupitt, *Christ and the Hiddenness of God* (London, 1971), pp. 67 and 74–80, and J. Hick, *Evil and the God of Love* (London, 1985), pp. 148–54.
19 See for example Waterland's *Advice to a Young Student* (Cambridge, 1730 but in fact first compiled in 1706), pp. 8–9. See C. Wordsworth, *Scholae Academicae. Some Account of Studies at the English Universities in the Eighteenth Century* ([1877], London, 1968).
20 M. Hoskin, '"Mining all within": Clarke's notes to Rohault's *Traité de physique*', *The Thomist*, 24 (1961): 353–63.
21 I have not, however, found any annotations in the copies I inspected in Cambridge college libraries or in the Cambridge University Library.
22 Cambridge 1732, pp. 1–3. These two *quaestiones* are under the natural philosophy heading. In this edition of the *Quaestiones*, *The Origin of Evil* is also referred to for 6 *quaestiones* in ethics, and for 22 out of 36 in

metaphysics; in the second edition, Cambridge 1735, for 2 in natural history, 1 in mechanics, 3 in logic, and several in metaphysics.
23 The *Defence of Dr. Clarke's Demonstration* appeared anonymously in London in 1732.
24 On Law's career see Gascoigne, *Cambridge in the Age of Enlightenment*. On Law's philosophical work and the debates surrounding it see L. Stephen, *History of English Thought in the Eighteenth Century* (London, 1902), 2nd edn; E. Cassirer, *Das Erkenntnisproblem in der Philosophie und Wissenschaft der neueren Zeit*, vol. II (Berlin, 1907), Book 7, ch. 2; more recently J. W. Yolton, *Thinking Matter: Materialism in Eighteenth-Century Britain* (Oxford, 1983), ch. 4, and *Perceptual Acquaintance from Descartes to Reid* (Oxford, 1984), ch. 4; also J. Stephens, 'Edmund Law and his circle at Cambridge: some philosophical activity of the 1730s', in G. A. J. Rogers and S. Tomaselli (eds.), *The Philosophical Canon in the 17th and 18th Centuries: Essays in Honour of John W. Yolton* (Rochester, 1996), pp. 163–73; V. Nuovo, 'Introduction' to his edition of *The Collected Works of Edmund Law* (Bristol, 1997), vol. I, pp. vii–xl; and Young, *Religion and Enlightenment*.
25 To John Clarke's *Defence* there followed, between 1733 and 1734, a string of anonymous publications (authored by, among others, John Clarke and John Jackson, and by Law himself and Joseph Clarke, a pupil of Waterland at Magdalene College, and Thomas Johnson, on the opposite front): on aspects of this dispute see the works cited above, n. 23, esp. J. Stephens, 'Edmund Law and his circle at Cambridge', and Young, *Religion and Enlightenment*.
26 *An Apology for Certain Gentlemen in the University of Oxford* (Oxford, 1756), p. 17.
27 *A Letter to Adam Smith on the Life, Death and Philosophy of his Friend David Hume. By one of the People called Christians* (Oxford, 1777).
28 William Jones, *Memoirs of the Life, Studies, and Writings of the Right Reverend George Horne, D.D., Late Bishop of Norwich* (London, 1795), pp. 92–3. Also copied out in the same copy of *The Origin of Evil*.
29 For King's objections to Locke's *Essay* (especially in connection with the rejection of innate ideas), their transmission to, and offhand dismissal by Locke, and then their abolition by Law, see J. W. Yolton, *Locke and the Way of Ideas* ([1956] London, 1996), pp. 5–6 and 49–50, and J. Stephens, 'Edmund Law and his circle at Cambridge', pp. 165–6.
30 'res extensa, immobilis, capax recipiendi seu continendi materiam et ab ea penetrabilis', p. 10. Law is a reasonably faithful, sometimes an insightful translator.

Further reading

M. M. Bakhtin, 'Discourse in the novel', in M. Holquist (ed.), *The Dialogic Imagination. Four Essays by M. M. Bakhtin*, trans. C. Emerson and M. Holquist (Austin, 1981), pp. 259–422

G. W. Bowersock, 'The art of the footnote', *The American Scholar* (Winter 1983/84): 54–62

A. Compagnon, *La Seconde main ou le travail de la citation* (Paris, 1979)
P. W. Cosgrove, 'Undermining the text: Edward Gibbon, Alexander Pope, and the anti-authenticating footnote', in Stephen A. Barney (ed.), *Annotation and Its Text* (New York and Oxford, 1991), pp. 130–51
B. Cronin, *The Citation Process. The Role and Significance of Citations in Scientific Communication* (London, 1984)
C. O. Frost, 'The use of citations in literary research: a preliminary classification of citation functions', *The Library Quarterly*, 49 (1979): 399–414
A. Grafton, *The Footnote. A Curious History* (London, 1997)
M. Hoskin, '"Mining all within": Clarke's notes to Rohault's *Traité de physique*', *The Thomist*, 24 (1961): 353–63
E. E. Kellett, *Literary Quotation and Allusion* (Cambridge, 1933)
J. P. Pittion, 'Hume's reading of Bayle: an inquiry into the source and role of the *Memoranda*', *Journal of the History of Philosophy*, 15(4), (1977): 373–86
H. F. Plett, 'The poetics of quotation. Grammar and pragmatics of an intertextual phenomenon', in S. Wyler (ed.), *Linguistik und literarisch Text – Linguistique et texte littéraire* (Neuchâtel, 1988), pp. 66–81
H. F. Plett (ed.), *Intertextuality* (Berlin and New York, 1991)
M. Sternberg, 'Proteus in quotation-land: Mimesis and the forms of reported discourse', *Poetics Today*, 3(2) (1982): 107–56

10 On the bureaucratic plots of the research library

> The clarification of the basic mysteries of humanity – the origin of the Library and of time – was also expected. It is credible that those grave mysteries can be explained in words . . .
> Jorge Luis Borges, 'The Library of Babel', *Ficciones*, trans. A. Kerrigan.

Into the Baroque era, colleges and universities, like private collectors with their *Wunderkammern*, did not consistently distinguish between libraries, museums, cabinets, laboratories, and often not between those and archives and treasuries. For a good part of the early-modern era, spaces we tend to separate were commonly not. The differentiation and articulation of a space for books and their 'simulacra' (such as catalogues and other 'virtual books') – the library in the modern sense – was a feat of the Enlightenment and the Romantic era. In an academic context, this transformation in the collection of books went hand in hand with a transformation of the interrelation of books, that is, the system of knowledge: the emergence of the modern research library is correlative with the transformation of the pursuit of academic knowledge from erudition to research. There were two essential moments in the conception of the modern research library. First, the Enlightenment bureaucratised library practices, especially of collection and registration. Second, the Romantic era concocted the ideology of 'culture', which fundamentally affected practices of acquisition.

Like many things bureaucratic, the modern research library traces its origins to the Germanies.[1] So I shall restrict attention to them in the sections below: the Baroque library, *bibliotheca universalis*, the Enlightenment library, *bibliotheca virtualis*, the Romantic library. The first two sections are meant generally, so the German case is exemplary only. The next two concern specific origins of the research library as part of the German Enlightenment. The final section is again meant as but exemplary.

The Baroque library

Johann Puschner's *Amoenitates Altdorfinae* (*circa* 1715), from which Figure 10.1 is drawn, depicts a tour of the small but wealthy

On the bureaucratic plots of the research library 191

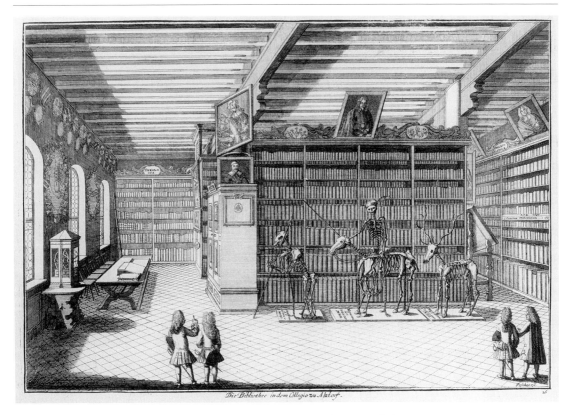

Figure 10.1 Johann G. Puschner, *Amoenitates Altdorfinae* (Nuremberg, ca. 1715), plate 16. (Reproduced with permission of the Universitätsbibliothek Göttingen)

University of Altdorf. As *Wunderkammer*, the library is one of the sites or sights to which one takes visitors, as the humans here indicate.[2] The books, like everything else, are 'monstrosities': things for display. Though not the most interesting, as monstrosities the books partake of an economy of the rare. Their materiality, including their covers (from which they are well judged), has a nature and a history independent of their contents, authors and ends.

The dynamic centre of Figure 10.1 is carved by the three fossils and the portrait of Johann Christoph Wagenseil hovering above them. (An isosceles triangle has its vertices at Wagenseil's forehead and points between the feet of the two pairs of observers.) Altdorf acquired Wagenseil's *Wunderkammer* between 1705 and 1708. This included the three fossils: a bear to the left, a stag to the right, a 'Croat' on his mount in the middle. Part of the Wagenseil collection, the cabinet to the left of the bear, probably contains his and other curiosities acquired by the university: a box with holy relics, a Lapp sorcerer's drum, a large dagger with an engraved calendar, objects from a synagogue, assorted coins, medals and so on. In Wagenseil's cabinet or elsewhere, the library also boasted a mineral collection. To the far left rests a valuable armillary sphere. On the table right of the sphere lies *Hortus Eychstettensis*, an expensive botanical work, as a token of the really rare. Besides Wagenseil's, portraits of other benefactors hover like saints here. To the far right is the portrait of Johann Stöberlein, who bequeathed his library of medical and philosophical books to the library in 1696. His medical books were shelved with the medical, while his philosophy collection, at first shelved in the philosophy lecture hall, here occupies separate shelves under his portrait, a plaque midway down declaring these as Stöberlein's 'testament'.

Also shelved intact, Wagenseil's books probably occupy the bookcase under his portrait, facing the observer. Wagenseil had become a full professor at Altdorf in 1667. In 1699 he became librarian too – a typical early modern practice, whereby professors acquired auxiliary offices to supplement salaries. Wagenseil wished his collection to the library, but willed it to his heirs in 1705. They first undertook what inspired fear in colleges and universities: extramural alienation of academic effects. From 1705 to 1706, the heirs tried to sell his library to the University of Leipzig. On-going war made Leipzig poor, and the heirs resisted dismembering Wagenseil's corpus. In 1708, Altdorf raised funds satisfying the heirs, thus bringing 'Wagenseil' back to the library.

The Baroque academic library grew like that, largely retrospectively, as an aggregated accumulation of the already accumulated.[3] The modern capital wonder, the instrument of the budget for regular discrete acquisitions (essential for transforming knowledge into 'research'), did not exist. Composed of monstrous materials, the pre-modern library grew by extraordinary events. Best were bequests and endowments of books or funds, the latter mostly used to buy collections of deceased academics who thus lorded over libraries beyond the grave. For early modern books often came unbound from the publisher and collectors frequently had all 'their' books bound in the same colour and style of binding. So you could tell a book by its cover – not the author but rather the collector was key. Important 'juridical personae', estates or testaments of a Wagenseil or Stöberlein, governed whole library shelves, often visible by their bindings. Other collectors, as we shall see, suffered subordination and dismemberment.

Though lacking the modern notion of the annual budget, pre-modern libraries sought some regular acquisitions, but usually made a miserable showing. Typical techniques included channelling student fines or fees to the library. Alas, fees of this sort were low or claimed elsewhere, and fines were hard to collect and usually replaced by incarceration. Many colleges and universities expected the faculty to donate their works (as did many princely or national libraries of their subjects), a policy also hard to enforce, and making collection essentially intramural. Beyond such techniques, interest from endowments earmarked for purchases of discrete volumes was all there was.

As said, acquisition en bloc of estates formed the chief pillar, with such books displayed intact, as in the case of Wagenseil and Stöberlein, or broken up and subjected to the rubric of the disciplines, visible in the sign 'Theologia', above the books to the left and rear of Figure 10.1, and in the sign 'Philosophia', to the left of Stöberlein's portrait. Estates and disciplines gave then the primary principles at play in shaping the Baroque and pre-modern library. The tendency of the Enlightenment library will be to give most play to the disciplines: collection via epistemic and ultimately bureaucratic systems, as opposed to the aggregate of 'juridical' estates and plots.

In the physical disposition and cataloguing of books, the Baroque library offered resistance to the hegemony of epistemic system. During the early modern era, save the few monstrously big or small, books typically came in folio, quarto or octavo. After the primary division of estate or discipline (for example, Stöberlein or theology), the size of the book formed the secondary principle. Shelf-units in Figure 10.1 have from seven to nine horizontal shelves. Each shelf-unit has shelves of various heights to accommodate all sizes, from folio at the bottom to octavo at the top. The book's materiality further refined the collection's articulation. And catalogues furthered that.

For a long time, catalogues, if existent, were often only shelf-lists. That seems logical; but one of the heroic feats of the Enlightenment and Romantic library will be prising the catalogue apart from the shelf-list. For, until the latter event, the materiality and history of the book, embodied in its size and binding, governed the catalogue, that is, the book of books – pre-modern catalogues usually were books, as opposed to cabinets of cards or 'Zetteln'. The pre-modern catalogue at times existed only virtually: in the librarian's memory. As the transformation of memory or 'history' into epistemic systems was a hallmark of modernity, it is no surprise that the catalogue partook of that development. But at its first manifestation beyond memory, the catalogue at a library such as Altdorf's exhibited the history of the collection. When a college or university acquired a private collection, it often came with a catalogue, having been kept by the collector or made by the seller. Like the books, the general library catalogue often existed then as an aggregate – a catalogue of catalogues.

And books would be shelved accordingly. If the collection of a Professor X was not to be shelved separately, but should the collection have a catalogue, one would dismember the collection only in terms of the disciplines, if at all. Academics often collected primarily in view of narrow interests. So if Professor X had been a theology professor, his books would tend to be largely theology. One would shelve them intact, divided by size, on the next open shelves in the theology section. One could then use the extant catalogue for the collection with minimal annotations to reflect the shelf-listing. If Professor X had also bought arts and philosophy books, he might have separated them in his catalogue, enabling them to be shelved intact in the philosophy section, rendering his catalogue usable. When collections came without a catalogue, it was still easier to shelf thus to catalogue them en bloc, but dismembered by the four disciplines or faculties: theology, law, medicine, and arts and philosophy (the latter including the sciences). Catalogues were at first usually shelf-lists.

Inscribed in the sphere of the *Wunderkammer*, composed as a collection of collections, bequeathed or bought, the Baroque academic library ultimately embodied estates, an archive or mausoleum of academic 'monstrosities', a juridical plot of private personae, an aggregate of idiosyncratic interests accumulated by extraordinary events – a library of libraries, its catalogue a book of books, reflecting the nature, materiality and history of the collection, despite the disciplines, resisting the systematic.

Bibliotheca universalis

[T]he systematic shelving of books . . . appeared as logical and essential to all who thought that a collection of universal scope ought to mirror the universe of knowledge and the order of the sciences. For such mirroring, the hall libraries of the time, with the collection arranged along the walls of one room, were well suited. One could take in the entire bibliographical universe in a single sweep of the eyes . . . [Thus] there was less need for detailed catalogues.[4]

So wrote Hugo Kunoff, the great historian of the eighteenth-century German library. As *Wunderkammer*, the Baroque library had aspired to embody and epitomise the universe. The eighteenth century took this aspiration earnestly and systematically but, while wishing to separate the 'library' from the 'archive' and the 'museum', adhered to the centrality of the gaze. The *bibliotheca universalis* remained in the sphere of the visible and the real. Ocular perspicuity came first; rational order second. The mastery of knowledge, like the sovereignty of power, abided in the realm of 'spectacle'. Kunoff noted further:

If necessary, the classification sequence was broken to please the eye. The expensive and rare items occupied the most prominent place. To make all books appear the same size, folios were cut, the bindings of quartos extended, and smaller items encased. The visual effect was not to be marred by ugly call numbers.[5]

In the theory and practice of rational order, the princely collections in Hanover and Wolfenbüttel lay at the forefront. Also known for work in mathematics and philosophy, Gottfried Wilhelm Leibniz became librarian in Hanover in 1676 and in Wolfenbüttel in 1690.[6] In the latter at least, he set to work immediately. Shelf by shelf, he had entries for all books written on sheets of paper, thirty-two books per sheet. The sheets were then cut into the thirty-two entries, producing cards or *Zetteln* that were alphabetised as the work proceeded. Once every book had acquired its *Zettel*, and all had been alphabetised, Leibniz had a universal alphabetical catalogue copied from the *Zetteln*, per custom, as a *Bandkatalog*: a series of volumes, with empty space for future entries. The *Zetteln* were finally re-shuffled in terms of disciplines, then pasted – with empty space left for future entries – into a *Bandkatalog* as a *catalogus materiam*, or *Realkatalog*, or 'systematic' or 'scientific' or 'subject' catalogue, all terms used then.

The dilemma of rational cataloguing lay in the problem of finite, material representation of the infinite order of universal knowledge, at once logical and historical: to make the order of books reflect the 'order of things'. Leibniz appreciated the difference between the shelf-list, which catalogued in linear, 'mechanical' order bound to the materiality of objects, versus alphabetic and subject catalogues, which might represent polydimensional and 'organic' interrelations and cross-references of contents, authors and ends. A shelf-list showed where a book was physically and really; alphabetic and subject catalogues localised it virtually and ideally. The latter relations resemble the order Leibniz envisaged in his 'monadology': while the physical world was like a machine, the 'metaphysical' was more like a well-catalogued library.[7]

The alphabetic catalogue embodied the historical moment of knowledge; the systematic catalogue the logical. Representing the historical by the author (and not the collector), composition of the alphabetical catalogue became in principle easy and universal for all libraries. Composition of systematic catalogues was the hard and, alas, parochial part but, since the eighteenth century favoured logic over history, systematic catalogues typically took precedence over alphabetic. Like others, Leibniz began with the four faculties as disciplines: theology, law, medicine, and arts and philosophy (and sciences). Further articulation of knowledge thus of books meant refining those disciplines into all relevant and necessary sub-disciplines, and sub-sub-disciplines and so on. Since no universal systematic division of knowledge had come to hold sway, Leibniz's 'bibliotheca universalis', in view of its systematic catalogue, like most, looked particular, non-universal, idiosyncratic.[8] That was the scandal of the universal library.

Help came from a developing genre called 'historia literaria'.[9] This emerged from Baroque projects envisaging a 'universal bibliography': knowledge of all books and authors ever. From the seventeenth to the eighteenth centuries, such projects often split on the historical versus logical classification of books – some organised by authors, others by disciplines. Organised by disciplines, *historia literaria* held the promise of a universal systematic catalogue, if one author's scheme could triumph. During the eighteenth century, whether organised primarily by authors or disciplines, works within *historia literaria* became less polymathic and descriptive, and more specialised and critical. So in the 1770s, for example, the physician Albrecht von Haller published critical bibliographies on anatomy, surgery and 'practical' medicine – each was called a 'bibliotheca'. Assembling all such specialised 'libraries', *historia literaria* as a discipline became a literary library of libraries, a catalogue of the ideal universal library, a basis of the newly emerging *Bibliothekswissenschaft*: library science.

The Enlightenment library

'In first place stands the library. Perhaps no public library has ever accomplished as much as the Göttingen. The whole university owes a large part of her celebrity to it . . . Many Professors may thank the library for their literary fame, as it supports them with all desirable resources for their learned works' – so, with the truth, did Prussian minister Friedrich Gedike begin his enumeration of Göttingen's 'institutes' in 1789.[10]

Opened in 1737, oriented on a pragmatic, rationalising view of knowledge, the University of Göttingen became the university of the German Enlightenment. The visible hand behind the foundation, the Hanoverian minister Gerlach Adolph von Münchhausen, had not only a plan for the university but also money for the library. Up to the 1780s, only 250–300 Thaler annually was officially budgeted; Münchhausen, however, transferred huge surpluses to the library each year. No records were kept. But judging from the growth, the funds must have been immense. In the 1790s a budget was set, averaging 3,000–4,000 Thaler per year. By 1800, at around 200,000 volumes, Göttingen's library was the third largest in the Germanies, behind only the princely libraries in Vienna and Dresden, and ahead of those in Berlin and Munich. It was the largest academic library in the Germanies and probably in the world.

Figure 10.2 shows the library in 1747 as a typical hall library. Captions indicate the room as 100 by 40 (German) feet and the collection as based on books willed the university by Johann Heinrich von Bülow. The next year, expansion began a process whereby the library took over other rooms on this floor then the whole floor in 1764 (later the whole building and more). Figures 10.3 and 10.4 depict the layout in 1765. History, ethnography and

Figure 10.2 Georg D. Heumann, *Wahre Abbildung der Köngl. Groß-Britan. u. Churfürstl. Braunschweigisch-Lüneburgische Stadt Göttingen*, (Göttingen, 1747) (E:GöttUB, gr. 2° H.Hann. V, 29 *rara*.). (Reproduced with permission of the Universitätsbibliothek Göttingen)

related books occupy the largest and (new) entry hall, 'A'. Wing 'B' has theology and 'C' law. The smallest hall houses medicine, philology, philosophy, mathematics, natural sciences, politics, economics, applied sciences and arts – all designated 'miscellaneous'. The place given historical and ethnographical works bespeaks Göttingen's (or Münchhausen's) idiosyncratic view of enlightenment.

The sheer size of the collection threatened chaos. Under Münchhausen's aegis, the library responded with a system of three catalogues. First came an accession catalogue, with a full bibliographic entry and the accession number of each book. By 1789 a complete alphabetical catalogue also existed, having been revised for the third time and now – with a page for every author and able to receive inserted pages – able to encompass future authors 'forever'. Finally, finished in its first revision 1743–55, the third and most famous was the systematic catalogue, at first with categories: (i) theology, (ii) law, (iii) medicine, (iv) philosophy, natural sciences, politics and art, (v) ancient philology, history and ancillary disciplines, (vi) *historia literaria*.

Perfection of this system was the work of Christian Gottlob Heyne, philology professor and director of the library, 1763–1812. The greatest librarian of the Enlightenment, Heyne saw to the rational interconnection of the catalogues. Most importantly, he turned the systematic catalogue into the shelf-list, which had long been but the memory of the book-fetchers. Heyne saw that the systematic catalogue's classification of a book, along with the page number in that catalogue, could offer a basis for a 'signature' – shelf or call number – of the book. A book's virtual location in the system of knowledge became its physical location. Göttingen's three

Figures 10.3 and 10.4
Johann Stephan Pütter, F. Saalfeld and G. H. Oesterley, *Versuch einer academischen Gelehrten-Geschichte der Georg-Augustus Universität zu Göttingen*, 4 vols. (Göttingen/Hanover, 1765–1838), vol. I. (Reproduced with permission of the Universitätsbibliothek Göttingen)

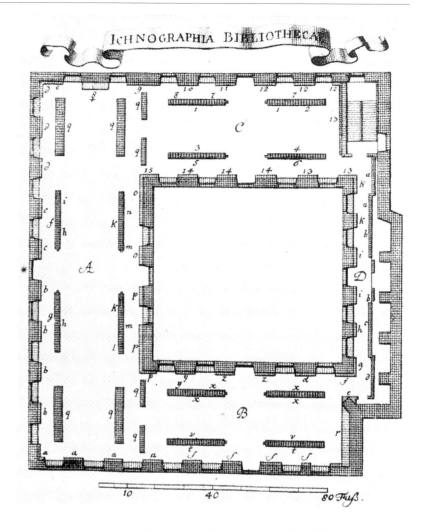

catalogues instantiated a cross-referencing, unified historical register and epistemic system.

Under Münchhausen's ministry, Heyne rationalised collection as well. Like most, he bought libraries en bloc as opportunity arose. But that no longer gave the primary principle of acquisition. The bureaucratic instrument of the budget changed everything. Administering one of the largest collections on earth, he had renounced the fantasy of the 'universal library': an enlightened librarian confronted finitude and thus made selections. Diametrically opposed to typical private and princely collectors, he avoided rare or costly works unless essential. He also avoided commonly owned, popular works, and multiple editions. In the Enlightenment's spirit, he collected rather the 'useful' for the public. And there only a book's contents mattered. Along with the systematic catalogue, this moved the library's centre from the material and

Figure 10.4

Erläuterung des Grundrisses von der Bibliothek.

A. *Der nördliche Saal hauptsächlich mit Historie besetzt.*
a a a a. Einleitung zur Geschichte, Geographie, nebst den Reisebeschreibungen, Chronologie, Genealogie, Heraldik, Diplomatik.
b b b b. allgemeine Weltgeschichte und Statistik.
c c. Geschichte u. Sammlungen der Friedens- u. anderer Tractate: c. auch vermischte historische Werke.
d d d. Schriftsteller von alten Denkmälern und Gebräuchen.
e. Geschichte der alten Reiche und Staaten.
f. – – – von Italien.
g. – – – von den nördlichen Reichen, Ungarn, Türkey, Asia, Africa, America.
h h. – – – von Teutschland.
i. – – – von der Schweitz.
k k. – – – von Engelland.
l. – – – von den Niederlanden
m m. – – – von Frankreich.
n. – – – von Portugall und Spanien.

* * *

o o. die alten Griechischen Schriftsteller.
p p p. die alten Lateinischen Schriftsteller.

* * *

q . . . q. Rastelen u. Tische unten mit Locaten versehen, und mit volumineusen Sammlungen besetzt.
♀ Das Bildniß des Curators Erc.

B. *Der westliche Saal, ganz mit Theologie besetzt.*
I. Bibelsammlung, und philologia und critica sacra.
s s s s. Ausleger der H. Schrift.
t t. Dogmatische Theologie.
v v. Patres und Kirchenscribenten.
x x x x. Allgemeine und besondere Kirchengeschichte nebst den Concilien, und Kirchengebräuchen.
y y. Streittheologie.
z z. Sittenlehre, Casuistik, Pastoraltheologie.
Homiletik, Ascetik.

C. *Der östliche Saal, enthält auf denen in der Mitte stehenden Repositorien.*
1. 1. Die bürgerliche Rechtsgelahrheit.
2. Das Lehnrecht, und Privatrecht.
3. Das Kirchenrecht.
4. Das Land- und Stadtrecht.
5. Das allgemeine, und Teutsche Staatsrecht.
6. Die Samlung von Deductionen.
7. 7 Die juristische Praxin, nebst Decisionen, Responsis, Consiliis, Obseruation., Disceptat. etc.
8. Zusammengedruckte Werke der Juristen.
   An der Ostseite.
9. Die Quellen der Arzneygelahrheit, nebst der Anatomie und Physiologie.
10. Die allgemeine Pathologie Semiotik, Diätetik.
11. Die allgemeine Therapie, Materia medica, Pharmacie, Chymie.
12. 12. 12. Die medicinische und chirurgische Praxin, Medicinam forensem, vermischte Werke.
An der Querwand, und Westseite.
13. 13. 13. Die Gelehrtengeschichte.
14. 14. 14. Die Philologie u. Etitik.
15. Die Werke der neuern Philologen und vermischte Schriften in allerley Sprachen.

D. *oder der schmale Saal enthält*
a a. die Mathematik.
b b. die allgemeine, und besondere Naturlehre.
c. die Philosophischen, alte und neue, Secten.
d. Logik, Metaphysik, natürliche Gottesgelahrheit, Recht der Natur, Moral.
e. Politik. f. Oeconomie.
g. Handlung und Gewerbe.
h. Schreibkunst, Mahlerey, Bildhauerey, Musik, Tanz-Fecht-Kriegskunst.
i i. Dichtkunst.
k k. Redekunst.

§. 117

visible towards the formal and rational. It dematerialised books, whose covers became incidental, and whose contents essential. Although it too put some on display, Göttingen's books were not 'monsters', not meant to be looked at, but rather to be read. The library became justly as famous for its reader-friendly atmosphere as for its catalogues – indeed the latter were part of the former. In Göttingen dead collectors no longer lorded over the library; future readers were rather served in their research.[11]

Acquisition of monstrous materials via extraordinary events largely gave way to regulated collection of the normal and useful: 'research' materials. The systematic catalogue, alongside *historia literaria*, gave an overview for rational planning of collection in respect of older literature – one saw what was missing. In 1789 Gedike praised the library for its 'well-conceived plan' of acquisition: not the subjective whim of the librarian, but rather the impersonal, objective dictates of research disciplines – determined by consultation with the relevant Göttingen professor – decided what would be bought. As the libraries in Alexandria and Vivarium had epitomised the ancient and medieval worlds, so Göttingen's was destined for the modern world: Göttingen rationalised the 'universal library' by bureaucratising collection.

Bibliotheca virtualis

Historia literaria had envisaged a universal library of the past. In *Allgemeine Deutsche Bibliothek* – 'ADB' – (1765–1806), Friedrich Nicolai conceived a review journal as a virtual (German) library of the present. He eschewed the subjective choices of most review journals in favour of 'comprehensive coverage' of all German academic and literary works. ADB reviews were anonymous. Short reviews followed the order of the disciplines and gave an 'objective' summary of the book's contents. ADB would be a systematic catalogue and virtual library of current German works – possessing it, one might forgo buying most books. This then furthered the dematerialisation of the book, as the virtual library reflected only a simulacrum of contents. By 1770, Nicolai had, however, abandoned the fantasy of a universal German library, and ever more books meant ever shorter summaries, thus ever more thin 'simulations' of virtual books.[12]

In the Germanies, book production was ten times greater in 1763–1803 than in 1721–63 – all the more need for virtual libraries such as Nicolai's. Many had arisen, more would, and most called themselves 'library' or 'journal for learned things'. Two were long most important: ADB and 'GGA' – *Göttinger gelehrten Anzeigen*. And the latter aimed to review not just German but rather all useful academic and literary works. Heyne edited GGA after 1770 and one saw his roles as librarian and editor as synergetic. All books fit for the library were reviewed in the journal, and all books reviewed in the journal went into the library.[13] For works published after a certain point, the journal became the virtual double or simulation of the real collection, though not in logical space but rather as the trace of its history – its yearly growth.

Because the Göttingen faculty was (supposedly) responsible for deciding what books to buy, and because all books bought were reviewed in the GGA (and vice versa), and because (supposedly) the faculty would write all reviews in the GGA, then the library served as a sort of archive of what the university as a collective had read. Or, as Heyne saw it, the GGA was a sort of diary of what the collective had thought and researched. While collectors – in view of books' covers – had lorded over the Baroque library, the collective – in view of books' contents – would come to embody itself in Göttingen's Enlightenment library, an archive of collective reading. In this sense the Enlightenment is best seen not so much as the great age of speaking or writing but rather of reading – the essential enlightened society being a reading group.

As contemporaries saw it, thanks to its collective review journal, as well as its other and individual virtual 'libraries' and works in *historia literaria* edited by the faculty, not to mention the many textbooks the faculty had written, the University of Göttingen emerged, in the Germanies, as a sort of 'supreme court' within the Republic of Letters. Heyne himself lamented the 'lack of policing [*Policey*] in the Republic of Letters' and hoped, through the Göttingen libraries, real and virtual, to instil a 'normal-law' (*Normalgesetz*) in the republic for rational order and good taste and style too.[14] Göttingen's enlightened budgetary acquisitions transformed the nature of the library from a juridical and historical plot or mausoleum into a bureaucratic and disciplinary system. Its system of catalogues transposed the order of books from a visual physical space into a rational virtual one. Its review journal, like a Baroque shelf-list, catalogued the history of the contemporaneous collection and also attested to the collective reading and research of the faculty. The Göttingen library in its own way also became an archive, a mausoleum of the faculty and its bureaucratic plots.

The Romantic library

'More light', said Johann Wolfgang von Goethe on 6 November 1817 (and not for the last time in life). After appearing unexpectedly on that day in the university library in Jena, so seriously did this poet and scientist take his new position as Supreme Co-Supervisor of the Library, the next day, without consulting the Jena Town Council, or even anyone for that matter, he ordered part of the ancient city wall torn down to allow more light into the library. Goethe's ministerial mission, which he discharged in person and with gusto to boot, was to turn the archaic library into a modern one.

He ordered the disjoint libraries integrated into one, leaving only the 'Buder' collection, per testament, separate. For the entire collection, he had an alphabetical catalogue begun, as well as a systematic one. Under Goethe's supervision, the library sought a 'pragmatic' division of knowledge, against what was seen as the overly 'scientific' bent at Göttingen: the arrangement of the systematic catalogue in Jena did not try to articulate a philosophical system of knowledge, but rather only used what were seen as commonplace thus user-friendly divisions of knowledge. Jena thus did not adopt the disciplinary divisions of the Göttingen systematic catalogue; but Goethe had the books reorganised so that the shelf-list embodied the systematic catalogue, à la Göttingen. So pressing did the task appear, he compelled the librarian, in order to devote more time to the library, to relinquish the editorship of Jena's review journal – the erstwhile *Allgemeine Literatur-Zeitung*, now the *Jenaische Allgemeine Literatur-Zeitung* and worthy rival to the GGA as a virtual library. A good sum was given for the reorganisation, but for a time the library's budget remained small, so Göttingen's heroic bureaucratisation of acquisition could not be imitated.[15]

In the nineteenth century, more and more took enlightened Göttingen as a model. Budgets eventually increased and allowed rational collection of current works, reviewed in ever more specialised disciplinary journals. The centre of collection moved everywhere from the rare and extraordinary to the regular and normal and 'serial'. Libraries differentiated themselves from museums, archives and treasuries. Most (Geman) stacks would eventually become closed. For a long time the systematic catalogue, Göttingen's supreme achievement, was seen as the librarian's ultimate duty.[16] But the Romantic era did witness the triumph of history and broke with the Enlightenment library on three points: the relation between the library and the faculty, the author versus systematic catalogues, and the triumph of 'culture' over utility.

Enlightened Göttingen probably embodied the last university whose faculty ideally had read everything in the library. Other universities with general review journals abandoned the fantasy that the faculty could review all books. The future belonged anyway to specialisation of journals and libraries. Romantic Germany became famous for specialised university institutes, seminars and laboratories, each eventually with its own budget, in part to buy books. The enlightened reading collective, vested in a central library and journal, found itself surrounded by a congeries of specialised journals and institute libraries, recapitulating the Baroque library as an aggregate, though now of epistemic disciplines and not juridical estates. The dynamics of research led to a dialectic of the centralised and collective versus the specialised and individual.

The Romantic library witnessed as well the triumph of the author catalogue, for which Albrecht Kayser's 'On the manipulation concerning the arrangement of a library and the composition of the catalogues' (*Ueber die Manipulation bey der Einrichtung einer Bibliothek und der Verfertigung der Bücherverzeichnisse . . .*, Bayreuth, 1790) was crucial. Romanticism advanced the dematerialisation of books still further, as visual and real presence lost all primacy to the conceptual and virtual. So Kayser insists 'the place where a book stands is most unimportant'. A systematic catalogue remains useful, he says, but is not essential. What is is a shelf-list and an author catalogue. The still parochial nature of systematic catalogues – witness Jena's rejection of Göttingen's classification of knowledge – meant that signatures of books had no extramural currency. The scandal went further, for different cataloguers using the same systematic catalogue might localise a book in different places. Without an author catalogue, a future reader might not find a book in a library, which itself might have accidentally acquired several copies, which various cataloguers had catalogued variously.

Even after librarians had read Kayser, and the author catalogue had emerged as their admitted first duty, investment in the systematic catalogue – the Enlightenment's legacy – often led to retardation of the author catalogue.[17] Nonetheless, in the name of the reader, Romanticism made the alphabet sovereign in the virtual library. The order of books no longer had to reflect the 'order of things', and surely not the history of the collectors, but rather the encyclopaedia of authors. The 'author-function' has varied and may be different in different books (thus play one role in literature and another in science).[18] But in the author catalogue, the Romantic author-function set a universal geneaology over a Baroque 'archaeology' (of collectors) and an Enlightenment 'topology' (of disciplines).

The Enlightenment had led to the disembodiment of the book by its transposition into a rational virtual space. Romanticism perpetuated and advanced this. For the Romantic author, though possessed of a juridical persona (vested in a copyright), had become most essentially a spiritual being through the notion of 'genius'.[19] And through the most spiritual of Romantic conceptions, the genius of a people, that is, 'culture', the Enlightenment's focus on the useful would be broken as well. Unlike the enlightened librarian, who only thought to collect materials for research which was 'useful', the Romantic librarian had to collect, in principle, everything, since everything, no matter how useless, is a manifestation of 'culture' and might provide grist for the mill of some future researcher. The Romantic research library fell into – or was condemned to – the Baroque fantasy of complete collection.

The Romantic library is the Library of Babel. As a great modern librarian noted of the latter, 'Everything is there: the minute history of the future, the autobiographies of the archangels, the faithful catalogue of the Library, thousands and thousands of false catalogues, a demonstration of the fallacy of these catalogues.'

Notes

1 On German libraries in general, see Hugo Kunoff, *The Foundation of the German Academic Library* (Chicago, 1982); Ladislaus Buzas, *Deutsche Bibliotheksgeschichte der Neuzeit (1500–1800)*, Elemente des Buch- und Bibliothekswesens, vol. II (Wiesbaden, 1976); id., *Deutsche Bibliotheksgeschichte der neuesten Zeit (1800–1945)*, Elemente des Buch- und Bibliothekswesens, vol. III (Wiesbaden, 1978).
2 On the university and collections, see Johann J. Baier, *Wahrhaffte und gründliche Beschreibung der Nürnbergischen Universität-Stadt Altdorf* (Altdorf, 1714), pp. 91–5. On the library, G. Werner and E. Schmidt-Herrling, *Die Bibliotheken der Universität Altdorf*, Beiheft zum Zentralblatt für Bibliothekswesen, vol. LXIX (Leipzig, 1937), pp. 6–22, 38–58. On museums and libraries, Paul Raabe (ed.), *Öffentliche und Private Bibliotheken im 17. und 18. Jahrhundert. Raritätenkammern, Forschungsinstrumente oder Bildungsstätten?*, Wolfenbüttler Forschungen, vol. II (Bremen/Wolfenbüttel, 1977). The histories of other academic libraries are similar – see *Geschichte der Universitätsbibliothek Jena*, Claves Jenensis, vol. VII (Weimar, 1958); Gottfried Zedler, *Geschichte der Universitätsbibliothek zu Marburg von 1527–1887* (Marburg, 1986); Otto Handwerker, *Dreihundert Jahre Würzburg Universitätsbibliothek 1619–1919* (Berlin 1932).
3 Buzas, *Bibliotheksgeschichte der Neuzeit*, pp. 35–6; Kunoff, *The Foundation*, p. 98.
4 Kunoff, *The Foundation*, p. 122. On the universal library, see also Roger Chartier, *The Order of Books*, trans. Lydia Cochrane (Cambridge, 1992), ch. 3.
5 Kunoff, *The Foundation*, p. 127.
6 On Leibniz, Otto von Heinemann, *Die Herzogliche Bibliothek zu Wolfenbüttel, 1550–1893*, 2nd edn. (Amsterdam, 1969), pp. 111–15, 128. In theory, the Electoral-Royal-State Library in Berlin was advanced in cataloguing: see Eugen Pauntel, *Die Staatsbibliothek zu Berlin . . ., 1661–1871* (Berlin, 1965), pp. 12–13, 85, 93–5.
7 Hans Schulte-Albert, 'Gottfried Wilhelm Leibniz and literary classification', *Journal of Library History*, 6 (1971): 133–52, at pp. 143–4. Hans Blumenberg, *Die Lesbarkeit der Welt*, 2nd edn. (Frankfurt a. M., 1983) ch. 10.
8 Kunoff, *The Foundation*, pp. 120–6.
9 Martin Gierl, 'Bestandsaufnahme im gelehrten Bereich: Zur Entwicklung der "Historia literaria" im 18. Jahrhundert', in id. (ed.), *Denkhorizonte und Handlungsspielräume. Historische Studien für Rudolf*

Vierhaus zum 70. Geburtstag (Göttingen, 1992), pp. 53–80; Kunoff, *The Foundation*, p. 76; Chartier, *The Order of Books*, ch. 3.

10 R. Fester (ed.), *Der Universitäts-Bereiser. Friedrich Gedike und sein Bericht [1789] an Friedrich Wilhelm II*, Archiv für Kulturgeschichte, supp. 1 (Berlin, 1905), p. 26. Generally, Arnold H. L. Heeren, *Christian Gottlob Heyne, biographisch dargestellt* (Göttingen, 1813), pp. 292–305; Werner Seidel, *Baugeschichte der Niedersächsischen Staats- und Universitätsbibliothek in Göttingen*, Hainbergschriften, vol. XI (Göttingen, 1953); Georg Schwedt (ed.), *Zur Geschichte der Göttinger Universitätsbibliothek. Zeitgenössische Berichte aus drei Jahrhunderten* (Göttingen, 1983); Karl J. Hartmann (ed.), *Vier Dokumente zur Geschichte der Universitätsbibliothek Göttingen* (Göttingen, 1937); Karl J. Hartmann and Hans Füchsel (eds.), *Geschichte der Göttinger Universitäts-Bibliothek* (Göttingen, 1937); Werner Arnold and Peter Vodosek (eds.), *Bibliothek und Aufklärung*, Wolfenbütteler Schriften zur Geschichte des Buchwesens, vol. XIV (Wiesbaden, 1988), pp. 17, 64–5.

11 See Heyne's report of 1810 reprinted in Hartmann, *Vier Dokumente*, pp. 14–18; also Heeren, *Christian Gottlob Heyne*, pp. 292–9; Fester, *Friedrich Gedike* (ref. 10) p. 27; Kunoff, *The Foundation*, ch. 6.

12 Ute Schneider, *Friedrich Nicolais Allgemeine Deutsche Bibliothek als Integrationsmedium der Gelehrtenrepublik* (Wiesbaden, 1995), pp. 76, 96, 270–2.

13 Herren, *Heyne*, pp. 259–62; Heinrich Oppermann, *Die Göttinger gelehrten Anzeigen . . .* (Hanover, 1844), pp. 13, 24, 30.

14 Oppermann, *Die Göttinger . . . Anzeigen*, pp. 27–28; Gustav Roethe, 'Göttingische Zeitungen von gelehrten Sachen,' in *Festschrift zur Feier des hundertfünfzigjährigen Bestehens des Königlichen Gesellschaft der Wissenschaften zu Göttingen . . .* (Göttingen, 1901), pp. 567–688, esp. p. 623; also the document in Emil Rössler (ed.), *Die Gründung der Universität Göttingen. Entwürfe, Berichte und Briefe der Zeitgenossen* (Göttingen, 1855), pp. 478–9, perhaps by Münchhausen, perhaps not.

15 Karl Bulling, *Goethe als Erneuer und Benutzer der jenaischen Bibliothek*, Claves Jenensis, vol. II (Jena, 1932); *Geschichte der Universitätsbibliothek Jena* (ref. 2), pp. 390–433; Kunoff, *The Foundation*, pp. 133–5.

16 Buzas, *Bibliotheksgeschichte der neuesten Zeit*, pp. 16, 31, 107, 135–6. On correlative change in archives, see William Clark, 'On the ministerial archive of academic acts', *Science in Context*, 9/4 (1996): pp. 421–86.

17 Buzas, *Bibliotheksgeschichte der neuesten Zeit*, p. 140.

18 See Michel Foucault, 'What is an author?', *Language, Counter-memory, Practice*, ed. Donald Bouchard (Ithaca, 1977); also Chartier, *The Order of Books*, ch. 2.

19 See Simon Schaffer, 'Genius in Romantic natural philosophy', in *Romanticism and the Sciences*, ed. Andrew Cunningham and Nicholas Jardine (Cambridge, 1990), pp. 82–98; id., 'Self Evidence', *Critical Inquiry*, 18 (1992): 327–62; William Clark, 'On the ironic specimen of the Doctor of Philosophy', *Science in Context*, 5/1 (1992): pp. 97–137.

Further reading

Ladislaus Buzas, *German Library History, 800–1945*, trans. W. D. Boyd and I. H. Wolfe (Jefferson, N.C., 1986)

John W. Clark, *The Care of Books, an Essay on the Development of Libraries from the Earliest Times*, 2nd edn. (Cambridge, 1909)

Philip Rowland Harris, *A History of the British Museum Library* (London, 1998)

Oliver Impey and Arthur MacGregor (eds.), *The Origins of Museums: The Cabinets of Curiosities in Sixteenth- and Seventeenth-Century Europe* (Oxford, 1985)

Hugo Kunoff, *The Foundation of the German Academic Library* (Chicago, 1982)

D. F. McKenzie, *Bibliography and the Sociology of Texts. The Panizzi Lectures, 1985* (London, 1986)

David McKitterick, *Cambridge University Library. A History: The Eighteenth and Nineteenth Centuries* (Cambridge, 1986)

J. C. T. Oates, *Cambridge University Library. A History: From the Beginnings to the Copyright Act of Queen Anne* (Cambridge, 1986)

Ian Philip, *The Bodleian Library in the Seventeenth and Eighteenth Centuries* (Oxford, 1983)

James W. Thompson (ed.), *The Medieval Library* (Chicago, 1939)

11 Encyclopaedic knowledge

After explaining the content and purpose of his *Lexicon Technicum* of 1704, John Harris (1667–1719) hoped that he had satisfied 'the Reader that it is a Book useful to be read carefully over, as well as to be consulted like other Dictionaries occasionally'.[1] In making this claim, he sought to distinguish this dictionary of arts and sciences from a range of other alphabetical compilations by affirming its status as a coherent work. This was a point made by most subsequent compilers of similar dictionaries during the eighteenth century. Ephraim Chambers (1680?–1740), the author of the influential *Cyclopaedia* (1728), trumped Harris by asserting that this possibility had not been fully exploited. In his preface, Chambers acknowledged Harris and others, but claimed to go beyond them by providing the option of a systematic reading: 'Former Lexicographers have not attempted any thing like Structure in their Works; nor seem to have been aware that a Dictionary was in some measure capable of the Advantages of a continued Discourse' (preface, vol. I, p. i). Prior to the second edition of 1738, Chambers boldly described his work as 'the best Book in the Universe'.[2]

Given our twentieth-century habit of using dictionaries and encyclopaedias as reference works of quick consultation, these remarks seem strange. This is partly because modern encyclopaedias have largely abandoned one of the aspirations of their Enlightenment predecessors – to show the connections between subjects in the circle of sciences. Although most encyclopaedic works of the eighteenth century were alphabetical, they still deferred, at least in principle, to the tradition of the earlier works arranged on some explicit plan – such as the seven liberal arts, or the cosmological chain of being with the Divinity as its apex. They promised to retain the integrity of each subject, and to offer paths that followed an appropriate round of learning. Accordingly, the prefaces of these eighteenth-century works were often devoted to the difficult task of explaining how this was possible given the scattering of material by the alphabet. When Harris and Chambers called their works 'books', this is the issue to which they alluded. I examine the assumptions associated with such statements and, in doing so, consider the ways in which dictionaries of arts and sciences sought to give a systematic account of knowledge. My focus is on Chambers' *Cyclopaedia* – the model for subsequent works of this kind, including the famous, and much larger, *Encyclopédie* (1751–72), edited by Denis Diderot and Jean D'Alembert.

It is important to recognise that the modern, multi-volume encyclopaedia derived from the smaller dictionaries of arts and sciences that appeared in the early 1700s. Prior to Chambers, this genre was pioneered by Antoine Furetière with his *Dictionnaire Universel* (2 vols., 1690; 2nd edn, 1694) and, more influentially, by John Harris with the *Lexicon Technicum: or, an Universal English Dictionary of Arts and Sciences* (Figure 11.1). Harris' work appeared in one folio volume in 1704, and a second edition followed in 1708. In 1710 he published a supplementary volume, including additional material; he called this volume two, even though, like the first volume, it covered the entire alphabet. These dictionaries offered short accounts of scientific and technical 'terms', arranged in alphabetical order. As such, they were closer in form to the 'hard word' language dictionaries such as Thomas Blount's *Glossographia* of 1656, than to the Renaissance encyclopaedic works, such as Johann Heinrich Alsted's *Encyclopaedia* of 1630 (in four folio volumes). This work consisted of lengthy treatises, in Latin, on the full range of liberal arts and sciences, reflecting the university curriculum but also going beyond it to provide what Alsted called 'the methodological understanding of everything than man must learn in this life'.[3] By contrast, Harris and Chambers aimed to give concise entries, in the vernacular, on the various terms used in the arts and sciences. Although the new dictionaries of arts and sciences included subjects such as law, music and grammar – regarded as 'sciences' – their acknowledged strong point was the physical and mathematical sciences and the practical subjects associated with them, such as navigation, architecture and shipbuilding. The latter were so prominent in the *Lexicon* that Harris earned the tag of 'technical Harris'; and he admitted that he had been 'designedly short' in grammar, logic, ethics and metaphysics (preface, vol. I). Chambers promised that his work would make good the omissions of previous dictionaries, especially in the advances 'of Natural Knowledge made in these last Years' (preface, vol. I, p. i). During the eighteenth century, at least in Britain, works such as this came to be called, simply, 'scientific dictionaries'.

These works must be distinguished from that another genre, the historical dictionary. The two key examples are Louis Moreri's *Grand Dictionnaire Historique, ou mélange curieux de l'histoire sacrée et profane*, first published in Lyon in 1674, and Pierre Bayle's two-volume *Dictionnaire Historique et Critique* of 1697. Both went through many editions during the eighteenth century; they were translated into English and inspired similar works such as the

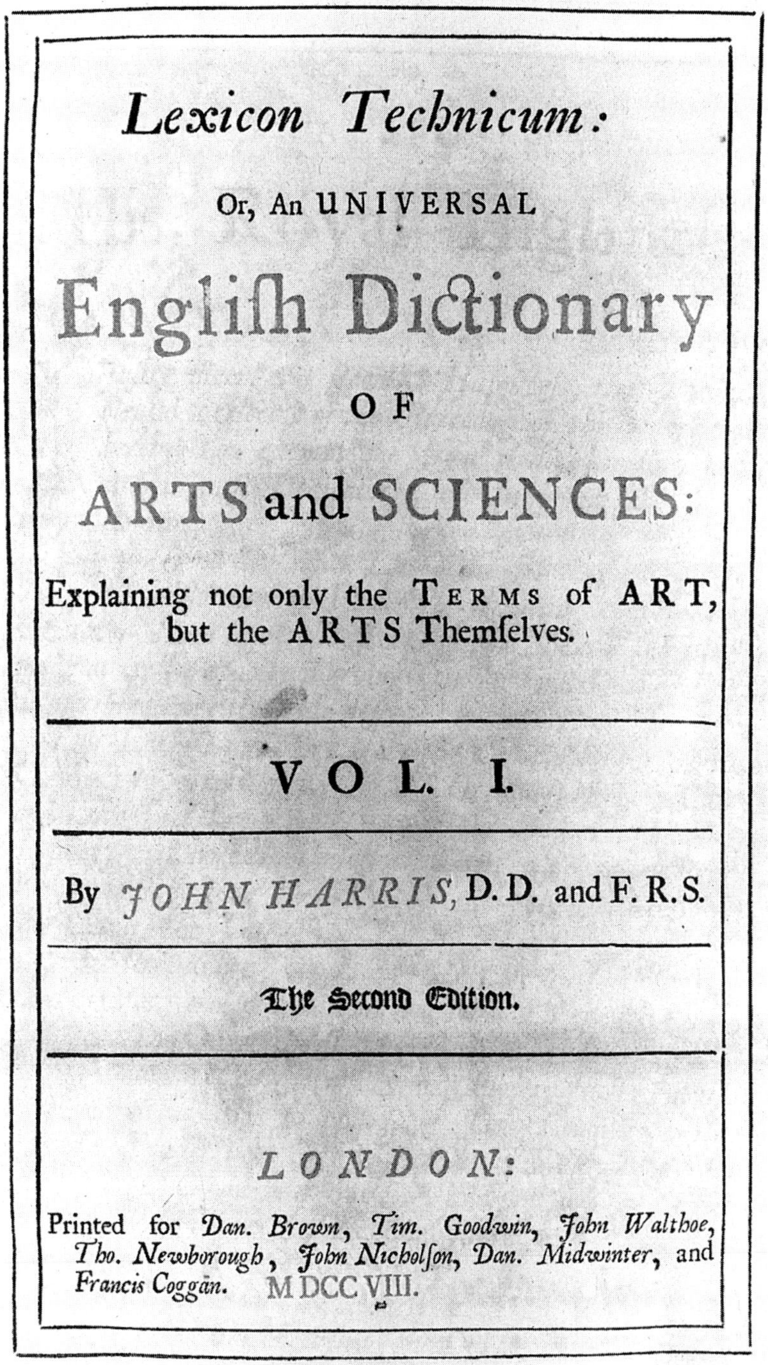

Figure 11.1 Title-page of John Harris' *Lexicon Technicum*, 2nd edn (London, 1708). (Courtesy of the Fisher Library, University of Sydney)

Biographia Britannica edited by William Oldys and then by Andrew Kippis (1st edn, 1747–66; 2nd edn, 1778–93). Their province was history, geography and biography rather than the arts and sciences. In contrast, the earliest dictionaries of arts and sciences did not have biographical entries and treated history and geography only in so far as this was relevant to the account of some technical term.[4]

The dictionaries of arts and sciences sought to carve out a special niche in the market. They rightly claimed to be more accessible than the neo-scholastic encyclopaedias written in Latin for university-educated readers. The ease of access provided by the alphabet was certainly a key selling point, but this was shared by the various philosophical, chemical, and medical lexicons that were common by the end of the seventeenth century. For example, Harris mentioned Stephen Chauvin's *Lexicon Rationale* (1692), Jacques Ozanam's *Dictionnaire Mathematique*, the 'Chymical and Physical Dictionaries of Johnson, Castellus and Blanchard', and the hard word dictionary by Edward Phillips, *The New World of Words*.[5] Presumably, these were the kinds of works he had in mind as being consulted 'occasionally'. By contrast, his own work sought to explain not just the terms of the arts and sciences, 'but also those *Arts themselves*' (preface, vol. I). One review opened with this point, noting that 'the design of this Dictionary is different from that of most others'.[6] This supported the more ambitious proposition already noted: namely, that unlike other dictionaries, the work could be read as a book. This was more than a matter of the simple denotation of 'book' – a label equally applicable to the other lexicons and dictionaries. Especially as elaborated by Chambers, it became an assertion that dictionaries of arts and sciences were judicious summaries of knowledge, informed by a respect for the systematic features of sciences and their relations to each other. This affirmation of the unity and coherence of knowledge justified his title – with its variant of the word 'encyclopaedia' – implying that the work embraced the circle of arts and sciences, and provided the opportunity to study 'a Course of Antient and Modern LEARNING' (Figure 11.2).

How could compilers of such works regard them as 'books' of this distinctive kind? I think there were two different, but related, ways in which this claim was made: (1) these dictionaries promised to replace other books, condensing knowledge of the arts and sciences into two large folio volumes; (2) they were conceived as having a structure or design, planned by an author.

A book in place of other books

The second edition of the *Cyclopaedia* carried the first significant entry on 'Book' in an English dictionary. Running to thirteen

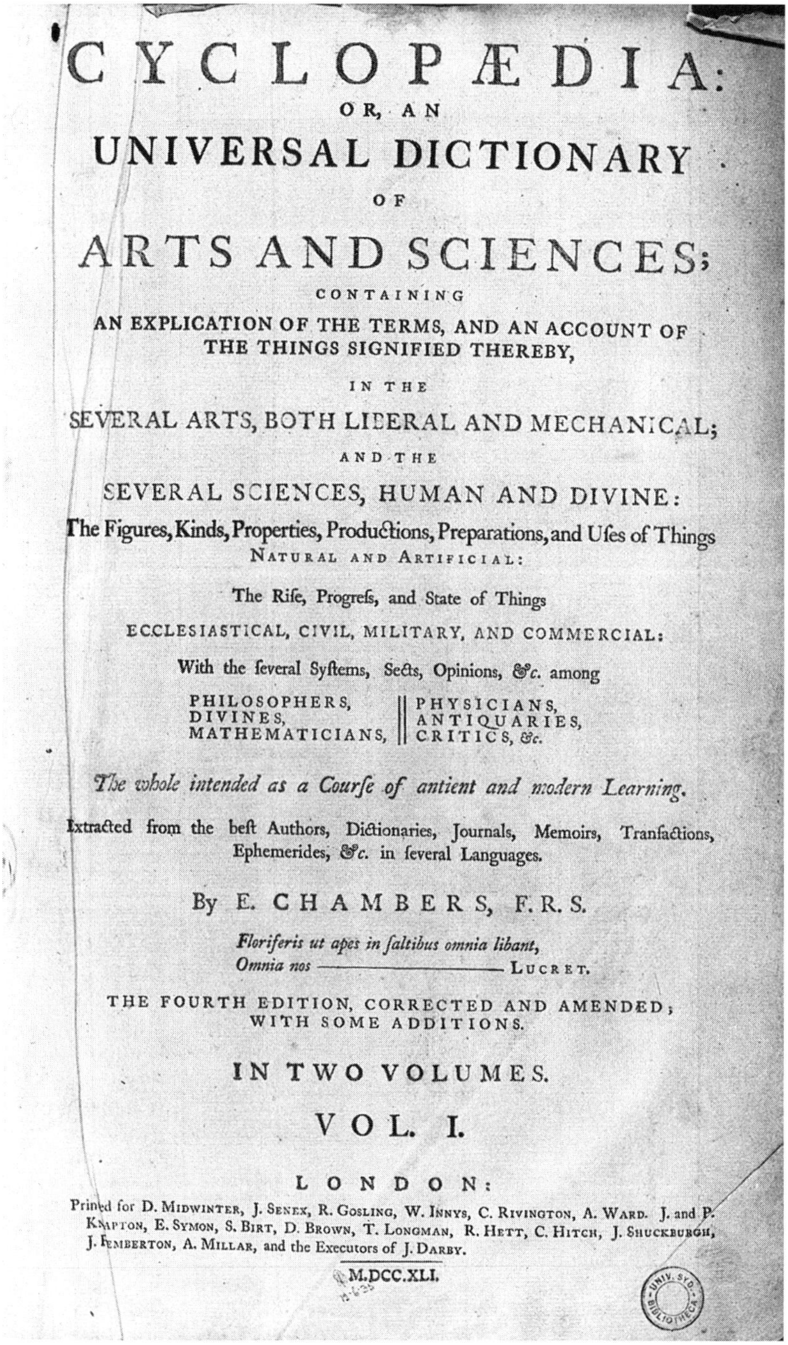

CYCLOPÆDIA:
OR, AN
UNIVERSAL DICTIONARY
OF
ARTS AND SCIENCES;
CONTAINING
AN EXPLICATION OF THE TERMS, AND AN ACCOUNT OF THE THINGS SIGNIFIED THEREBY,
IN THE
SEVERAL ARTS, BOTH LIBERAL AND MECHANICAL;
AND THE
SEVERAL SCIENCES, HUMAN AND DIVINE:
The Figures, Kinds, Properties, Productions, Preparations, and Uſes of Things
NATURAL AND ARTIFICIAL:
The Riſe, Progreſs, and State of Things
ECCLESIASTICAL, CIVIL, MILITARY, AND COMMERCIAL:
With the ſeveral Syſtems, Sects, Opinions, &c. among
PHILOSOPHERS, DIVINES, MATHEMATICIANS, || PHYSICIANS, ANTIQUARIES, CRITICS, &c.

The whole intended as a Courſe of antient and modern Learning.

Extracted from the beſt Authors, Dictionaries, Journals, Memoirs, Tranſactions, Ephemerides, &c. in ſeveral Languages.

By E. CHAMBERS, F.R.S.

Floriferis ut apes in ſaltibus omnia libant,
Omnia nos ——— LUCRET.

THE FOURTH EDITION, CORRECTED AND AMENDED; WITH SOME ADDITIONS.

IN TWO VOLUMES.

VOL. I.

LONDON:
Printed for D. MIDWINTER, J. SENEX, R. GOSLING, W. INNYS, C. RIVINGTON, A. WARD. J. and P. KNAPTON, E. SYMON, S. BIRT, D. BROWN, T. LONGMAN, R. HETT, C. HITCH, J. SHUCKBURGH, J. PEMBERTON, A. MILLAR, and the Executors of J. DARBY.
M.DCC.XLI.

Figure 11.2 Title-page of Chambers' *Cyclopaedia*, 4th edn, 2 vols. (London, 1741). (Courtesy of the Fisher Library, University of Sydney) columns (making it one of the longest entries in the work), this was one of the new entries singled out in the 'Advertisement concerning the Second Edition'.[7] The entry offered an impressive historical survey of the different kinds of books. It also went beyond descriptive information, allowing Chambers to canvass attitudes to books, especially to the increasing and unmanageable 'multitude' of them in his own time, thereby positioning his own work as part of this book culture, but also as an answer to some of its problems. Alluding to this issue in the preface, Chambers identified some inappropriate uses of books: bibliophilia was a great sin; many libraries were merely opportunities for acquisition and display. The *Cyclopaedia*, he said, would 'answer all the Purposes of a Library, except Parade and Incumbrance', and would be more useful 'than any, I had almost said all, of the Books extant' (preface, vol. I, p. ii).

The problem of the increasing number of books had been noted by thinkers of the previous generation. In 1680 Gottfried Wilhelm Leibniz spoke of that 'horrible mass of books which keeps on growing', so that eventually 'the disorder will become nearly insurmountable'; and in the meantime it would become a disgrace rather than an honour to be an author. This made it more difficult for the republic of letters and the academies to achieve and communicate any consensus on fundamentals. So Leibniz recommended that royal academies arrange for the 'the quintessence of the best books' to be extracted, and to add to them the observations, not yet recorded, of the best experts of each profession.[8] Pierre Bayle supported this need for judicious summary by somewhat different reasoning. For him, the problem was not just too many books, but limited access to those he needed. In the preface (dated 23 October 1696) to his *Dictionnaire* of 1697 he explained that 'the prodigious scarcity of books, very necessary to my design, stopped my pen a hundred times a day'.[9] Bayle's aim was, in part, to serve the republic of learning by reviewing books, as he did in his journal, *Nouvelles de la République des Lettres*, which he edited between 1684 and 1687. This required reliable and succinct abridgement, something that carried over into his *Dictionnaire*. Admitting that some topics were adequately treated by Moreri, he assured readers that he had not included superfluous information because he did not want them to buy 'the same things twice'. Pitching to those who could not afford many books, Bayle offered the *Dictionnaire* as a book to stand in place 'of a library to a great many people' (preface, pp. 3, 6). Thus, for Leibniz and Bayle, both the large number of books *and* their uneven availability, made abridgement imperative.

Chambers regarded his *Cyclopaedia* as a response to these problems. He called for 'a reduction of the vast bulk of universal knowledge into a lesser compass'; this was 'growing every day more and more necessary' because the objects of inquiry were increasing, books were becoming more numerous, and new points of dispute more frequent. Some convenient summary gave the only chance of making the essential knowledge of a range of sciences accessible to all, including specialist scholars. The *Cyclopaedia* therefore took words or *terms*, not *disciplines*, as its basic unit, and aimed to distil knowledge of the arts and sciences in a more manageable form, as a way of mapping the '*terra cognita*'. In this way, it would be a book to replace other books, a library in its own right. Moreover, it was a book composed by a single individual to meet a need not properly supplied by the learned academies of Europe.

It is well known that Bacon and his followers during the seventeenth century made derogatory remarks about the bookish culture of scholastic learning. They regarded this as one of the reasons for the alleged poverty of natural knowledge. As a remedy, they asserted that men must study the great book of Nature rather than the little books of men. Yet the new knowledge collected in nature had to be recorded in new books, only now more of them than ever before. When he pointed to the limits of memory in the face of expanding knowledge Bacon declared that 'up to now thinking has played a greater part than writing in the business of invention, and experience has not yet become literate. But no adequate inquiry can be made without writing.'[10] Chambers responded to the consequences of this dilemma: judging that empirical and theoretical knowledge of nature was expanding, that the number of books was beyond the capacities of individual readers, he declared that a single book was required to record the present state of knowledge.

The notion of a book distilling other books was accompanied by claims about authorship. Both Harris and Chambers regarded themselves as authors, not just compilers. On their own testimony, reliable authorship of a dictionary involved carefully selecting information, pruning errors, abridging texts and avoiding repetition – the virtues also affirmed by Bayle, even though he said that he had not intended to put his name on the title-page. Harris was anxious to distance himself from some sources which he relied on, but whose reputation or veracity he doubted. Thus in the case of Blanchard's (i.e. Blankaart's) dictionary he noted that

tho' many things are well enough done in him, yet some can hardly be said to be so; so that in many Places I have been obliged to put his Name to what my *Amanuensis* or *Assistant* transcribed from him, lest the Reader shou'd mistake it for my own Words (preface, vol. I).[11]

Moreover, Harris insisted that '*much the greater part*' of his *Lexicon* was 'collected from *no Dictionaries*, but from the best Original Authors I could procure in all Arts and Sciences, and is the Result

Figure 11.3 Portrait of John Harris in the frontispiece of his *Lexicon Technicum*, 2nd edn (London, 1708). (Courtesy of the Fisher Library, University of Sydney)

of some Years Labour and Consideration' (preface, vol. I; original emphasis). The portrait of Harris in the frontispiece identifies him with his book, echoing the images by which Renaissance authors, architects and painters began to assert their intellectual property (Figure 11.3). Similarly, Chambers was an active promoter of his authorship – something recognised by the Royal Society of London in 1729 when it accepted him as a Fellow: the nomination describes him as the 'Author of the Universal Dictionary of Arts and Sciences'.[12] Chambers depended on other books, but contended that not only had he made a reliable summary of the arts and sciences, but that this was informed by awareness of the integrity of subjects and their place in the circle of sciences. Indeed, in his preface, he spent twenty-five dense folio pages explaining how the work should be used. At one point he stopped to make this significant apology:

The reader begins to feel this preface grow tiresome; and yet several things are still behind. When so large a work was to follow, he perhaps imagines he should have been excused from a long introduction . . . But . . . several matters were purposely waved in the course of the BOOK, to be treated of in the PREFACE; which appeared the most proper place for such things as have a regard to the whole work. (Preface, 1738, vol. I, p. xvi; the original preface has thirty pages)

In other words, the preface, and the alphabetical entries, were all part of the one coherent book. The *Cyclopaedia* was a collation of knowledge; but it also had a design.[13]

A well-ordered Common-Place Book

The publication of the *Cyclopaedia* was announced in an advertisement of 17 January 1726 entitled 'Proposals for printing by Subscription'. Here Chambers explained that:

The Character of this Work is to be a DICTIONARY, and a SYSTEM at the same time. It consists of an infinite Number of Articles, which may either be consider'd separately, as so many distinct *Parts* of Knowledge; or collectively, as constituting a *Body* thereof.

Chambers promised to display the 'System' of knowledge but explained that it would be

a System taken asunder, and its Members re-compos'd in a different Order, to render 'em more obvious, and easy of access. Instead of the Natural Order, wherein the vast Assemblage would be unwieldy, and the Complication of numerous Parts make each difficult to come at and conceive; they are here presented *singly*, and in the familiar Order of the *Alphabet*.

He thus assumed the advantages of alphabetical order, but was not content to make this the major selling point. The *Cyclopaedia* acknowledged the existence of a 'Natural Order' – in spite of the way in which this systematic arrangement was disrupted by the alphabet – and it enabled the parts of knowledge to be 'put together, and the Whole set to view'. Chambers assumed that readers would wish to do this, and promised that the relations between parts of knowledge would be indicated:

their Natural Order and Dependance is here pointed out, and how and where they join to each other. Each Head is pursu'd into the Borders of the neighbouring ones; where the Pursuit is resum'd, and carried on by a Chain of References to others; and thence in an orderly Progression to the rest.

In this way, he assured potential buyers, it would be possible to rise from particulars to 'the Generals' of any art or science, or vice versa.[14]

This notion of a reader pursuing a subject in some logical or systematic fashion, rather than merely checking the meaning of a term, was central to the claim that the *Cyclopaedia* was a book to be read. The corollary of this was that it was also *composed* as a book, informed by some design. But given that it was an alphabetical dictionary, how can this be? I think this question demands a qualification of the historical perspective in which Chambers is usually placed. From the mid-eighteenth century his reputation as the 'father' of the encyclopaedic enterprise was established, and subsequently acknowledged by most commentators, from famous *philosophe*s such as Diderot and D'Alembert to London booksellers with an interest in publishing, or imitating, his work. Indeed, Diderot admitted that the rationale of the *Encyclopédie* owed much to Chambers.[15] But this should not conceal the ways in which Chambers depended on older traditions of both reading and storing information. In particular, the Renaissance tradition of commonplace books is evident in his remarks about the organisation and retrieval of knowledge.

Some clues to this link are apparent in the words Chambers used in the proposal and preface. The mention of 'Heads' and the distinction between 'General' and specific topics, suggest the terminology of the commonplace tradition that began in classical Greek culture and flourished in the Renaissance. This was manifested in the practice widespread among students and scholars of keeping a commonplace book for recording phrases, argument and factual information noted in the course of reading. This practice supported the rhetorical demands of Renaissance schooling that encouraged students to memorise quotations on literary and moral themes from classical authors. Arranged under 'Heads', or recorded as 'common-places', this material could be embellished in speeches and essays, as well as serving as a reminder of the location in the original source.[16] The commonplace method was thus a technique for storing and retrieving knowledge.

How does this help us appreciate Chambers' claims about the coherence of his *Cyclopaedia*? This can be approached in two steps: first, the relevance of a method of organising knowledge; and second, the implications of this for reading a dictionary of arts and sciences.

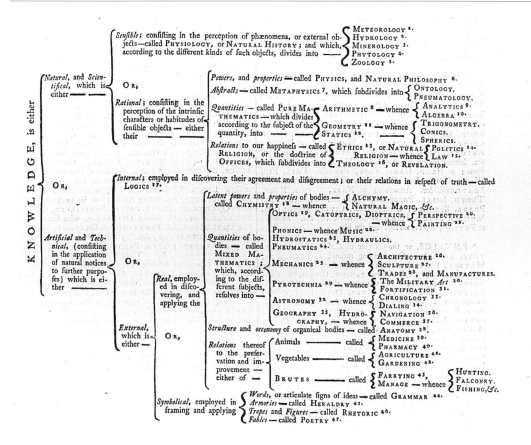

Figure 11.4 The 'View of Knowledge' in the Preface of Chambers' *Cyclopaedia*, 4th edn, 2 vols. (London, 1741), vol. I, p. iii. This appeared in all editions. (Courtesy of the Fisher Library, University of Sydney)

Consider the diagram of the arts and sciences in the preface of the *Cyclopaedia* (Figure 11.4). Such schemas (often based on the Porphyrian tree, but with horizontal inclination) were common in earlier encyclopaedias and were often arranged in a similar manner, proceeding through a series of dichotomies. This was especially the case following the impact of the pedagogic 'method' of Petrus Ramus (1515–72) from the late sixteenth century, but such diagrams had also been compatible with Aristotelian frameworks.[17] In encyclopaedic works of the middle ages and Renaissance, these diagrams usually indicated subjects covered in distinct treatises within the work. For this reason, their presence in a dictionary of alphabetical terms was odd: Furetière and Harris did not include such representations of subjects.[18] But Chambers' boast of going beyond previous dictionary makers in offering the chance of a 'continued discourse' was closely linked with this graphic display of the sciences. Each subject shown in the diagram was linked to a footnote containing a list of the terms belonging to it, so that, with cross-references, the reader could reconstitute a science that had been scattered alphabetically. How was this meant to work?

One way to approach this is to imagine the *Cyclopaedia* as a large

Figure 11.5 Note to the 'View of Knowledge' in the Preface of Chambers' *Cyclopaedia*, 4th edn, 2 vols. (London, 1741), vol. I, p. iv. Note 6 shows the various terms under the Head of 'Physics'. (Courtesy of the Fisher Library, University of Sydney)

⁶ PHYSICS, or the doctrine of *CAUSES*; as NATURE, LAW, &c. Occasions or means; as PRINCIPLE, MATTER, FORM, &c. Their composition, or constitution, in ELEMENT, ATOM, PARTICLE, BODY. CHAOS, WORLD, UNIVERSE, SPACE, VACUUM, &c. Properties of body; as EXTENSION, SOLIDITY, FIGURE, DIVISIBILITY, &c. Powers thereof; as ATTRACTION, COHESION, GRAVITATION, REPULSION, ELASTICITY, ELECTRICITY, MAGNETISM, &c. QUALITIES; as FLUIDITY, FIRMNESS, DUCTILITY, HARDNESS, VOLATILITY, DENSITY, POLARITY, LIGHT, HEAT, COLD, &c. Operations or effects thereof; as MOTION, RAREFACTION, DILATATION, CONDENSATION, DISSOLUTION, EBULLITION, FREEZING, EVAPORATION, FERMENTATION, DIGESTION, EFFERVESCENCE, &c. VISION, SEEING, HEARING, FEELING, SMELLING, &c. Modifications or changes; as ALTERATION, CORRUPTION, PUTREFACTION, GENERATION, DEGENERATION, TRANSMUTATION, &c. Systems or hypotheses hereof; CORPUSCULAR, EPICUREAN, ARISTOTELIAN, PERIPATETIC, CARTESIAN, NEWTONIAN, &c. — Occult and fictitious qualities, powers, and operations; ANTIPERISTASIS, SYMPATHY, ANTIPATHY, ARCHÆUS, &c. MAGIC, WITCHCRAFT, FASCINATION, VIRGULA DIVINA, LIGATURE, TALISMAN, CABBALA, &c. DRUID, BARD, BRACHMAN, GYMNOSOPHIST, MAGI, ROSICRUCIAN, and the like.

commonplace book. In the preface, Chambers described the contents of the work as 'extracts and accounts from a great number of books of all kinds', and went on to say, in effect, that he had sought to condense and order this collection: 'So that the difficulty lay in the form, and oeconomy of it; so to dispose such a multitude of materials, as not to make a confused heap of incoherent Parts, but one consistent Whole' (preface, 1738, vol. I, p. ii). Thus his 'View of Knowledge' shows forty-seven Heads – in this case various arts and sciences – numbered simply according to their position on the diagram, from Meteorology to Poetry. The notes attached to each art or science show the cognate terms belonging to it – as they might have been listed under one Head in a commonplace book (Figure 11.5). However, in the text of the work these terms are thrown into alphabetical order. The purpose of the chart, and its accompanying forty-seven footnotes listing terms under each subject, was to show that this had not been done without method, because the terms had first been collected under an appropriate Head. This deference to the respected tradition of commonplaces was a way of asserting the pretensions of the *Cyclopaedia* above those of a mere lexicon of terms.

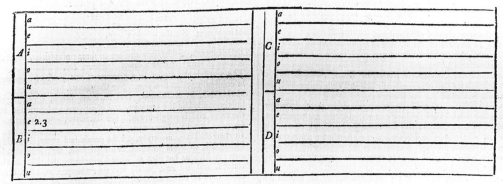

Figure 11.6 Illustration of John Locke's Index in the entry for 'Common-Place-Book' in Chambers' *Cyclopaedia*, 4th edn (London, 1741), vol. I. (Courtesy of the Fisher Library, University of Sydney)

The *Cyclopaedia* carried an entry for 'COMMON-PLACES' in the first edition, and this reappeared in the second as 'COMMON-PLACE-Book', with some additional sentences. The first edition gave this definition:

COMMON-PLACES, *Adversaria*, among the Learned, are a Register, or orderly Collection of what things occur worthy to be noted, and retain'd in the Course of a Man's reading, or Study; so despos'd, as that among a Multiplicity of Heads, and Things of all Kinds, any one may be found, and turn'd to at pleasure. (1728, vol. I)

As well as describing a commonplace book, the entry recommended a method for keeping one – namely the procedure advanced by John Locke, 'the great Master of Order', as Chambers called him. In his *A New Method of a Common-Place-Book*, first published in French in 1687, and then in an English translation in 1706, two years after his death, Locke explained this as follows:

If I would put any thing in my COMMON-PLACE-BOOK, I find out a Head to which I may refer it. Each Head ought to be some important and essential Word to the matter in hand, and in that Word regard is to be had to the first Letter, and the Vowel that follows it; for upon these two Letters depend all the use of the *Index*.[19]

In fact, this was a method for making an index to a commonplace book (Figure 11.6). After assigning a passage or quotation to an appropriate Head, one first considered the first letter/vowel combination of this Head, then checked the index to see if a page of the commonplace book had already been allocated. If so, then this passage was entered on that page. Material was therefore recorded in the book on pages occupied by other Heads sharing the same initial letter and first vowel. For example, Locke listed 'Ebionitae, Episcopus, Echinus, Edictum, Efficacia' as Heads that might be found in the same part of his commonplace book. The index simply recorded the page (or pages) on which these Heads had been entered. Locke's index allowed for 100 Heads which, he said, had always been adequate for his own purposes; but he did note that 500 could be accommodated by taking account of a second vowel (*New Method*, pp. 319, 321–2). The prospect of this larger number of Heads is signalled by the examples given in Chambers' entry: 'Beauty, Benevolence, Bread, Bleeding, Blemishes etc'. Of these, the last three do not belong to the standard Renaissance repertoire of moral, theological and philosophical topics, but rather to medicine or natural history – subjects that Locke no doubt anticipated when he suggested using 'a Book for every Science'. But some of the topics instanced by Chambers do not represent *general* categories from the relevant fields. If others at this specific level were added, the notion of collecting material under a manageable number of Heads would be threatened. Unless, of course, notice of the more general Heads was taken into account as the work was collated. This, I think, is a way of interpreting Chambers' emphasis on the plan of the dictionary. He urged readers to use the 'View of Knowledge' and its notes to guide their reading so as to regroup the scattered articles that belonged to a particular Head: namely, a certain art or science. In this way, the integrity of particular sciences might be restored. These considerations still exercised Chambers in the paper of 1738 where he accepted that the second edition could still be improved upon: 'The Oeconomy of the Articles might also be better adjusted, by distributing the larger ones under certain Heads, and throwing the Particularities into *Notes*' ('Considerations' [1738], p. 2). In other words, it was desirable to make the *Cyclopaedia* an even better commonplace book by consolidating more topics under major Heads.

What did this mean for the way the *Cyclopaedia* might be used, or read? To begin with, we should note that Chambers said that his work was not primarily designed for 'men of the first-rate erudition' who, like the ancient writers themselves, 'study their authors at first hand' (preface, 1738, vol. I, p. xxiv). But the analogy with commonplace books provided him, and later editors and publishers of his work, with a way of imagining two types of reader, thus maximising the potential audience. Abraham Rees, the editor of a revised and enlarged edition published serially from 1778, made this point in his new preface. Rees spoke of the scholar and the less learned reader. Referring to the former, he suggested that:

> those, who are proficients in science, will find it useful, on many occasions, to consult a Dictionary, as they would refer to a common-place book, in order to assist their memories, without the labour and the loss of time, which it would require to recur to a great number of distinct treatises, whence their knowledge was originally derived.[20]

Thus scholars could use the *Cyclopaedia* as a commonplace book to prompt their memory for earlier reading, or perhaps to explore topics outside their own field. Other readers, however, could study it as a single point of reference, accepting the warrant of the author that it was based on reliable abridgement of the major subjects. This might imply that Chambers' map of knowledge was more crucial to the person who did not already possess a good sense of the parameters of the various subjects and their cognate terms. For such readers, the *Cyclopaedia* functioned as a ready-made commonplace book, and an encyclopaedia, to be consulted and read for almost all their needs.

Conclusion

In the second edition (1701) of his *Dictionnaire*, Bayle remarked that 'books of this kind' are not usually 'read through, from the beginning to the end'.[21] I have tried to show that the authors of dictionaries of arts and sciences did not want to concede this. Unlike historical (and biographical) dictionaries, the works of Harris and Chambers professed to treat technical and scientific subjects whose content could be set out by way of principles, theories, experiments and observations that demanded systematic study. Here we seem to have the contention that summaries of this kind of knowledge demanded a certain kind of book. So while these dictionaries of arts and sciences promoted the advantages of alphabetical arrangement, their compilers still insisted on the possibility, and desirability, of respecting the integrity and interrelations of the various subjects. Chambers presented his two-volume work as an efficient summary of knowledge, and as a book informed by a concern for the coherence and order of the sciences. In explaining that many of its entries could be regrouped under the Heads of major subjects, Chambers implied that he had also compiled the *Cyclopaedia* in this way. So while his work was undoubtedly a crucial stimulus for the *Encyclopédie* and most subsequent encyclopaedias, it also looked back to an earlier tradition that placed more responsibility on readers to organise their own reading and knowledge.

Notes

I acknowledge the support of the Australian Research Council.

1 J. Harris, *Lexicon Technicum: or, an Universal English Dictionary of Arts and Sciences* (London, 1704), preface; emphasis in the original. No pagination in the preface.
2 E. Chambers, *Cyclopaedia; or, an Universal Dictionary of Arts and Sciences*, 2 vols. (London, 1728), vol. I, p. i. Also his *Some Considerations offered to the Publick, preparatory to a second edition of Cyclopaedia: or, an Universal Dictionary of Arts and Sciences* (no date or place of publication [1738]), p. 4.
3 J. H. Alsted cited in P. R. Percival, *A Neglected Educator: Johann Heinrich Alsted. Tranlations from the Latin of his Encyclopaedia* (Sydney, 1910), p. 23.
4 See R. Yeo, 'Alphabetical lives: scientific biography in historical dictionaries and encyclopaedias', in M. Shortland and R. Yeo (eds.) *Telling Lives in Science: Essays on Scientific Biography* (Cambridge, 1996), pp. 139–69, at pp. 140–3.
5 Harris, *Lexicon*, vol. I, preface. The work by Phillips first appeared in 1658; there was a new edition by John Kersey in 1706.
6 'An Account of a Book', *Philosophical Transactions of the Royal Society of London*, vol. 24, no. 292 (1704): 1699–1702, at p. 1699.
7 E. Chambers, *Cyclopaedia*, 2 vols., 2nd edn (London, 1738), 'Advertisement'. All subsequent references to this edition in the chapter are indicated by the date of 1738.
8 G. W. Leibniz, 'Precepts for advancing the sciences and arts', in P. Wiener (ed.), *Leibniz Selections* (New York, 1951), pp. 29–30, 32.
9 P. Bayle, 'Preface to the first French edition', in *The Dictionary Historical and Critical of Mr Peter Bayle, the Second Edition*, trans. P. Des Maizeaux, 5 vols. (London, 1734–8), vol. I, p. 4. On Bayle's role, see A. Goldgar, *Impolite Learning. Conduct and Community in the Republic of Letters, 1680–1750* (New Haven and London, 1995).
10 F. Bacon, *Novum Organum*, trans. and ed. P. Urbach and J. Gibson (Chicago and La Salle, 1994), Aphorism 101, p. 109.
11 Stephen Blankaart's *Lexicon Medicum* was published in Amsterdam in 1679 and translated into English in 1684.
12 Royal Society of London, Journal Book copy XIII, 358, 16 October 1729.
13 These dictionaries were composed by 'authors' who registered their 'copy', or assigned it to a bookseller/publisher, as a new book deserving the legal protection described in the Act of 1710. See A. Johns, *The Nature of the Book: Print and Knowledge in the Making* (Chicago, 1998), ch. 3, and for the particular case of encyclopaedias, see R. Yeo, *Encyclopaedic Visions: Scientific Dictionaries and Enlightenment Culture* (Cambridge, 2001).
14 E. Chambers, 'Proposals for printing by Subscription, Cyclopaedia' (London, 1726), no pagination. The copy I consulted is in the National Library of Scotland: Mss 1.14a. Some words on this copy have been partly erased.

15 See R. Yeo, 'Modèles d'outre-Manche', *Cahiers de Science et Vie*, no. 47 (October 1998): 24–6.
16 See J. M. Lechner, *Renaissance Concepts of the Commonplaces* (Westport, Conn., 1962); A. Blair, 'Humanist methods in natural philosophy: the commonplace book', *Journal of the History of Ideas*, 53 (1992): 541–51; A. Blair, q.v., ch. 4, and A. Moss, *Printed Commonplace-Books and the Structuring of Renaissance Thought* (Oxford, 1996). In the early modern period, the usual form was 'Common-Place book'. When not quoting, I use the modern convention: commonplace book.
17 N. Gilbert, *Renaissance Concepts of Method* (New York, 1960); C. Schmitt, *Aristotle and the Renaissance* (Cambridge, Mass., 1983); J. S. Freedman, 'Diffusion of the writings of Petrus Ramus in Central Europe, c. 1570–c. 1630', *Renaissance Quarterly*, 46 (1993): 98–152.
18 At the end of vol. II (1710) Harris added an 'Alphabetical Index' listing the terms treated under twenty Heads; but it did not show relations between subjects. See R. Yeo, 'Classifying the sciences', in R. Porter (ed.), *Cambridge History of Science: Eighteenth Century* (Cambridge, forthcoming).
19 J. Locke *A New Method of a Common-Place-Book*, in *The Posthumous Works of John Locke* (London, 1706), pp. 312–22 at pp. 316–17. The original version (in French) was in Jean Le Clerc's *Bibliothèque Universelle et Historique de l'année*, vol. II (1687), pp. 315–19, published in Amsterdam.
20 E. Chambers, *Cyclopaedia . . . With the Supplement and Modern Improvements Incorporated in One Alphabet. By Abraham Rees*, 4 vols. (London, 1786–88), vol. I, new preface, p. i.
21 Bayle, 'Advertisement concerning the second French edition', in *Dictionary Historical and Critical*, vol. I, p. 14.

Further reading

A. Blair, 'Bibliothèques portables: les recueils de lieux communs dans la Renaissance tardive', in M. Baratin and C. Jacob (eds.), *Le Pouvoir des bibliothèques* (Paris, 1996), pp. 84–106
R. Chartier, *The Order of Books*, trans. L. Cochrane (Cambridge, 1994)
R. Darnton, 'Philosophers trim the tree of knowledge', in *The Great Cat Massacre and Other Essays in French Cultural History* (London, 1985), pp. 185–207
R. De Maria, *Johnson's Dictionary and the Languages of Learning* (Chapel Hill, 1986)
C. Donato and R. M. Maniquis (eds.), *The Encyclopédie and the Age of Revolution* (Boston, 1992)
A. Goldgar, *Impolite Learning. Conduct and Community in the Republic of Letters, 1680–1750* (New Haven and London, 1995)
A. Grafton, *Defenders of the Text: The Traditions of Scholarship in an Age of Science, 1450–1800* (Cambridge, Mass., 1994)
A. Johns, *The Nature of the Book: Print and Knowledge in the Making* (Chicago, 1998)

F. Kafker (ed.), *Notable Encyclopedias of the Seventeenth and Eighteenth Centuries: Nine Predecessors of the Encyclopédie*, Studies in Voltaire and the Eighteenth Century, vol. 194 (Oxford, 1981)

A. Moss, *Printed Commonplace-Books and the Structuring of Renaissance Thought* (Oxford, 1996)

R. Yeo, 'Reading encyclopaedias: science and the organisation of knowledge in British dictionaries of arts and sciences, 1730–1850', *Isis*, 82 (1991): 24–49

'Ephraim Chambers's *Cyclopaedia* (1728) and the tradition of commonplaces', *Journal of the History of Ideas*, 57 (1996): 157–75

12 Periodical literature

In 1698, Pierre-Daniel Huet, the Bishop of Soissons in France, found himself concerned not with his flock's state of grace or lack thereof, but with the worsening state of scholarship. 'You would be appalled,' he confided to a friend, 'if you knew what decadence letters have fallen into in France.' So dreadful had things become, Huet continued, that he had even heard that 'if you took a Latin book to be printed in the rue St. Jacques, people would laugh in your face'. The barbarians disdained by this latter-day Boethius were not Goths, but something much worse: journalists. The abridgements of books carried in the journals coming out of Paris, Rotterdam and Leipzig were, in Huet's words, the 'indubitable proofs' of the corruption of letters in Europe.[1]

Shortly less than a century later, the barbarians had not only breached the gates, they appeared to have overrun the terrain. So evidently was this so that Karl Philipp Moritz, whose own strikingly modern career shifted between the roles of classical pedagogue, novelist and journalist, could proclaim almost trivially in 1784 that a newspaper or periodical was 'the best vehicle . . . by means of which useful truths can be spread among a people'. True to his own word, Moritz's journalistic undertakings squarely aimed at promoting the general good. Most prominent among these was his *Magazin zur Erfahrungsseelenkunde* (magazine of empirical knowledge of the soul), in which provincial Lutheran pastors and Prussian government counsellors, private tutors and salon habitués eagerly elbowed each other aside in the rush to advance the cause of Enlightenment by baring in print the secrets of their innermost mental lives.[2]

The change between Huet's era and Moritz's was dramatic indeed, measurable in part by the ever-increasing volume of journal production. Whereas Huet could have quite possibly enumerated and individually condemned all the publications that he regarded as shaking the pillars of scholarship, by Moritz's time their numbers had swelled beyond all reasonable measure. In German-speaking countries alone, more than 2,000 periodicals of all kinds were launched between 1765 and 1800.[3] And while it might be an exaggeration to elevate this phenomenon to the status of a 'second printing revolution', there can be no question that periodicals dramatically altered both the form of scholarly publication and its content. Over the long term, periodicals would replace longer monographic treatises as the focal point of scientific communication. In 1687, Isaac Newton found the price of scientific immortality to be the long and intricately argued *Principia mathematica*, and even as late as 1859, when Charles Darwin rushed into print to gain credit for his theory of evolution, his 'sketch' of the theory, *The Origin of Species*, ran to some 490 pages. By contrast, James Watson and Francis Crick found it possible to make their reputations in 1953 with a laconic two-page announcement in the journal *Nature*. Huet, it seems, had not been so very far from the truth.

The task of this chapter will be to account for the rise of periodicals as a form of literature and to describe their early history. It will examine first the origins of periodicals as an outgrowth of established correspondence networks among scholars and as part of the growing demand for 'news' among readers in the seventeenth century. Then we will turn to the explosion of the literary market in the eighteenth century and the boom in journal production that it fed. Finally, we will examine the specialisation of journal audiences that began taking place and its connection to the formation of specialised research communities.

Before commencing, however, we need to specify just what a periodical is. A 'periodical' or 'journal' is distinguished from the authored monograph first by presenting itself deliberately as part of an ongoing series of such publications that appear at regular intervals. Secondly, periodicals typically do not present a single authorial voice. Instead, each issue contains a multiplicity of voices that sometimes speak to each other, sometimes to other writings, and sometimes to no one in particular. On the other side, periodicals are distinguished from newspapers in their relationship to topicality. Whereas newspapers are wholly topical and transfer their attention to the next matter of interest with every issue, periodicals use topicality as the occasion for more sustained discussion and reflection. Thus, if a scientist reports an experimental finding in a journal, that report has a certain topical interest and is 'news'. But the report also presents itself as contributing to a larger project, discussion or body of literature.[4]

The demand for news and the birth of the periodical press

By most accounts, the origins of newspapers and periodicals lie back in the sixteenth century. During the political and religious upheavals that followed the onset of the Lutheran Reformation in 1517, there developed a network of postal routes in the territories of the Holy Roman Empire, including much of what is now Austria, Germany, Northern Italy and the Netherlands. At first the imperial post carried diplomatic correspondence, but it soon assumed responsibility for regular private correspondence as well.

The establishment of regular postal routes in Europe made it possible for networks of correspondents to form and remain in contact with each other. Writers could exchange news of the latest political developments and commercial activities in ports or important trading centres. More than a few rulers made it their business to employ correspondents in different parts of Europe to send them reports on current events. Moreover, well-placed men of letters, such as the Lutheran reformer Philip Melanchthon, regularly obtained and passed on bits of news in their correspondence. Indeed the production of entire newsletters could grow into a major undertaking, as demonstrated by the Fugger Newsletter, a handwritten collection of news reports produced between 1568 and 1604 by the powerful banking and merchant house of Fugger in Augsburg. Postmasters were especially well positioned to perform this service, because they had the most direct access to the postal networks, and consequently to information from abroad. These early newsletters, which were copied by hand, focused overwhelmingly on political developments and commercial news, but it was not long before they began incorporating reports of a more diverse sort.

What is evident in all this activity is a growing demand for political and commercial 'news'. Handwritten newsletters like the Fugger newsletter, it should be noted, did not circulate openly; their subscribers were recruited from those people most directly involved in trade, diplomacy and government. Yet already in the sixteenth century there were traces of a more broadly based form of news reporting, which appeared in the form of broadsides, so-called *newe Zeitungen*, and other printed matter. These publications reproduced news reports from far-off places for a less exclusive readership than the subscribers to handwritten newsletters. Yet these early news sheets appeared too sporadically to be called newspapers in our understanding of the word.

It would not be long, however, for regular newspapers to start appearing. The earliest known example was published by one Johann Carolus, a bookseller and printer who applied to the city council of Strasburg in 1605 for permission to print a weekly newspaper. Postmasters too were involved in many of the early undertakings, a role reflected in the titles given to them, such as the 'Post' or the 'Courier'.[5] Although the appearance of regularly published newspapers might appear to have followed naturally from the unprinted newsletters discussed above, we should note at least one major difference: newsletters went out to people, such as government ministers and bankers, whom we might expect to have a direct interest in news from foreign sources. The new printed newspapers, on the other hand, circulated to a broader audience of readers whose eagerness to learn about the latest goings-on in Amsterdam or Stockholm is a little harder to comprehend. As difficult as it is to comprehend, the growing demand for news marked a change in literate culture, the significance of which can scarcely be exaggerated.

The extension of the communication network in Europe in the sixteenth and seventeenth centuries also supported the transmission of scholarly 'news'. As Anne Goldgar has shown, scholars used their correspondence to secure and bestow patronage on one another, report on their researches and request aid in securing information that could not be obtained locally, and communicate news of the latest books appearing in important publishing centres. Indeed, well placed scholars at the centre of such correspondence networks became, like postmasters, virtual clearinghouses of information for their contacts in diverse corners of Europe.[6]

One person who enjoyed such an advantageous position was Henry Oldenburg. Born in the German port city of Bremen and widely travelled throughout Europe, Oldenburg had an admirable facility with modern languages and a way of ingratiating himself with other educated men. During the 1650s, Oldenburg twice found himself in England, representing the interests of Bremen before the parliamentary regime of Oliver Cromwell. There he came into contact with prominent English scholars and natural philosophers, among them John Wilkins and Robert Boyle, and he secured employment as a tutor in Boyle's family. When the Royal Society was organised in 1660, Oldenburg was proposed as a member, whereupon he assumed an influential role as the Society's secretary.[7]

Oldenburg's position as secretary rapidly made him the conduit of information between the Royal Society and the rest of the scholarly community in Europe. The management of such a position could be a delicate business, for complaints occasionally surfaced that Oldenburg's reports gave scholars elsewhere in Europe a chance to publish work that English scholars felt properly belonged to them.[8] Quite possibly for reasons of recognising priority, but also to simplify his correspondence and to make a little money, Oldenburg inaugurated in March of 1665 the *Philosophical Transactions*, a repository of natural scientific inquiry that exists to this day. The sub-title given the journal – *Giving Some Accompt of the Present Undertakings, Studies and Labours of the Ingenious in Many Parts of the World* – offers a clear sense of the role it was to play in formalising Oldenburg's correspondence network.

The early *Philosophical Transactions* was a wide-ranging collection of chemical investigations, astronomical and meteorological reports, monstrous births and other wonders of nature, and descriptions of contrivances intended for artisanal use. It represented the Royal Society's conviction – or at least its profession – that the proper business of natural philosophy was not the construction of elaborate theoretical systems, but the careful and systematic collection of useful facts. The journal's standing in the scholarly world was secured by the prestige of the early Royal Society, and by the extensive network of contacts maintained by Oldenburg. In short order, the *Philosophical Transactions* became a leading source of 'news' from the world of natural science.

A counterpoint to the kind of news offered in the *Philosophical Transactions* also began running in 1665. The *Journal des Sçavans*, edited by the jurist and aristocrat Denys de Sallo, favoured notices and reviews of recently published books from all corners of the scholarly world over empirical research reports. The *Journal*'s early days were bumpy. De Sallo's free-wheeling style of criticism, along with the suspicion that he was taking sides in the struggle between the Jansenist and Jesuit factions in the French Catholic Church, led to the non-renewal of the journal's privilege from the French government after only thirteen weeks. When the *Journal* resumed publication in 1666, it had a new editor, the Abbé Jean Gallois, a less combative tone, and a link to the newly organised *Académie Royale des Sciences*.[9]

In the decades that followed the inauguration of these two learned periodicals, a host of others would appear, among them the *Acta Eruditorum* in 1682 and the *Nouvelles de la République des Lettres* in 1684. Yet the basic contours of periodical literature followed the patterns set by the *Philosophical Transactions* and the *Journal des Sçavans* in three ways. First, journals became repositories of research and vehicles for ensuring priority of discovery. Second, by reporting the latest scholarly works and delivering to readers digests of recent books, journals created an entirely new medium for transmission of ideas. No longer did someone have to read a long and ponderous treatise to learn what it contained, or wait to be informed of it in private correspondence. Journals gave readers a chance to learn, more or less, what a given book had to say without actually reading it.

The third innovation, the introduction of critical judgements into reviews of books, at first glance seems modest enough, especially in light of the trouble De Sallo quickly fell into with his critical efforts in the *Journal des Sçavans*. Yet the consequences of this innovation, as we shall see in the next section, were significant both for the cultural role of periodicals and for the public authority of scientific knowledge. For judgements published in journals had a public character that did not pertain to opinions and judgements contained in unpublished letters. This is not to say that private letters never had public consequences; in 1667, Henry Oldenburg found himself tossed into the Tower of London for a brief time because, it seems, he had imprudently conveyed information deemed of military importance to one of his continental correspondents. But judgements made in print became not just one person whispering in a correspondent's ear, but instead a new kind of public and authoritative voice. Those voices would assume a dominant role in the culture of the Enlightenment.

Journalism and criticism in the Enlightenment

Pierre Bayle was someone who well appreciated what it meant to live on the margins. Born in Languedoc in extreme southern France, the son of a Calvinist minister in a Catholic country, Bayle spent much of his adult life in exile (twenty-five years) in Rotterdam, where he occupied himself as a journalist and man of letters. It was perhaps for this reason that Bayle created his enormously successful *Nouvelles de la République des Lettres* in 1684 for an audience who were not themselves members of the scholarly correspondence networks that had given birth to the earliest journals. Instead, Bayle tailored the *Nouvelles* for a new group of consumers of learned culture, people whose wealth, education and leisure gave them a taste for what was going on in the scholarly world, but who were not disposed to conduct research themselves.[10]

As with all such efforts, Bayle's *Nouvelles* succeeded because it caught the beginnings of a wave that would continue forming and growing throughout the eighteenth century. For many years, historians have pointed to a set of social and cultural changes that took place during this period which had a profound impact on reading and publishing, and more generally on modern culture. The first such change involved the reading practices of educated people. During the eighteenth century, reading habits shifted from intensive study of a few, often devotional and religious texts, to a much broader assortment of fictional and non-fictional reading.[11] This was accompanied by the emergence of what has often been called 'new forms of sociability' during the later seventeenth and eighteenth centuries, represented by the appearance of gathering places such as coffee-houses, salons, masonic lodges and reading societies. These institutions, although quite distinct from each other, nevertheless shared the quality of minimising the traditional hierarchies that had shaped European society.[12]

The third important change, closely connected with the others, was the growing awareness of participants of themselves as part of a new collectivity, the 'public'. It is significant that this 'public' was understood by contemporaries as having a distinctly different character than the republic of letters that had given rise to the *Philosophical Transactions* and the *Acta Eruditorum*. The republic of letters was understood as a community having a more or less circumscribed membership. The eighteenth-century public, by contrast, was simply thought to be 'everyone', in so far as everyone was a member of civil society and at the same time no one in particular.[13] Whereas it was possible to take a census of the republic of letters – a task performed by the scholarly lexicons of the period – no one would have even conceived of taking a similar census of the public. The public was the idealised projection of enlightened society, an image of society having little in common with 'the people' so often disparaged by writers as vulgar and governed by passion and superstitious beliefs instead of reason.

In this booming literary market, Bayle's *Nouvelles* and its many imitators and successors found their eager audience, feeding its unquenchable demand for novelty and information. The consequences for the sciences of this new cultural and social environment were considerable. Scientific knowledge ceased being the preserve of a few scholars communicating among themselves over minutiae of ancient chronologies or the motion of pendulums, and became instead a matter of public interest. Just as importantly, the role of science as a form of public knowledge was decisively shaped by the market for periodical literature and by the distinctive structure of the publications.

The first and most obvious way that science became public knowledge during the Enlightenment occurred via innumerable writings aimed at spreading useful knowledge. The range of advice was remarkable, covering such topics as agricultural improvement and estate management, proper methods of child-rearing, management of public health, and stimulation of domestic industry and technological improvement. Periodicals were well suited to this role, because the knowledge being proffered could easily be packaged in an article and inserted in between other fare of interest to the general reader. Accordingly, articles containing practical medical and scientific knowledge appeared in a huge range of publications, ranging from the *Gentleman's Magazine* in England, to the various *Affiches* in France, to the *Journal des Luxus und der Moden* in Germany, a fashion magazine appealing to readers who wanted to keep up on all that was happening in Paris and London.

While periodicals undoubtedly contributed significantly to the diffusion of scientific knowledge during the Enlightenment, other institutions, such as public lectures and academies of useful knowledge, performed the same function.[14] In another and less obvious way, however, magazines and journals were uniquely responsible for making science a form of public knowledge – as conduits of criticism. Criticism was a discourse that evaluated a particular object, for example, a book, social practice, work of art or technical artifact, against a theory or model (such as a model of beauty) that was presumed to inform or justify the object. In the hands of Enlightenment reformers, criticism became a means of adjusting laws, social habits, technologies, or just about anything to what the reformers claimed was a more scientific and natural basis.

Criticism reverberated back and forth through the periodical media. As early as 1709 in England, Joseph Addison and Richard Steele began publishing the *Tatler* and the *Spectator*, two weeklies that set themselves up as anonymous voices emanating from within the public to instruct readers on proper social behaviour and elevate their taste. So successful were these two publications that they spawned literally scores of imitators throughout the century. A second kind of critical discourse, also anonymous, was the book review. As we saw above, book reviews began appearing as early as the 1660s, in the *Journal des Sçavans*. Yet from such modest seventeenth-century beginnings, the practice of book reviewing flourished to become a literature of its own, and gave rise to a distinctive kind of periodical, the literary review.

More perhaps than in any other institution of Enlightenment culture, literary reviews embodied the ideology of the public as a rational and enlightened collectivity. All forms of scholarship and *belles lettres* came under their scrutiny, because all knowledge 'belonged' to the enlightened public. Thus treatises on natural philosophy and mathematics found themselves being discussed and dissected in the literary reviews right alongside the latest collections of fables and treatises on political economy. Just as important as their comprehensiveness, however, was the fact that the literary reviews voiced criticism in a way that seemed to speak both to and for the enlightened public. Like the anonymous 'spectator' in Addison and Steele's weeklies, the anonymous critics in literary reviews began positioning themselves as proxies for the public, instructing it on matters of taste, judging the adequacy of the newest published works in the natural and moral sciences, and evaluating how particular pieces of art and literature measured up to canons of beauty. Here the anonymity of criticism was especially crucial, for this voice became the projection of the public's – that is, it said what the public would say if only it had the opportunity to do so. Thus the anonymous review, while instructing the public, simultaneously stood in for it: this generalised, anonymous voice permitted critical reviews to speak 'for' the public.[15]

Of course, literary reviews were not just tools in an altruistic campaign to improve the public's taste and morals. No less than other publications, they too were products of the marketplace, and at root their role consisted of telling readers which books to buy. By making scholarly knowledge public, the reviews made it newsworthy, and by making it newsworthy they made it an object of consumption. Friedrich Nicolai, a young bookseller in Berlin in the 1750s, recognised the existence of this market, and perceived how criticism would have to respond to it.

> A scholar who, from the innermost recesses of his study, seeks to introduce a new taste and way of thinking, and who takes no notice of the habits, beliefs, and indeed the prejudices of his countrymen, is like a natural scientist who seeks to prove a thesis by means of deductions, without taking the trouble of gathering the observations necessary for its proof.[16]

Nicolai's essential insight, like Bayle's before him, was to recognise the public's twin potential as consumers of books and of news and criticism about books. Critical reviews would elevate taste and provide news of the latest books, while at the same time they would stimulate a bookseller's business. Together with his friends Moses Mendelssohn and Gotthold Ephraim Lessing, Nicolai launched a succession of literary reviews, the most successful of which was the *Allgemeine deutsche Bibliothek*, a huge, forty-year undertaking that attempted (but failed) to review every book published in German.

Thus the emergence of a booming literary market during the eighteenth century worked along two distinct but closely connected paths that took science beyond the interests of a narrow group of scholars and transformed it into a matter of public interest. One path consisted of the provision of practical scientific and medical advice to readers in signed articles. Rendering advice in signed articles enhanced the authoritativeness of both the advice – by demonstrating the reliability of the source – and the author, by displaying the author as someone who cultivates useful scientific knowledge. At the same time, however, the authority of scientific knowledge in Enlightenment culture was not merely a matter of diffusion, for critical discourse made the possession and evaluation of such knowledge something belonging to the public itself. By casting themselves as voices emanating from *within* the public, and not as people speaking *to* the public, critics presented readers with knowledge that was, in a certain sense, already in the public domain.

Market segmentation and specialised journals

From one point of view, the emergence of the public sphere during the eighteenth century dissolved the republic of letters that had given birth to the earliest periodicals. Not in a strict sense, of course: scholars continued to recognise and acknowledge each other's research and areas of expertise, and to speak of a scholarly community. But the emergence of public discourse in the print media gave scholarship a new set of references, some cultural and intellectual, others economic. Most importantly, scholarship became public knowledge and was required to submit to the public's scrutiny through criticism.

Yet the very same economic circumstances that had given rise to the general-interest literary reviews also provided the framework for what would become a specialised scientific press. To understand this, we need to see that in the eighteenth century journals could afford to have small circulations. This is because the cost of materials, especially paper, was high in comparison to periodicals today, while the costs related to producing the journal's contents were low. As circulation increased, the costs related to paper and printing did too, but of course so did income from subscribers. Thus, one reason that a lot of journals could be launched during the period, even if they folded after just a few issues, is that such undertakings did not require investment of a lot of capital and were not terribly risky.[17]

The relative ease with which smaller circulation journals could be supported encouraged the creation of journals targeted for specific audiences. Those new journals did not cease speaking as if their audience was the enlightened public. They continued to address readers as if on the one hand they were instructing them and spreading enlightenment and useful knowledge, and on the other hand as if they were representing truths that were, in some sense, already part of the public sphere. But although the more narrowly focused publications shared the same discourse of public knowledge as the more broadly based literary reviews, they offered a more restricted palette of topics.

One such publication was the *Chemisches Journal für die Freunde der Naturlehre, Arzneygelahrtheit, Haushaltungskunst und Manufacturen*, begun in 1778 by chemical lecturer and journalist Lorenz Crell. The audience for the new publication was unmistakably signalled in the title: 'for friends of natural philosophy, medicine, domestic economy and manufactures'. The community of reader/contributors invoked by this journal – along with a subsequent publishing project, the *Chemische Annalen*, the latter of which circulated to some 400 subscribers in Germany and elsewhere in Europe – was not a clearly delineated professional group. Nor was it quite identical to the members of the republic of letters who had furnished the audience for the earliest seventeenth-century periodicals. Instead, Crell's subscribers were an assortment of scholars, physicians, pharmacists, mining engineers and amateurs who performed chemical experiments themselves. Although this group supported the pursuit of what could be labelled 'chemical science' in Germany, it was scarcely an exclusive occupation carried out by highly trained specialists, who circulated their experimental results in reports that were only comprehensible to a small, expert community.[18]

Another journal, less focused topically than Crell's but claiming a more restricted audience, was the *Observations sur la Physique, sur l'Histoire Naturelle, et sur les Arts*, edited by the clergyman and agricultural improver Jean Baptiste Rozier. The preface to the journal's first issue claimed that, in contrast to other publications, Rozier's journal would be a serious forum for rapid publication of scientific work. 'We will not offer to idle amateurs works which are merely agreeable', he intoned, 'nor the pleasant illusion of believing oneself an initiate in those sciences of which one is ignorant.'[19] Yet Rozier's journal was hardly more a repository of specialised scientific work than the *Philosophical Transactions* or other such journals. It succeeded largely because it offered scholars a rapid way to 'prepublish' work that would appear in more extensive form elsewhere. Thus Rozier's journal too responded to the market demand for scholarly news.

General-interest science journals would continue going strong throughout the nineteenth century, and indeed new ones such as *Nature* (founded in 1869) would continue to appear. But after 1800 there developed a clear trend towards the publication of journals having both a more focused audience and a more limited selection of contents. This can be seen, for example, in the *Deutsches Archiv für die Physiologie* (founded in 1815), a successor to a general-interest journal of medical theory and criticism, the *Archiv für die Physiologie*. In contrast to its predecessor, the *Deutsches Archiv* contained no book reviews or wide-ranging discussions of the requirement for a science of physiology and its relationship to the rest of medical theory. Instead, it published almost exclusively the results of experimental and comparative anatomical studies of animal form and function. Just as significantly, the *Deutsches Archiv* claimed on its title-page to be a collaborative project by the most prominent physiologists in Germany. In journals like the *Deutsches Archiv*, we detect the beginnings of the symbiosis between scientific research communities and specialised periodicals, a pattern of mutual support that has proven so productive during the past two centuries.[20]

What we have seen in the evolution of periodical publishing, therefore, are three important developments that would remake the form and ultimately the content of scholarly work. First, periodicals arose in the seventeenth century amidst a growing demand for news, and, once established, fed that demand by offering notices of new books and opportunities for rapid publication. Second, the flourishing of the periodical press contributed fundamentally to establishing the cultural authority of science during the eighteenth century. And finally, the economic circumstances of the periodical market permitted the creation of smaller circulation journals, through which discrete communities within the larger reading public could coalesce and pursue their interests. Of course, we would not want to conclude that the advancing specialisation of science was wholly created by the literary market. But it would be fair to say that the market provided the indispensable basis for it, as well as the conduits of research exchange. Modern science would not be the journal-based entity we know it as without the foundations that developed in both the scientific community and the broader public sphere of the eighteenth century.

Notes

1. Cited in A. Goldgar, *Impolite Learning: Conduct and Community in the Republic of Letters 1680–1750* (New Haven and London, 1995), p. 54.
2. On Moritz and the Enlightenment ideal of journalism, see W. Martens, 'Die Geburt des Journalisten in der Aufklärung', in *Wolfenbütteler Studien zur Aufklärung*, Bd. 1 (Wolfenbüttel, 1974), pp. 84–98, quoted on p. 85.
3. For a general history of the German periodical press, see J. Kirchner, *Das Deutsche Zeitschriftenwesen: seine Geschichte und seine Probleme*, vol. I (Wiesbaden, 1958), esp. pp. 72–3, 115.
4. For a more exhaustive discussion of the periodical as a genre, see D. A. Kronick, *A History of Scientific and Technical Periodicals*, 2nd edn. (Metuchen, NJ, 1976).
5. M. Lindemann, *Geschichte der deutschen Presse, Teil I. Deutsche Presse bis 1815* (Berlin, 1969), pp. 15–21, 44–9; J. Weber, '"Die Novellen sind eine Eröffnung des Buchs der gantzen Welt": Entstehung und Entwicklung der Zeitung im 17. Jahrhundert', in K. Beyrer and M. Dallmeier (eds.), *Als die Post noch Zeitung machte* (Frankfurt am Main, 1994), pp. 15–25.
6. Goldgar, *Impolite Learning*, ch. 1.
7. For biographical information on Oldenburg, see the entry by A. R. Hall in C. C. Gillispie (ed.), *Dictionary of Scientific Biography*, vol. X (1974), pp. 200–3.
8. This was the charge levelled against Oldenburg by Thomas Sprat, the early historian and apologist for the Royal Society, and by the biographers of the architect Christopher Wren. See D. Stimson, *Scientists and Amateurs: A History of the Royal Society* (New York, 1948), pp. 66–7.
9. B. T. Morgan, *Histoire du Journal des Sçavans depuis 1665 jusqu'en 1701* (Paris, 1929), ch. 1.
10. Goldgar, *Impolite Learning*, p. 55.
11. On the new reading habits, see R. Engelsing, *Der Bürger als Leser: Lesegeschichte in Deutschland, 1500–1800* (Stuttgart, 1974); and R. Chartier, 'Urban reading practices, 1660–1780', in id., *The Cultural Uses of Print in Early Modern France*, trans. Lydia G. Cochrane (Princeton, NJ, 1987), pp. 183–239.

12 For a general overview, see M. C. Jacob, 'The mental landscape of the public sphere: a European perspective', *Eighteenth-Century Studies*, 28 (1994): 95–113.
13 On the emerging ideology of the public, see A. J. La Vopa, 'Conceiving a public: ideas and society in eighteenth-century Europe', *Journal of Modern History*, 64 (1992): 79–116; and D. Goodman, 'Public sphere and private life: toward a synthesis of current historiographical approaches to the old regime', *History and Theory*, 31 (1992): 1–20.
14 On this aspect of Enlightenment culture, see H. Lowood, *Patriotism, Profit and the Promotion of Science in the German Enlightenment: The Economic and Scientific Societies, 1760–1815* (New York, 1991); and L. Stewart, *The Rise of Public Science: Rhetoric, Technology, and Natural Philosophy in Newtonian Britain, 1660–1750* (New York, 1992), ch. 4.
15 On this point, see T. Broman 'On the epistemology of criticism: science, criticism and the German public sphere, 1760–1800', in J. Schönert (ed.), *Literaturwissenschaft und Wissenschaftsforschung* (forthcoming).
16 Quoted in P. E. Selwyn, 'Philosophy in the Comptoir. The Berlin Bookseller-Publisher Friedrich Nicolai, 1733–1811' (Ph.D. diss. Princeton University, 1992), p. 18.
17 For discussion of the economics of periodical publishing, see H.-M. Kirchner, 'Wirtschaftliche Grundlagen des Zeitschriftenverlages im 19. Jahrhundert', in J. Kirchner, *Das Deutsche Zeitschriftenwesen*, vol. II, pp. 379–85.
18 On Crell's publications, see K. Hufbauer, *The Formation of the German Chemical Community (1720–1795)* (Berkeley and Los Angeles, 1982).
19 Quoted in Kronick, *History of Scientific and Technical Periodicals*, p. 109.
20 T. H. Broman, 'J. C. Reil and the "journalization" of physiology', in P. Dear (ed.), *The Literary Structure of Scientific Argument* (Philadelphia, 1991), pp. 13–42.

Further reading

C. Bazerman, *Shaping Written Knowledge: The Genre and Activity of the Experimental Article in Science* (Madison, WI, 1988)
T. Broman, 'The Habermasian public sphere and "science *in* the Enlightenment"', *History of Science*, 36 (1989): 123–49
J. Censer, *The French Press in the Age of Enlightenment* (London, 1994)
F. Donoghue, *The Fame Machine. Book Reviewing and Eighteenth-Century Literary Careers* (Stanford, 1996)
A. Goldgar, *Impolite Learning: Conduct and Community in the Republic of Letters 1680–1750* (New Haven and London, 1995), esp. chs. 1 and 2
J. Habermas, *The Structural Transformation of the Public Sphere: An Inquiry into a Category of Bourgeois Society*, trans. Thomas Burger with the assistance of Frederick Lawrence (Cambridge, MA, 1989)
D. A. Kronick, *A History of Scientific and Technical Periodicals*, 2nd edn (Metuchen, NJ, 1976)

A. A. Manten, 'Development of European scientific journal publishing before 1850', in A. J. Meadows (ed.), *Development of Science Publishing in Europe* (Amsterdam, 1980), pp. 1–22

L. Nyhart, 'Writing zoologically: the *Zeitschrift für wissenschaftliche Zoologie* and the zoological community in late nineteenth-century Germany', in P. Dear (ed.), *The Literary Structure of Scientific Argument* (Philadelphia, 1991), pp. 43–71

13 Natural philosophy for fashionable readers

At the end of the seventeenth century, natural philosophers aspired to participate in the fashionable culture of the cosmopolitan cities of Europe and Britain. The fashion for the physical sciences over the next hundred years fuelled demand for natural philosophy books accessible to non-specialists. Conversely, the appeal these subjects held for readers in polite society contributed to the growing cultural legitimacy enjoyed by the sciences. Men of science wrote for a heterogeneous 'public' comprised variously of ladies and gentlemen, provincial amateurs, bourgeois householders, men and women of letters, journal subscribers, government officials, and other men of science. Many of these authors sought the approbation of readers of discerning taste, partly as a way of selling books, but also as a way of making reputations in the rarefied atmosphere of the social and cultural elite. A book's success was measured by how rapidly an edition disappeared from the booksellers' shelves, but also by how much talk it stimulated in the press and in polite conversation. This chapter deals with two complementary historical processes: the dissemination of scientific ideas and ideals into polite and sociable circles, and the legitimation of natural philosophy by social elites.

Philosophical conversations

In genteel European society, conversation became the hallmark of politeness, governed by standards of taste instantly recognisable by insiders and opaque to the vulgar lower classes. Whig culture at the turn of the eighteenth century, as Lawrence Klein has shown, expressed politeness in 'intelligent and stylish conversation about urbane things, presided over by the spirit of good taste'.[1] Across the Channel in France, the self-styled quarrel of the ancients and the moderns revolved around the contrast between the solitary learning of the pedant and the sociable wit of the conversationalist, liberated from the restrictions of classical form. Conversations about a broad range of subjects, from poetry to philosophy to astronomy, took place in salons, coffee-houses, gardens and boudoirs, often in mixed-gender groups. Reading complemented conversation as a mode of sociability, since reading was often done in company and talk focused on written texts. The fashion for natural philosophy and for sociable conversation coexisted in England and France. But although books, letters, periodicals and even some people travelled back and forth between London and Paris, sharing certain conventions of civility, other cultural differences emerged in subtleties of style and moral tone. These nuances shaded the landscape of polite culture through which natural philosophers moved, stimulating conversations about discoveries, instruments, theories and books.

The reasoned exchange of civil conversation also appeared in print, sometimes explicitly in the form of dialogues between genteel characters, sometimes more implicitly in an author's (or editor's or translator's) addresses to readers. Bernard de Fontenelle's *Conversations on the Plurality of Worlds* (1686) became the canonical example of natural philosophy packaged for a polite audience. A key figure on the Parisian literary scene in the last decades of the seventeenth century, Fontenelle used the dialogue form to display Cartesian natural philosophy to genteel readers. He modernised the classical Platonic dialogue by ornamenting his arguments with conversational banter. In the text, the male narrator undertakes to explain the intricacies of cosmology to his philosophically naive but intelligent female companion, combining accounts of the latest telescopic discoveries with speculations about the inhabitants of other worlds. He presents the universe as an impresario presents a spectacle, unveiling the heavens for her delectation. She responds with her own conversational gambits, showcasing her cleverness. Fontenelle may well have imagined his ideal reader as an acquaintance from the salon circuit, who could identify with fictional aristocratic characters conversing in an elegant garden. He certainly struck a chord with his contemporaries, resonating far beyond his initial target audience in the French capital. Not only did he personally supervise ten separate revised editions over the next fifty years, but the book appeared in many other editions from Amsterdam and elsewhere.[2] English interest was sufficient to warrant four different translations (in numerous editions) between 1687 and 1715, in spite of the disjuncture between Fontenelle's worldly Cartesianism and the Newtonian natural theology coming into ascendancy in Britain. Over the years, the book's conversations echoed even further, as nearly every French, Italian or English book published in the eighteenth century for a polite or a popular audience harkened back in some manner to Fontenelle.

Fontenelle prefaced his conversations with instructions about how to read them. The learned should expect to be amused, the ignorant should expect to be instructed, eased along by a style based on the 'natural liberty of conversation'. Women should find reading his book to require no more effort than a careful reading of

The Princess of Clèves. Invoking this immensely successful novel by Mme de Lafayette, herself a legendary figure in salon culture, Fontenelle claims that Lafayette's readers can appreciate natural philosophy as well as romance and intrigue. The physical attributes of the book also reflect its intended use. The first edition does not name the author, suggesting Fontenelle's uncertainty about its success. He had the work printed as a small duodecimo volume on heavy paper, with wide margins that make reading comfortable and give an impression of elegance and luxury. The size and layout made it easy to handle, and presumably pleasant to read more than once. The first edition contains only one illustration, a folding plate that extends well beyond the dimensions of the pages, representing the cosmos with its many worlds. Every planet is encircled by its satellites and every sun looks like a small face at the centre of its vortex (Figure 13.1). The fine quality of the engraving, clearly visible in the large foldout, marked the book as an appropriate addition to an aristocratic library.

Geoffrey Sutton has shown how Cartesian astronomy and natural philosophy became part of literary culture in France at the end of the seventeenth century, 'a subject of polite conversation in *le monde*'.[3] Fontenelle's own place in this comfortable world gave him the insight into conversational and literary taste that made his book such a success. By the 1690s, Fontenelle was well known, in France and abroad, as the author of plays, poems, novels and criticism. He also became the spokesman for the Paris Academy of Sciences, by virtue of his appointment as perpetual secretary. In this capacity, he presented the work of the Academy to the same genteel audience who read and discussed his books.[4] At the turn of the eighteenth century, the overlapping and intermingling of the literary, sociable, philosophical and scientific made possible Fontenelle's many roles in the cultural matrix of Paris. Fashionable worldly readers were essential to Fontenelle's project to legitimise the sciences as worthy of the attention of the 'modern', and soon to be 'enlightened', person.

English conversations on natural philosophy

Fontenelle's *Conversations* took on a somewhat different meaning when brought to England. One translation appeared in 1688 from the pen of the prolific novelist and playwright Aphra Behn.[5] A woman making her living as an author, Behn was sensitive to the market for translations of French works. She advertised her own gender as one of her selling points: 'I thought an English Woman might adventure to translate any thing, a French Woman may be supposed to have spoken' (n.p.). In her translator's preface, she disparaged the English fascination with French culture, even as she

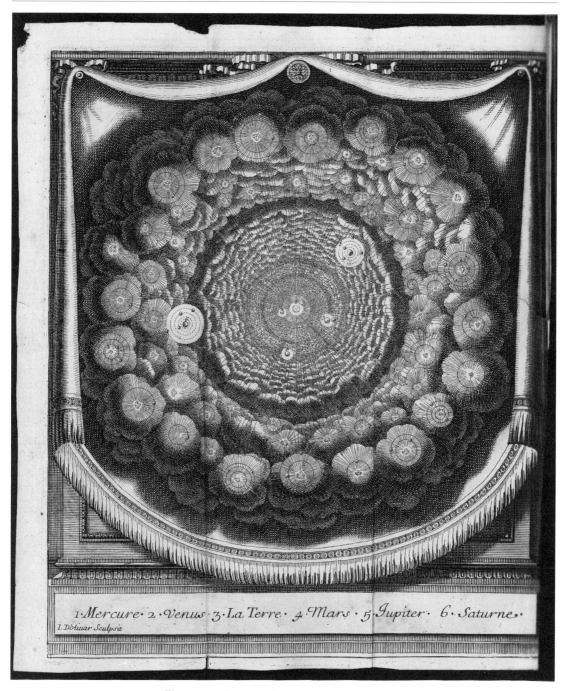

Figure 13.1 Bernard de Fontenelle, *Entretiens sur la pluralité des mondes* (Paris, 1686), frontispiece. The many worlds, each in its own cloud-like vortex, are displayed on a theatre curtain. (By permission of Houghton Library, Harvard University)

capitalised on it, and explained the difficulties of translating from French to English. 'But as the French do not value a plain Suit without a Garniture, they . . . confound their own language with needless Repetitions and Tautologies; . . . If one endeavours to make it English Standard, it is no Translation. If one follows their Flourishes and Embroideries, it is worse than French Tinsel' (n.p.). With her long preface, the translator put herself into the book as an author in her own right, setting her words apart in italic print. Although she admired Fontenelle's 'ingenuity', Behn showed a decided ambivalence about his book, apologising for 'what some may understand to be a Satyr against him'. Judging the success of his attempt to make natural philosophy understandable to everyone, she decides he has failed, 'for endeavouring to render this part of Natural Philosophy familiar, he hath turned it into Ridicule; he hath pushed his wild Notion of the Plurality of Worlds to that heighth [sic] of Extravagancy, that he most certainly will confound those Readers, who have not the Judgment and Wit to distinguish between what is truly solid (or, at least probable) and what is trifling and airy'. Finally, she suspects Fontenelle of irreligion. 'He ascribes all to Nature, and says not a Word of God Almighty, from the Beginning to the End; so that one would almost take him to be a Pagan' (n.p.). She objects to his frivolity, to his tendency to fancy, to his ornamental style – to his Frenchness, in fact. At the same time, she approves of his Copernicanism, and ends her preface with a long peroration on the inappropriateness of reading Scripture as astronomy, recapitulating the Catholic Church's debate with Galileo.

Behn reflects the ambivalence of the English response to French aristocratic gallantry, which sold books even if it was theologically suspect and logically flawed. Descartes' vortices, part of Fontenelle's truth, become in the English version just another fiction, 'trifling and airy'. But English readers apparently found Fontenelle to their taste. Behn's translation went through five more editions in the next twenty years.[6] Presumably the readership also expanded down the social scale, since English aristocrats, many of them fluent in French, would not have been the primary market for translations. The translations are in fact less elegant books than the French, mostly without frontispieces and printed on thinner paper with narrow margins.

In early-eighteenth-century England, natural philosophy became an activity suitable for ladies as well as gentlemen, and the audience for natural philosophical displays grew along with the authority of Newtonianism. The audience for books about astronomy, electricity, magnetism and natural history extended far beyond the Royal Society to include women, children and consumers of related products of all sorts, especially instruments. Newtonian writers used the intelligibility of their physics as a way of arguing for its validity. They saw women, who lacked formal education, as the ideal receptive audience for knowledge relying on the testimony of the senses rather than the obfuscating pedantry of the schools.[7] Fontenelle had made a similar claim for Cartesianism, though the English Newtonian audience included distinctly middle-class consumers of culture. The morally virtuous overtones of Newtonianism, especially because of its links to natural theology, set it apart from Fontenelle's more aristocratic gallantry.

English authors imitated Fontenelle as well as translating him. By 1719, when John Harris wrote his *Astronomical Dialogues Between a Gentleman and a Lady*, he could assume at least a passing familiarity with the *Conversations* on the part of his English readers, and he has his heroine, Lady M., begin her initiation into astronomy by reading Fontenelle. Although the *Astronomical Dialogues* take place in an aristocratic country residence, the style and content of Harris' work reflect the consumer culture of London. Harris, a fellow of the Royal Society and a Boyle Lecturer, taught mathematics to 'gentlemen' and gave public lectures on natural philosophy. His course of lectures at the Marine Coffeehouse prompted him to publish a book on how to use celestial and terrestrial globes, as a handy reference for his auditors.[8] This work went through six editions by 1725, benefiting from the expanding market for books pertaining to the globes, orreries, barometers and other instruments being turned out by London workshops and bought by prosperous city dwellers, as well as by aristocrats.[9] Harris incorporated his explanation of globes into the *Astronomical Dialogues*, where he laid out the basic principles of astronomy in conversational form.

Harris dedicated his book to a real Lady Cairnes, locating himself and his natural philosophy squarely in the world of sociable conversation and elegant objects. Taking Fontenelle as his model, he avoids a 'crabbed and abstracted' style in favour of a 'pleasing and agreeable' one (v). But where Fontenelle strove to make learning pleasurable by presenting telescopic discoveries alongside speculations about the inhabitants of other worlds, Harris wants to 'Improve [the] Minds' of his readers, whom he assumes to be 'Persons of Birth and Fortune' (iv). This kind of knowledge will certainly enhance polite conversation and distract readers from dissolute or frivolous amusements, but it will also turn them into eager customers for the instruments portrayed in the text and accompanying plates. Instead of the cosmic plenum illustrating Fontenelle's *Conversations*, Harris has plates of commercially available objects: the armillary sphere, terrestrial and celestial globes, and the orrery. The images represent instruments designed to demonstrate accepted truths, not to produce new discoveries or facilitate speculations about other worlds. Though Harris makes a stab at emulating Fontenelle's literary style by quoting poetry and inserting gallant remarks into his dialogue, the Englishman never cultivates the fanciful dimension of his subject, and does not encourage his female character to do so. He rather disapproves of these features of Fontenelle's work, though he only says so elliptically: 'I embarrass her with no clumsy Epicycles, or imaginary and indeed impossible Vortices: But I shew her at first the Coelestial World just as it is; and teach her no Hypotheses . . .' (v).

A pair of celestial and terrestrial globes, newly arrived in her drawing room, spark Lady M.'s philosophical conversations. These objects fascinate and perplex her more than the evocation of other worlds in Fontenelle, and possibly even more than the stars themselves. The globes also allow the conversation to take place indoors, in daylight, rather than in the compromising night air, rendering them safer from a moral point of view. Harris' book is more than an instructional manual, but less than a fiction, starting with simple definitions and building up gradually to observations. Rather than asking her to use her imagination, he teaches the lady to use compasses to determine positions on the globes, and then to relate the representations of stars on the globe to the stars visible in the night sky. Ultimately, her knowledge of the cosmos leads her to a new appreciation of the wisdom and power of God: 'What a glorious idea doth it give us of the Almighty Power! Of the Wisdom and Goodness of the Divine Nature!' (78). The consumption of philosophical learning and goods thus ends with a moral purpose, making Lady M., and the reader, a better person.

English writers, recognising the value of appealing to a female audience, developed the popularising dialogue along bourgeois lines. *The Young Gentleman and Lady's Philosophy* (1759), by the London instrument maker and lecturer Benjamin Martin, provides the best example of this particularly English form of moralised empiricism. Martin replaced the subdued eroticism of Fontenelle with a safely domesticated conversation between brother and sister (Figure 13.2). The brother brings the new physics home from college to his sister, showing her how to think philosophically without leaving home. She already has a liking for reading: 'Your remarkable Disposition to reading . . . has given you an elegant Taste, and rendered you capable of understanding, and conversing with Persons on such Subjects as come but too rarely on the Carpet in any Conversation, especially that of your Sex' (1). It turns out that 'the most delightful Part of this Science' is conveyed by experiment, rather than abstract reasoning, and that experiment is compatible with domestic modesty and virtue. Martin illustrated his dialogues with plates of his own instruments, as well as those of other London instrument makers (Figure 13.3). In spite of its dialogue form, Martin's book is more of a systematic textbook than a

Figure 13.2 Benjamin Martin, *The Young Gentleman and Lady's Philosophy* (London, 1759), frontispiece illustrating the didactic conversation between brother and sister. (By permission of Special Collections, Young Research Library, UCLA)

Natural philosophy for fashionable readers 247

Figure 13.3 Benjamin Martin, *The Young Gentleman and Lady's Philosophy* (London, 1759). Working planetarium marketed by Martin, showing motions of planets around the sun, against background of fixed stars. The instrument is both instructive and decorative. (By permission of Special Collections, Young Research Library, UCLA)

diversion; it takes up three volumes and covers not only astronomy, but pneumatics, optics, sound and hearing, meteorology, electricity and mechanics. The conversational form is designed to be particularly appropriate for 'the fair Sex', and marks natural philosophy as virtuous as well as tasteful; but this work could not be read lightly.

This virtuous tinge coloured much of the natural philosophy presented to English polite society in books and lectures. On the Continent, Newton's physics evoked different resonances. Far from being entrenched as the reigning orthodoxy, in France and Italy Newtonianism represented the enlightened, slightly subversive avant-garde. Francesco Algarotti, a Venetian nobleman and poet, aspired to the kind of literary success that Fontenelle had enjoyed a generation earlier, and he saw Newton's physics as the vehicle to that success. Under the title *Newtonianism for the Ladies* (1738), Algarotti presented gallant conversations as a spectacle for the reader, who should be diverted as if she were at the theatre. His dialogues negotiated a transformation in his noble heroine, from Cartesian to Newtonian, as she learned about optics and gravity without diagrams or equations. The reference to Fontenelle was obvious, made more so in the first Italian edition with a handsomely engraved frontispiece depicting an aristocratic couple conversing in a garden.[10] Indeed, Algarotti posed as an admiring disciple in a prefatory 'Letter to M. Fontenelle', where he acknowledges his debt to the master of the conversational genre. Admiration is tempered with irony, however, since his text demolishes the Cartesian physics that formed the centrepiece of Fontenelle's book. 'The Light of Newtonianism has dissipated the Cartesian Phantoms which deluded your sight', the lady learns by the end of the book.[11] Algarotti's readers would know that Fontenelle himself had never accepted the reality of Newton's gravitational force, and that the octogenarian's philosophy had become superannuated by the 1730s, even if he kept publishing new editions of his book. This irony gives a twist to the ideal of politeness originally represented in the *Conversations*. In practice, the canonical reasoned exchange governed by evenhanded fairness often involved witticisms at the expense of antagonists and barbs cloaked in politeness. Seasoned readers knew how to read this sort of exchange.

Reading lecture demonstrations

Fontenelle had certainly discovered a formula for success in his witty conversations about the diverting aspects of astronomy and natural philosophy, and others adapted it to their own ends. All these books explicitly identified their intended audience either as ladies or as 'young gentlemen and ladies', and sought either to amuse or to instruct, or both. Whether for the purposes of moral uplift or simply for entertainment, their authors set the dialogues apart from 'scholarly' or 'serious' works. These dialogues were not the only books aimed at polite readers, however. The public lecture courses fashionable in London, Paris and Leiden often spawned books based on the spoken presentations and demonstrations.[12] Some of the English dialogues overlapped with this category, as in the books by John Harris and Benjamin Martin. Other lecturers, such as J. T. Desaguliers, 'sGravesande and Jean-Antoine Nollet, marketed books to accompany their courses, or to interest potential auditors, or to extend the reach of these lectures beyond their immediate locations. Since lecturers built their performances around their demonstration apparatus, some of which was for sale as well, the books were liberally illustrated with plates of experiments specifically designed to teach students about physical principles.[13]

In all these books, the key that distinguishes the true 'savant' from the broader category of readers is a knowledge of mathematics. Authors steered clear of the 'dry' and 'abstract' qualities associated with equations, or even diagrams. For example, the Abbé Nollet, the most successful French lecturer at mid-century, explained in his *Lectures in Experimental Philosophy* (1748) that he avoided mathematics so that he would not demand 'more pain and application than can be expected' of his readers. The unspoken assumption is that mathematics requires training, special aptitude and hard solitary work: it is not a sociable activity, and certainly not a feminine one. On the other hand, 'natural effects', such as how bodies behave in collisions or what happens to animals in the receiver of an air pump, can be displayed to spectators armed only with 'the good taste of the age' (xv). These demonstrations, Nollet notes, are 'suitable to all Conditions', including nobility, and to 'young persons of both sexes' (xxiv).

A fashion for theory

While experiments lent themselves to public display and spectacular effects, controversial theoretical matters also engaged polite audiences. In France, the fashion for learning geometry tied into public awareness of rather esoteric disputes about mechanics and cosmology. Men of science recognised the potential value of taking such disputes to the literate public. When P.-L. de Maupertuis (1698–1759), an up-and-coming mathematician in the Paris Academy of Sciences, undertook to sell Newton's theory of gravity to a French audience in the 1730s, he made a strategic decision to publish a book that might be read more widely than a technical paper presented to the academy. In his *Discourse on the Different Shapes of the Heavenly Bodies* (1732), he situated mathematical results about the shapes of rotating fluid bodies in the context of an overview of the Cartesian and Newtonian 'systems', written not for those who could have read Newton, but for anyone intellectually adventurous enough to consider the ramifications of his physics.[14] Even though Maupertuis did not adopt the dialogue form, Voltaire claimed to have read the book 'with the pleasure of a girl reading a novel, and the faith of a devout believer reading the gospel'.[15] Voltaire's remark indicates that the book's style and controversial subject matter could engender the kind of intellectual pleasure appropriate to unlearned readers.

By judiciously considering both sides of the Cartesian–Newtonian divide in measured but engaging tones, Maupertuis effectively brought gravity (or 'attraction') into civil conversation, where it was to become a hot topic. In spite of the rhetoric of impartiality, the discerning reader could detect the author's proclivities, and Maupertuis made his name as a Newtonian on the strength of this book. Ten years later, after the most virulent controversy over gravity had died down, Maupertuis published a second edition of his *Discourse*. Elegantly produced, this volume displayed a confident command of astronomical phenomena for genteel readers. A frontispiece was added, representing the stars as centres of gravitational force and light, each with its orbiting planets and comets (Figure 13.4). Reminiscent of Fontenelle's many worlds, the image is more geometrical and less fantastic, suggesting that solid Newtonian explanations have replaced the speculations of Fontenelle's characters. The frontispiece adds to the elegance of the volume (in some copies it was coloured a blue-green) along with the heavy paper and wide margins. Then the preface reassures the reader that all equations and diagrams have been newly segregated in an appendix 'so that this book can be read easily by everybody'. The text stands independent of the calculations, which are nevertheless there for 'those who want to take the trouble'. Thus the mathematics signals the special knowledge of the author, not erased from the book entirely, but displayed separately.

A rather different example is Voltaire's *Elements of Newtonian Philosophy* (1738), a controversial book with a complicated publication history and a wide readership in the cosmopolitan world of letters.[16] Just by virtue of its author's identity, it was guaranteed substantial visibility. Voltaire undertook his exposition of Newton as a counterpart to Algarotti's book, particularly to its lightness of tone. Voltaire rejected the light and witty format of the polite dialogue in favour of a more didactic treatise aimed at 'those who know no more than the name of Newton and his philosophy'.[17] Voltaire set out to be a torch-bearer for Newton in France, following in Maupertuis's footsteps, but aiming at more comprehensive coverage of mechanics and optics than had been attempted for this audi-

Natural philosophy for fashionable readers 251

Figure 13.4 Pierre-Louis de Maupertuis, *Discours sur les différentes figures des astres*, 2nd edn (Paris, 1742), frontispiece. (By permission of Special Collections, Young Research Library, UCLA)

ence in an original French work. Voltaire assumed that his readers would be willing to work at their own enlightenment, as long as he was able to replace Newton's abstruse style with French clarity of expression, devoid of equations. Voltaire himself must have worked hard to synthesise his material, incorporating the experiments of the *Opticks* with the theory of universal gravitation, and especially with an attack on the physics of Cartesian vortices. He oversaw the printing of his manuscript in Amsterdam, where engravers produced a lavish allegorical frontispiece and numerous vignettes to decorate the chapter breaks.[18] He evidently intended the book to be aesthetically pleasing as well as instructive. When it did appear, in 1738, it provoked a number of attacks in pamphlets and periodicals. Editors and publishers promoted the controversy to build readership for subsequent editions. Voltaire, never one to avoid a battle, responded in kind, so that the book and the 'civil war' between Cartesians and Newtonians fuelled conversation and correspondence.[19] As a result, a reviewer in the *Mémoires de Trévoux* claimed that 'All Paris resounds with Newton, all Paris gabbles of Newton, all Paris studies and learns Newton.'[20]

In England, the fashion for Newtonian natural philosophy was linked to a flourishing consumer culture and to natural theology, as well as to polite sociability. In France, Newtonianism also became a fad of sorts, taken up by fashionable society (*le monde*). Imported into the discourse of ironic wit that characterised French philosophical controversies, its novelty and foreign origin gave it an aura of riskiness that drew the attention of the reading public. English books made natural philosophy virtuous, a safe subject for domestic conversation and the education of children. The moral value of philosophising is far less prominent in the French books we have considered. French writers cultivated an aristocratic audience, while also boasting that their works were suitable for people 'of both Sexes and all Conditions' (Nollet, x). In both England and France, books brought genteel readers, women and men, into the cultural arena where natural philosophy mattered. Authors saw these readers not just as a potential market for commodities, but also as crucial legitimators of the whole undertaking of natural philosophy. While polite conversation remained an ideal, emulated in many books on scientific subjects, readers and writers also engaged in what we might call impolite conversation: ad hominem attacks, accusations of impiety, or challenges to intellectual integrity. Books were only one of many formats for such exchanges, polite and impolite. They became part of an extended web of discourse made up of reviews, pamphlets, responses, letters, translations and revised editions. In all of these genres, we can read traces of conversations, between readers and authors, listeners and lecturers, teachers and students. Authors wrote dialogue into their texts, and books provoked real conversations in their turn. And so public interest in natural philosophy reflected the 'good taste that prevails in our Age'.

Notes

1 Lawrence Klein, *Shaftesbury and the Culture of Politeness* (Cambridge, 1994), p. 8.

2 See Robert Shackleton (ed.), *Entretiens sur la pluralité des mondes de Fontenelle* (Oxford, 1955).
3 Geoffrey Sutton, *Science for a Polite Society* (Boulder, 1995), p. 162.
4 Erica Harth, *Cartesian Women* (Ithaca, 1992); M. Terrall, 'Gendered spaces, gendered audiences', *Configurations*, 2 (1995): 207–32.
5 Fontenelle, *A Discovery of New Worlds*, trans. Aphra Behn (London, 1688).
6 Other translators included W. D. Knight (1688); John Glanvill (1688, 1695, 1702 and 1719); and William Gardiner, *A Week's Conversation on the Plurality of Worlds by M. de Fontenelle* (London, 1715), 2nd edn., 1728. Based on a revised edition, this version was reprinted at least four more times.
7 John Mullan, 'Gendered knowledge, gendered minds: women and Newtonianism, 1690–1760', in *A Question of Identity*, ed. Marina Benjamin (New Brunswick, NJ, 1993).
8 John Harris, *Astronomical Dialogues between a Gentleman and a Lady* (London, 1719), preface.
9 Alice Walters, 'Conversation pieces: science and politeness in eighteenth-century England', *History of Science*, 35 (1997): 121–54.
10 The image resembled an engraving used for the frontispiece of one of the editions of Fontenelle's *Conversations*.
11 Francesco Algarotti, *Sir Isaac Newton's Philosophy Explain'd For the Use of the Ladies* (2 vols.) trans. Elizabeth Carter (London, 1739), vol. II, p. 247.
12 Larry Stewart, *The Rise of Public Science* (Cambridge, 1992); Simon Schaffer, 'Natural philosophy and public spectacle in the eighteenth century', *History of Science*, 21 (1983): 1–43.
13 Sutton, *Science for a Polite Society*, ch. 6.
14 Maupertuis, *Discours sur les différentes figures des astres* (Paris, 1732).
15 Voltaire to Maupertuis, 20 November 1732, *Oeuvres complètes de Voltaire* (Oxford, 1968), vol. 86, p. 252.
16 For publication history, see introduction to *Eléments de la philosophie de Newton*, ed. R. L. Walters and W. H. Barber, *Oeuvres complètes de Voltaire* (Oxford, 1992), vol. 15.
17 Voltaire, *Eléments*, 1738 edn., Avant-propos 'A Mme la Marquise du Ch★★.
18 Robert Walters, 'The allegorical engravings in the Ledet-Desbordes edition of the *Eléments de la philosophie de Newton*', in *Voltaire and his World*, ed. P. Howells et al. (Oxford, 1985), pp. 27–49.
19 The phrase is Voltaire's.
20 *Mémoires de Trévoux* (August 1738), p. 1674.

Further reading

Ann Bermingham and John Brewer (eds.), *The Consumption of Culture, 1600–1800: Image, Object, Text* (London, 1995)
Roger Cooter and Stephen Pumphrey, 'Separate spheres and public places: reflections on the history of science popularization and science in popular culture', *History of Science*, 32 (1994): 237–67
Joan DeJean, *Ancients Against Moderns: Culture Wars and the Making of a Fin de Siècle* (Chicago, 1997)

Aileen Douglas, 'Popular science and the representation of women', *Eighteenth-Century Life*, 18 (1994): 1–14

Bernard LeBovier de Fontenelle, *Conversations on the Plurality of Worlds*, trans. H. A. Hargreaves (Berkeley, 1990)

Lawrence E. Klein, 'Gender, conversation and the public sphere in early eighteenth-century England', in Judith Still and Michael Worton (eds.), *Textuality and Sexuality: Reading Theories and Practices* (Manchester, 1993), pp. 100–15

Terry Lovell, *Consuming Fiction* (London, 1987)

John Mullan, 'Gendered knowledge, gendered minds: women and Newtonianism, 1690–1760', in Marina Benjamin (ed.), *A Question of Identity: Women, Science, and Literature* (New Brunswick, NJ, 1993), pp. 41–56

James Secord, 'Newton in the nursery: Tom Telescope and the philosophy of tops and balls', *History of Science*, 23 (1985): 127–51

Kathryn Shevelow, *Women and Print Culture* (London, 1989)

Geoffrey Sutton, *Science for a Polite Society: Gender, Culture and the Demonstration of Enlightenment* (Boulder, 1995)

Naomi Tadmor, "In the even my wife read to me": women, reading and household life in the eighteenth century', in J. Raven, H. Small and N. Tadmor (eds.), *The Practice and Representation of Reading in England* (Cambridge, 1996)

Mary Terrall, 'Gendered spaces, gendered audiences: inside and outside the Paris Academy of Sciences', *Configurations*, 2 (1995): 207–32.

'The uses of anonymity in the age of reason', in Mario Biagioli and Peter Galison (eds.), *What is a Scientific Author?* (forthcoming, Chicago, 2001).

Alice Walters, 'Conversation pieces: science and politeness in eighteenth-century England', *History of Science*, 35 (1997): pp. 121–54.

14 Rococo readings of the book of nature

Many twentieth-century readers are struck by the way in which shells are depicted in conchologies of the eighteenth century. Whilst such works make evident the ornamental qualities of shells, twentieth-century ones present an apparently more transparent relationship with nature, in which shells are not arranged in a decorative fashion. It appears to modern readers that the earlier works are belaboured, artificial and stilted, that their authors struggle to generate something which is not 'really' there by placing shells in patterns. These authors seem inexpert, as if they do not know how to go about treating natural objects. In effect, their books cannot be read as scientific. They seem no more than empty works of decoration, reflecting the frivolous concerns of the eighteenth-century European elite.

Our analyses, often unreflecting, are also asymmetrical, however. For we are still happy to live within a world of natural history books in which objects are represented to us in strictly geometrical relations. Thus Figure 14.1 immediately appears reassuringly informed by a scientific taxonomy. Geometricity and lack of ornamentation are silent signifiers for scientific authenticity, and hence for nature, even though these two qualities ostensibly have no bearing upon reality 'out there', just as books themselves are material artefacts of paper, ink, binding, words and illustrations which have no intrinsic and privileged relationship with either 'scientific truth' or 'nature'. The relation between a book and nature, or between a book and scientific authority, is thus one which is constructed at least at one level through the aesthetics of the book's presentation. How can agreement be reached about what constitutes the 'right' way of representing nature; how is that representation policed; and above all, how do people assent to the 'naturalness' of those modes of representation?

Clearly, the natural history book is involved in this process. However, we cannot afford to treat it as an isolated historical object. Historians of the book have suggested that it is in the process of interpretation that books acquire meaning. In this sense the book was one element in a network of relations between things, texts and people. Thus the book's interpretation was mediated by the

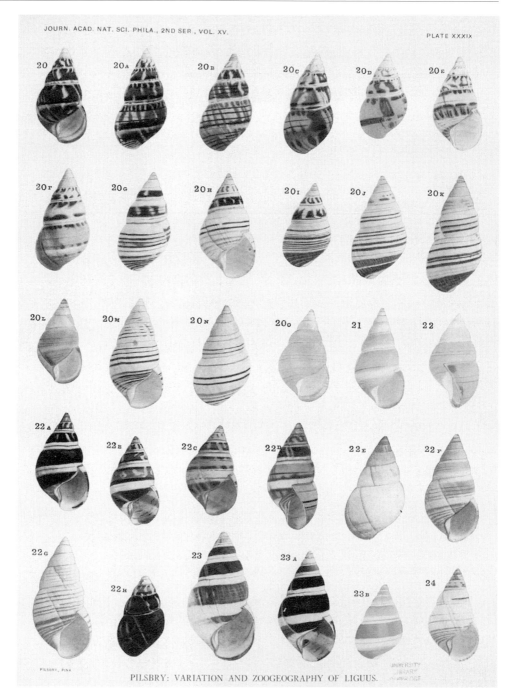

Figure 14.1 Henry Augustus Pilsbry, 'A study of the variation and zoogeography of Liguus in Florida', *Journal of the Academy of Natural Sciences of Philadelphia*, 15 (1912), plate 39. Such grid-like arrangements of shells on printed plates first appeared in early-nineteenth-century British publications, but by the late nineteenth century had become standard for most scientific conchologies. (By permission of the Syndics of Cambridge University Library)

Rococo readings of the book of nature 257

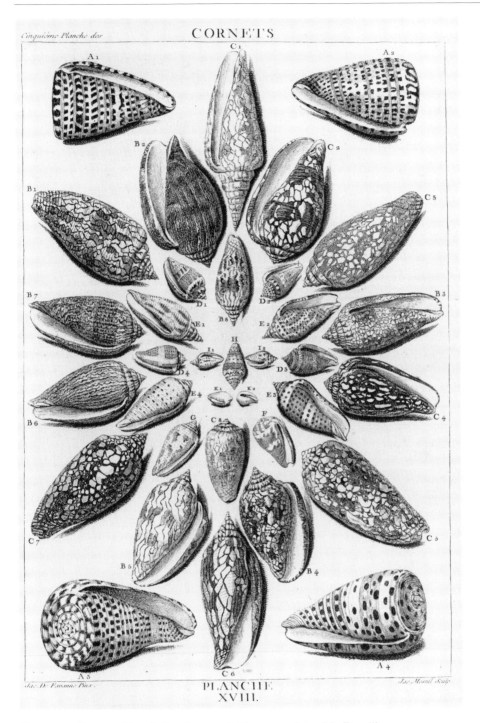

Figure 14.2 Jacques Mesnil after Jacques de Favanne, plate 18 in Dezallier d'Argenville, *La Conchyliologie*, 1780. Two artists, Guillaume and Jacques de Favanne, edited d'Argenville's book after his death. In adding plates, they emphasised symmetrical and decorative arrangements of depicted naturalia. (By permission of the Syndics of Cambridge University Library)

problems of power, social and learned status, wealth and self-fashioning which preoccupied its owners and readers. My essay is concerned with the interplay between the material object of the book and the means of making it 'work' – the form that its credibility takes in a particular culture.[1] For whilst books survive, understandings of the natural transmute. In order to show how the history of the book might allow us to develop new understandings of the history of nature, I will look in some detail at a particular genre of natural historical writings, the conchological literature of mid-eighteenth-century France.

Symmetry, contrast, variety

The 1685 *Historiae sive synopsis methodicae conchyliorum* of Martin Lister (1639–1712), traveller, FRS and medical practitioner, was widely used during the eighteenth century. His daughters Susanna and Anna prepared copperplate engravings of each specimen depicted in the book. This contained virtually no text – the merest indication of Lister's system, and a name under each engraving, were all that was supplied. Unlike other contemporary shell books, each specimen was separately engraved onto a small copperplate, several of which were then impressed upon a large page; separate pages were then bound in the usual way. This process enabled errors to be corrected and new engravings to be added during the production process (Figure 14.3). The book was thus a flexible entity, composed of fixed units which were combined and recombined, jettisoned and replaced. Accordingly, few copies were identical. As Johns argues, the early modern book cannot be seen as a stable object corresponding to a single fixed vehicle for transmitting knowledge. The lack of stability in Lister's work enabled multiple interpretations and readings by naturalists across Europe even whilst they attempted to fix its bibliographical identity.[2]

From the 1690s onwards, however, it became customary for shell books to be illustrated with specimens arranged in symmetrical patterns, from the relatively simple designs of Figure 14.4 to the sophisticated patterns of Figure 14.2. It was, in particular, the types of naturalia collected by the polite elite in the early eighteenth century which warranted such symmetrical arrangements, although different categories of naturalia became 'fixed' in particular modes of representation. Thus chameleons, lizards, shells, minerals, corals and figured stones or fossils all appeared in symmetrical designs in cabinets and books, and were explicitly admired because of their intricately patterned colours and shapes. Conversely, birds and many other quadrupeds warranted schematised landscape settings. Principles of symmetry, variety and contrast were apparent in the illustrations to many natural history

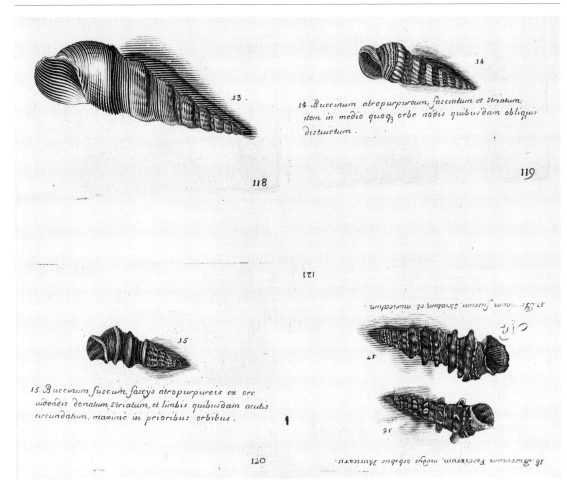

Figure 14.3 Martin Lister, *Historiae . . . conchyliorum*, 1685, plates 118–20. The Lister family's methods of manufacturing the shell book produced very variable outcomes. Specimens were placed in various orientations, and were even sometimes printed upside down. The conventions for shell depictions in conchologies from the eighteenth century are in sharp contrast. (By permission of the Syndics of Cambridge University Library)

works published during the first thirty years of the century. However, the authors of natural history works were virtually silent about such choices of representation.[3] How then might we understand the operation of such striking aesthetic choices within the natural history book?

Symmetry, variety, uniformity, contrast – this was a language widely employed from the Renaissance until the end of the eighteenth century, in philosophical and artistic treatises whose authors sought to define the nature of beauty and taste. The definition of symmetry in its modern sense, as 'the reflecting correspondence of points in a plane, or parts of a body, with respect to a common axis' was, however, a seventeenth-century invention.[4] Symmetry was a design principle which was most widely employed in literature concerning the philosophical underpinnings of the fine arts, particularly writings harking back to classical Palladian or Vitruvian architectural principles. Many eighteenth-century writers on taste and beauty pointed to natural productions as the material

Figure 14.4 An early symmetrical shell plate, showing typical features: arrangement of individual printed naturalia about a vertical central axis, and bilateral symmetry of form and orientation which supersedes simple horizontal alignments. F. Ertinger, in Claude du Molinet, *Le Cabinet de la Bibliothèque de Sainte Geneviève* (Paris, 1692), plate 44. (By permission of the Syndics of Cambridge University Library)

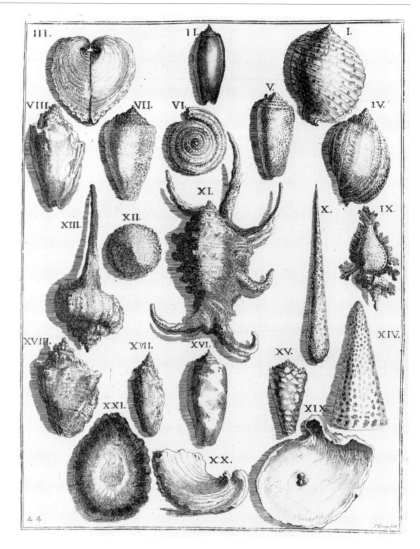

manifestation of fundamental aesthetic criteria. The fact that naturalia displayed symmetry *and* irregularity, variety *and* uniformity was evidence of the existence of a designing deity; most such treatises and many natural historical works, therefore, presented a natural theological account of the cosmos.

In describing the process of making order within the collection, European naturalists frequently appealed to arguments from design. But in France in particular, naturalists simultaneously used collections to legitimate their membership within polite society by demonstrating their mastery of taste. Taste as a criterion for self-fashioning spread from a small, highly wealthy elite of collectors in early eighteenth-century France to a wider, more anonymous public with the money to purchase some collectables but also to afford the growing number of printed works representing collections and specimens.

Selling the rococo

Shell collecting first became widespread in France partly thanks to the successful marketing tactics of Edmé-François Gersaint (d.1750), a merchant with a boutique on the Pont-Neuf in Paris. He began selling curiosities in 1718, specialising in objects of the fine arts. Over the next decades, his network of contacts expanded to include many expert Parisian collectors. These owners of prints and paintings by rococo artists such as Watteau, Lajoüe and Boucher were also Gersaint's readers, clients and friends, and when they died, he sold off their collections piecemeal, sometimes with regret but always with profit.

In the early 1730s, Gersaint made several trips to the Dutch Republic to purchase curiosities of art and nature, and even whole collections, for export to eager French connoisseurs. At around this time he began to buy the shells which ranked alongside paintings in Dutch collections, with a view to launching them as a new form of curiosity in Paris. The first of many auction catalogues was published in 1736. Gersaint entered the world of print to recruit connoisseurs to a new form of curiosity from which he could profit, and also to an unfamiliar, Dutch style of purchasing, the auction. He used his publications to offer guidelines as to the proper mode of appreciation of naturalia for those unfamiliar with this branch of curiosity. Few French collectors read Dutch, the language of many shell books; thus they depended upon travelling experts such as Gersaint to interpret specimens and books from the Dutch Republic.[5]

In the first decades of the eighteenth century, as several historians have noted, French collectors were few in number, extremely wealthy, and tended to restrict their collecting activities to antiquities and works of the fine arts in particular. The representation of collections in print was to produce considerable transformations in the nature and meaning of collecting which coincided with the advent of a widespread fashion for collecting natural objects.[6] The financial aspect of the virtual possession enabled by the publication of a collection characterised Dutch and French natural history books from the start of the eighteenth century. As a 1711 review of Rumpf's *Thesaurus imaginum piscium testaceorum* in the *Journal des Sçavans* explained: 'The purchase of this single volume will put [readers] in possession of a shell ... which cost its owner as much as 500 florins.' Putting collections into print was thus both a commercial enterprise and a mode of self-fashioning; it made what had been private into a public, purchasable commodity. But by the middle decades of the century, polite print culture increasingly emphasised the exercise of taste as a mode of self-conduct which set educated and well-to-do people aside from the impolite. Accordingly, increasingly vociferous criticisms were aimed at those who concerned themselves too obviously with the commercial value of specimens.[7]

Because most natural historical readers defined themselves as persons of taste first and foremost, most natural history authors, engravers, printers and publishers strategically presented their books as objects recognisably belonging within that world of wealth, fashion and instruction.[8] Natural history books were one category of luxury good, and artists were involved in their production. The commercial reproduction of works of art played a growing role in the financial support of artists over the course of the century. Books of rococo ornaments were bought for *découpage* by the well-to-do, and the same artists who worked for naturalists might publish books of flower engravings which were at once works of natural history, silk and embroidery design sourcebooks, and entertaining volumes for the polite reader.[9] A rococo work of natural history was thus part of a larger artistic style. To be understood, the author depended upon the reader's prior familiarity with certain aesthetic conventions; thus the book contained textual and pictorial devices to indicate its links with such stylistic traditions.

From the mid-1730s, there was a convergence of French natural history works and picturesque artistic styles. *Rocaille*, a picture of marine naturalia by François Boucher, a leading rococo artist, was the model for two frontispieces he supplied for Gersaint's 1736 and 1744 auction catalogues, and for the most widely read conchological work of the century, Dezallier d'Argenville's *Histoire naturelle éclaircie* (Figure 14.5). The same frontispiece appeared in a separate edition of the conchological part, *La Conchyliologie, ou Histoire naturelle des coquilles de mer, d'eau douce, terrestres et fossiles* (Paris, 1757); also, slightly retouched, in a third edition of 1780 which was considerably more lavish and ornately illustrated than the previous two. The presence of a Boucher frontispiece announced, before the reader had even reached the text, that the book in question belonged within a very specific visual tradition, and that its possession would denote one as a person of taste. Conversely, Boucher himself became a leading natural history collector whose own catalogue contained a eulogy by the auctioneer, Gersaint's colleague Pierre Remy, celebrating him as a man of taste.

Like most of Gersaint's collectors, Antoine-Joseph Dezallier d'Argenville (1680–1765) came to natural history from the standpoint of a connoisseur in the arts. D'Argenville, the son of a Paris *libraire* or guild publisher, wrote on both the fine arts and natural history, contributing to ongoing debates about whether the artist

Rococo readings of the book of nature 263

Figure 14.5 Quintin-Pierre Chedel after François Boucher, frontispiece for Dezallier d'Argenville, *L'Histoire naturelle éclaircie*, 1742. The presence of a Boucher frontispiece announced, before the reader reached the text, that the book belonged within a particular visual tradition. Here naturalia were tied to a world of taste through the intermediary of a multiply reproduced, commercial object, the print. (By permission of the Syndics of Cambridge University Library)

should portray *la belle nature*, enhanced nature. Trained by leading artists, he allotted nature a powerful role as source of all beauty and model for all lesser, human creativity. Thus works of art and the sciences could be subject to the same type of scrutiny and criticism. D'Argenville was also a correspondent of numerous scientific societies who published on the natural history of shells and minerals. Whilst distinguishing between the *curieux* and the *philosophe*, d'Argenville sought to woo existing connoisseurs to accept a marriage of method and beauty. Like Gersaint, he suggested that the two requirements of 'recreation for eye and mind' promised in the title of one widely read conchology could be reconciled by the collector.[10] Philosophical method could still translate into the language of admiration applied to flowers and miniatures: 'Although it seems that in arranging Shells by family, one loses the fine enamel of the colours, one can nevertheless approach the beauty of Borders in a numerous collection, such that each drawer is filled with a single family, then the colours which vary between different species of the family present a most agreeable perspective, and one only loses out in the diversity of shapes.'[11] Again, d'Argenville sought to reconcile *savant* and *curieux* concerns at the level of tastefulness in his recommendation that shell specimens be displayed in pairs, one polished and one brute. D'Argenville's *Histoire naturelle* set a standard for subsequent shell books, most of which displayed their naturalia arranged by family even as their authors celebrated the ornamental qualities of their subject.

The manner of representing nature favoured by polite collectors reflected a shared emphasis upon tastefulness, with its origins in expertise concerning objects of the fine arts (Figure 14.6). D'Argenville claimed that experience in the artificial representation of natural objects actually *assisted* proper classification. Certain treatments of shells were viewed as ways to demonstrate the true nature of the specimen. 'Helping nature and making her appear in all her lustre is the principal goal to which everyone should aspire: conversely, there is nothing more presumptuous than wanting to force nature and enhance her graces.' Thus rococo readers were advised to remove the '*drap marin*', a greyish outer coating on some species.

Although several Naturalists wish to have all Shells as they come from the sea, in other words, covered with mud and in a rough condition, one cannot completely agree with them . . .; a Shell is like a rough Diamond which one cannot enjoy until it is revealed, cut and polished; it is by that means that one acquires new species, and, so to speak, a second Shell.[12]

However, not all artful uses of nature were acceptable. D'Argenville criticised the Dutch habit of painting shells with different colours or engraving them with 'Historical subjects', 'stripes, circles, or

Figure 14.6 At the level of wealth, print and collecting, d'Argenville's book eroded the distinction between *savant* and *curieux*; every plate was labelled with the name of the sponsoring collector (here Bonnier de La Mosson) who had made its publication possible. Unknown artist, in Dezallier d'Argenville, *L'Histoire naturelle éclaircie*, 1742, plate 20. (By permission of the Syndics of Cambridge University Library)

stars . . . and a thousand other figures which Nature had not thought fit to give them . . . they impoverish, so to speak, the subject in attempting to make it unique . . .'.[13] However, d'Argenville never abandoned the language of artistic appreciation: the shell was described as 'the subject', the judgement was that of a connoisseur in the fine arts discussing a painting. It was thus the very fact that Nature *did* paint pictures on naturalia which made Dutch practices illicit, a rivalry or forgery rather than an enhancement. Since the late seventeenth century, natural historical writers had claimed that nature's own handiwork could never be surpassed. Technologies like the microscope exposed the gulf between the clumsiness of the human artist and the delicacy of nature, as an *Encyclopédie* article noted:

The finest and smallest writing, such as the whole of 'Our Father' written on a silver penny . . . when examined under the microscope, appears to be as deformed, coarse and barbarous as if it had been written by the heaviest hand; but the marks on the wings or bodies of moths, scarabs and other insects . . . are found to be precisely circular, and other adjacent lines and marks seem to be drawn regularly and delicately, with all possible exactitude.[14]

The visual world of these lovers of natural history demanded a language of workmanship and artistry. Natural theological accounts of natural history in effect provided a space in which artistic and philosophical concerns coincided.

The counter-aesthetic

In early eighteenth-century conchological publications, a balance of regular with irregular, symmetrical with asymmetrical, rational with imaginative was admired. But after 1750, the possibility of cohabitation of methodical and curious models which had been explored in earlier conchological works was challenged by some. 'That same beauty which caused attention to be paid to shellfish has become a powerful obstacle to the progress of this science', claimed the young naturalist Michel Adanson (1727–1806) in the preface to his natural history of the shells of Senegal. Problematising the nature–art relation, Adanson claimed that an aesthetic appreciation of shells had diverted naturalists' attention from the humble animal within. In fact, however, natural theological conchologists regularly called for more attention to the Providential design of the living animal.[15] Underlying Adanson's claims was something else: a radical account of the natural status of the classificatory category. For Adanson, only a combination of criteria drawn from all body parts could serve as the basis for a truly natural classification of shelled animals. He denounced earlier classifications of shells as superficial, the product of an hour of leisure, rather than the prolonged and complicated study 'filled with thorns and difficulties' of the true nature of the shell and its animal. 'This work will thus be less a method or system than a new way of considering Shellfish . . .'.[16]

The claim that the methods of one's predecessors were inadequate was widespread among conchological authors in this period. The juxtaposition at stake was not simply that between methodical and unmethodical, scientific and unscientific, as modern authors would have it.[17] Instead, Adanson rejected the study of conchology for aesthetic pleasure, a mode of understanding shells which had given purpose and meaning to their study since the late seventeenth century and within which Adanson's contemporaries made sense of their activities as conchologists. Similarly, Adanson's work contained no natural theological discussions, an approach increasingly rejected in his manuscripts dating from this period.[18]

Conchological contemporaries often failed to appreciate the basis for Adanson's dismissal of design as superficial, therefore the wrong source for a truly natural order. As the British collector Emanuel Mendes da Costa put it, 'the especial colours and forms of the Shells do not the less give a distribution or order, and proper characters to go by, than does the mechanism of the very fish themselves'.[19] Even when, in 1775, the conchological writer Favart d'Herbigny rejected 'puerile arrangements and symmetries of colour' in shell collections, he went on to set the study of shells firmly within the world of connoisseurship and the language of artful design, noting the 'variety' and 'regularity' of shell patterning and construction. His criticisms signalled not the abandonment of a tasteful understanding of nature but the rejection of aesthetic criteria which had predominated in the late seventeenth century, the age of tulipomania, shell grottoes and 'embroidered' flower borders. 'Borders' made of shells, common in early Dutch collections, were now perceived to have declined in status to the vulgarly commercial, 'permitted to the florists, and necessary to textile manufacturers and carpenters'.[20] Producers and consumers of luxury goods were not to be presented as equals. The French shell literature catered to the self-perceptions of readers by an appeal to good taste, and the wholesale study of the soft parts of shellfish was ultimately ruled out by several authors on that basis. A natural history of shells with their animals, on the lines Adanson suggested, would not predominate until well into the following century.

It is very significant, then, that Adanson's reforms of conchology extended to the conventions of shell display in the book, the interface at which the author's tasteful self-presentation and the reader's artistic judgement operated. Like many natural historical authors,

Adanson had chosen a renowned artist, in this case Marie-Thérèse Reboul, a regular Salon exhibitor, to prepare the shell drawings for his plates. But, he insisted, the plates in the *Histoire naturelle du Sénégal* deviated from 'the severity of the ordinary rules of drawing'. Shadows had been abolished from the illustrations 'after the custom of naturalists who suppose their objects to be detached from all neighbouring bodies'. Adanson's emphasis upon the living animal as the primary source of meaning for the individual specimen led to a radical design change. 'I have scrupulously avoided those orientations against nature which certain Authors have given their shells, in depicting them with the point at the top, situations as bizarre as ones representing men with their head at the bottom and feet at the top' (Figure 14.7). The remains of Adanson's collection of shells were unearthed earlier this century. Here he formed small groups of shells of different sizes and colours which, based upon his observations of the resident animal, he viewed as members of a single species differentiated by age and gender. Thus the species conferred an ontological unity which superseded the visual presentation of the specimens on each 'page' of the collection, where groups of real shells pasted onto card mingled with printed engravings excised from published works and Adanson's Indian ink sketches. This was a far cry from a leading rococo collector's ornate *coquillier* in which specimens were symmetrically arranged upon blue and white satin.[21] In the plates to Adanson's book, however, these display practices were not followed. Instead, Reboul's images largely conformed to the characteristic symmetrical presentation of shells. Likewise, the vignette to page 1 of the *Histoire naturelle du Sénégal* was typical of the rococo book in its floral garlands and shellwork. Between Adanson's collection and the printed work was a visual distance resulting from the implementation of certain aesthetic criteria which, as I have shown, were an integral part of the conchological genre (Figures 14.8 and 14.9).

But if natural history books were fashioned according to precise standards of beauty, it is interesting to reflect upon the ways in which they also served authors' statements about the nature of the natural world. It is arguably no more 'natural' to display engraved shells pointing vertically than to portray them in symmetrical arrangements or even garlands. Indeed, all the authors I have discussed claimed both to be scientific and to represent natural objects 'from nature' in their books. The distinction which Adanson was constructing operated, in effect, at the level of the aesthetics of display, *but it was ascribed to the intentions of nature*. Choices about which words and images may perform as signs of the natural in books can thus serve to define the characteristics and content of a nature which is then claimed to supply these principles as fundamental truths.

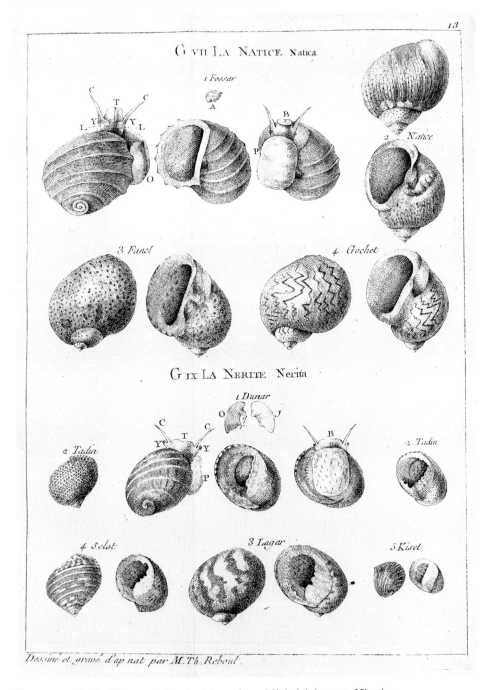

Figure 14.7 Marie-Therese Reboul, plate 13 from Michel Adanson, *Histoire naturelle du Sénégal*, 1757. Most of Reboul's plates reveal that symmetry structured the presentation of the objects upon the printed page. Here, however, symmetry has broken down and the balance between the two sides of the page is uneven. (By permission of the Syndics of Cambridge University Library)

Figure 14.8 The natural history cabinet of Bonnier de La Mosson, showing how the function of naturalia as specimens was inseparable from their role in the ornamentation of the room. This architect's elevation resembled countless other plans of interior design generated during the same period. From K. Scott, *The Rococo Interior: Decoration and Social Spaces in Early Eighteenth-Century Paris* (New Haven and London, 1995).

Conclusion

Since the publication of Fiske Kimball's *The Creation of the Rococo* in 1943, the rococo has been a respectable field for art historians. The term has been used of the bibliographic ornamentation familiar to any reader of eighteenth-century books, in the form of cartouches, frontispieces and vignettes. Such elements are generally viewed as peripheral to the 'real' meaning of the book. But in the interpretation of shell books in the eighteenth century, the style of ornamentation was crucial for readers, for it indicated whether the book belonged within a larger literature governed by particular aesthetic conventions, a literature which was the property of the polite elite.

The term 'rococo' was not a contemporary one. It was first used in the 1830s, of architecture and interior decoration 'having the

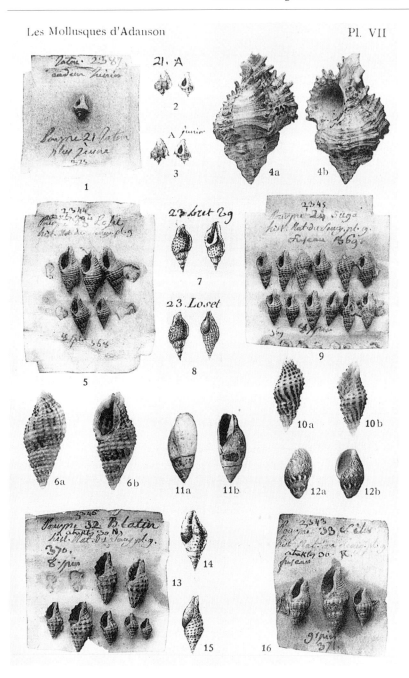

Figure 14.9 A 'page' from Michel Adanson's shell collection, showing the juxtaposition of specimens with ink drawings by Adanson and his annotated cuttings from published works. Here the interpenetration of collection and text was complete; at the same time, there were no concessions to the demands of rococo style. From E. Fischer-Piette, 'Les Mollusques d'Adanson', *Journal de Conchyologie*, 85 (1942): 103–377.

characteristics of Louis Quatorze or Louis Quinze workmanship, such as conventional shell- and scroll-work and meaningless decoration: excessively or tastelessly florid or ornate'.[22] The appearance of the word 'rococo' thus signified the circumscription and devaluation of a decorative style which assisted the interpretation of mid-eighteenth-century natural history books.

In the sense that the term 'rococo' might draw our attention to the existence of past *styles* operating in scientific books, which shaped readings and even underpinned the meaning of 'nature', it is perhaps a useful expression. The term acknowledges the irredeemably past nature of such styles, which are no longer accessible to us as readers and thus no longer convincing or meaningful. Subsequent historians have struggled to make the printed materials of d'Argenville, Gersaint and others fit the dichotomy of art versus science. The cabinet of one wealthy collector, Joseph Bonnier de La Mosson, is often taken to exemplify the rococo.[23] Jean Courtonne's drawing of the natural history cabinet reveals the conventional decorative use of the specimens, the bilateral symmetry which operated not only in each drawer of the cabinet but also in the juxtaposition of larger specimens outside the cupboards.

Such display strategies have proved invisible for one historian, who describes how 'the flora and fauna have slipped their cages to climb all over the panelling, so that it is often impossible to distinguish between sculpted representations of serpents, stags' heads, horns, coral, sponges and shells and the specimens themselves'. Such a muddle, she suggests, effectively prevented the implementation of a properly scientific classificatory system. 'For a brief moment, before being completely undone by science, these late *Wunderkammern* seem to have been host to a tense conviviality between fantasy and reason and fable and history.'[24] In a world of fundamental opposition between science and art, a style which seems to traverse the boundary between natural and artificial is unsettling. Scott's discomfiture demonstrates the ethical freight that the representation of nature subsequently acquired and which it has never lost.

Notes

1 Roger Chartier, *Cultural History: Between Practices and Representations*, trans. Lydia G. Cochrane (Cambridge, 1988), introduction; Stanley Fish, *Is There a Text in This Class? The Authority of Interpretative Communities* (Cambridge, Mass., 1980); Steven Shapin, 'Cordelia's love: credibility and the social studies of science', *Perspectives on Science*, 3 (1995): 255–75; id., *A Social History of Truth: Civility and Science in Seventeenth-Century England* (Chicago, 1994); and see Peter

Lipton, 'The epistemology of testimony', *Studies in History and Philosophy of Science*, 29 (1998): 1–32.

2 Adrian Johns, *The Nature of the Book: Print and Knowledge in the Making* (Chicago, 1998), introduction; G. Keynes, 'Dr. Martin Lister: a bibliography', *The Book Collector*, Winter 1979: 501–20, Spring 1980: 50–81.

3 Philip Stewart, *Engraven Desire: Eros, Image and Text in the French Eighteenth Century* (Durham, 1992); E. C. Spary, 'Codes of passion: natural history specimens as a polite language in late eighteenth-century France', in Peter Hanns Reill and Jürgen Schlumbohm (eds.), *Wissenschaft als kulturelle Praxis, 1750–1900* (Göttingen, 1999), pp. 105–35.

4 David Summers, 'Symmetry', in Jane Turner (ed.), *The Dictionary of Art* (London, 1996), vol. XXX, p. 171; Werner Szambien, *Symétrie, goût, caractère: théorie et terminologie de l'architecture à l'âge classique 1550–1800* (Paris, 1986), ch. 3; Charles de Secondat, baron de Montesquieu, 'Essai sur le goût dans les choses de la nature & de l'art', in Denis Diderot, Jean de La Rond d'Alembert (eds.), *Encyclopédie, ou Dictionnaire raisonnée des sciences, des arts et des métiers*, 35 vols. (Paris, 1751–1772), vol. VII, pp. 762–7.

5 Gersaint, *Catalogue raisonné de coquilles, insectes, plantes marines, et autres curiosités naturelles* (Paris, 1736), pp. 41, vi, 46–59; id., *Catalogue d'une collection considérable de Curiositez de differens genres* (Paris, 1737); Yves Laissus, 'Les cabinets d'histoire naturelle', in René Taton (ed.), *Enseignement et diffusion des sciences en France au XVIIIe siècle* (Paris, 1986), pp. 659–712.

6 Krzysztof Pomian, *Collectors and Curiosities. Paris and Venice, 1500–1800*, trans. Elizabeth Wiles-Portier (Cambridge, 1990), chs. 4 and 5; Francis Haskell, *The Painful Birth of the Art Book* (London, 1987).

7 *Journal des Sçavans*, 50 (July-December 1711), pp. 625–31; Emile Hublard, *Le Naturaliste hollandais Pierre Lyonet: sa vie et ses oeuvres (1706–1789) d'après des lettres inédites* (Bruxelles, 1910), chs. 6 and 7; Madeleine Pinault-Sørensen, 'Dezallier d'Argenville, l'*Encyclopédie* et la *Conchyliologie*', *Recherches sur Diderot et sur l'*Encyclopédie, 24 (1998): 101–48, esp. p. 104.

8 Anne Goldgar, 'The absolutism of taste: journalists as censors in eighteenth-century Paris', in Robin Myers and Michael Harris (eds.), *Censorship and the Control of Print in England and France, 1600–1910* (Winchester, 1992), pp. 87–110; A. Laign, 'French ornament engravings and the diffusion of the Rococo', in H. Zerner (ed.), *Le Stampe e la diffusione delle imagini e degli stili, Atti del XXIV Congresso Internazionale di Storia dell'Arte, 1979* (Bologna, 1983), vol. VIII, pp. 109–27.

9 Michel Faré, *La Nature morte en France. Son histoire et son évolution du XVIIe au XXe siècle*, 2 vols. (Genève, 1962); vol. I, part 2, ch. 1; Madeleine Pinault, *The Painter as Naturalist: From Dürer to Redouté* (Paris, 1991); Peter Thornton, *Baroque and Rococo Silks* (London, 1965), p. 25; Katie Scott, *The Rococo Interior: Decoration and Social*

Spaces in Early Eighteenth-Century Paris (New Haven/London, 1995), ch. 10; Giles Barber, 'The Parisian fine-binding trade in the last century of the *ancien régime*', in Robert Fox and Anthony Turner (eds.), *Luxury Trades and Consumerism in* Ancien Régime *Paris: Studies in the History of the Skilled Workforce* (Aldershot, 1998), pp. 43–62.

10 Filippo Buonanni, *Ricreatione dell'Occhio e della Mente, nell Osseruation' delle Chiocciole, Proposta a' Curiosi delle Opere della Natura* (Roma, 1681).

11 D'Argenville, *L'Histoire naturelle*, p. 196; Faré, *La Nature morte*, vol. I, p. 133.

12 D'Argenville, *L'Histoire naturelle*, pp. 117, 185–6.

13 Ibid., p. 190.

14 Louis de Jaucourt, 'Ouvrages de l'art et de la nature', *Encyclopédie*, vol. XI, pp. 722–724, p. 723. See, in particular, Lorraine Daston, 'Nature by design', in Caroline Jones and Peter Galison (eds.), *Picturing Science Producing Art* (New York, 1998), pp. 232–53.

15 Adanson, *Histoire naturelle du Sénégal. Coquillages* (Paris, 1757), p. v; compare Gersaint, *Catalogue* (1736), p. 9; d'Argenville, *L'Histoire naturelle*, p. 110; Georg Wolfgang Knorr, *Les Délices des yeux et de l'esprit, ou collection générale des différentes espèces de coquillages*, 2 vols. (Nürnberg, 1764–71), vol. I, part 1, p. 6.

16 Adanson, *Histoire naturelle*, pp. x, xij. The review in *Journal de Trévoux*, January 1758, vol. I, pp. 119–32, censured Adanson for his dismissive attitude towards his predecessors.

17 D'Argenville, *L'Histoire naturelle*, pp. 34–5, and review in *Journal de Trévoux*, February 1743, pp. 322–60, March 1743, pp. 423–66; review of *La Conchyliologie* in *Journal de Trévoux*, March 1758, pp. 702–22. Compare Antoine Schnapper, *Le Géant, la licorne et la tulipe: collections et collectionneurs dans la France du XVIIe siècle* (Paris, 1988), conclusion, with Giuseppe Olmi, 'From the marvellous to the commonplace: notes on natural history museums (16th–18th centuries)', in Renato Mazzolini (ed.), *Non-Verbal Communication in Science Prior to 1900* (Florence, 1993), pp. 235–78, esp. pp. 272–3.

18 Jean-Paul Nicolas, 'Adanson, the man', in George H. M. Lawrence (ed.), *Adanson: The Bicentennial of Michel Adanson's 'Familles des plantes'*, 2 vols. (Pittsburgh, 1963–4), vol. I, pp. 1–122, esp. pp. 55, 71.

19 Emanuel Mendes da Costa, *Elements of Conchology: Or, an Introduction to the Knowledge of Shells* (London, 1776), p. 19.

20 Christ-Elisée Favart d'Herbigny, *Dictionnaire d'Histoire Naturelle, qui concerne les Testacées ou les Coquillages de Mer, de Terre & d'Eau-douce*, 2 vols. (Paris, 1775), vol. I, p. xlvij.

21 Adanson, *Histoire naturelle*, pp. xxv-xxvj; E. Fischer-Piette, 'Les Mollusques d'Adanson', *Journal de Conchyliologie*, 85 (1942): 103–377; E. F. Gersaint, *Catalogue raisonné d'une Collection considerable de diverses curiosités en tous Genres, contenuës dans les Cabinets de feu Monsieur Bonnier de la Mosson* (Paris, 1744), p. 173.

22 *The Oxford English Dictionary*. 2nd edn, 20 vols. (Oxford, 1989), vol. XIV, s.v.v. 'Rococo', p. 27; Owen E. Holloway, *French Rococo Book Illustration* (London, 1969).

23 Christopher Hill, 'The cabinet of Bonnier de la Mosson (1702–1744)', *Annals of Science*, 43 (1986): 147–74; M. Roland Michel, 'Le Cabinet de Bonnier de la Mosson, et la participation de Lajoue à son décor', *Bulletin de la Société de l'Histoire de l'Art Français*, 1976 (1975): 211–21.
24 Scott, *Rococo Interior*, pp. 171–2.

Further reading

S. Peter Dance, *A History of Shell Collecting* (Leiden, 1986)
Lorraine Daston and Katherine Park, *Wonders and the Order of Nature 1150–1750* (New York, 1998)
Christopher Hill, 'The cabinet of Bonnier de la Mosson (1702–1744)', *Annals of Science*, 43 (1986): 147–74
Oliver Impey and Arthur MacGregor (eds.), *The Origin of Museums: The Cabinet of Curiosities in Sixteenth- and Seventeenth-Century Europe* (Oxford, 1985)
A. Laign, 'French ornament engravings and the diffusion of the rococo', in H. Zerner (ed.), *Le Stampe e la diffusione delle imagini e degli stili, Atti del XXIV Congresso Internazionale di Storia dell'Arte, 1979* (Bologna, 1983), vol. VIII, pp. 109–27
Giuseppe Olmi, 'From the marvellous to the commonplace: notes on natural history museums (16th–18th centuries)', in Renato Mazzolini (ed.), *Non-Verbal Communication in Science Prior to 1900* (Florence, 1993), pp. 235–78
Madeleine Pinault, *The Painter as Naturalist: From Dürer to Redouté* (Paris, 1991)
Krzysztof Pomian, *Collectors and Curiosities. Paris and Venice, 1500–1800*, trans. Elizabeth Wiles-Portier (Cambridge, 1990)
Katie Scott, *The Rococo Interior: Decoration and Social Spaces in Early Eighteenth-Century Paris* (New Haven/London, 1995)

15 Young readers and the sciences

Charles Sneyd Edgeworth was about ten years old when he was first 'delighted with the four volumes' shown in Figure 15.1.[1] They made a lasting impression, for a decade later his sister recorded that Sneyd 'has just finished a poem called the "Transmigrations of Indur"' based on a story in the same book.[2] *Evenings at Home* was typical of the 'instructive and amusing' genre which dominated children's literature between about 1780 and 1820. The instructive element of these books could be anything from geography to Scripture history, but it frequently involved the sciences.

Books written specifically for children were one of the new commodities of the eighteenth century, with their development usually attributed to John Newbery (1713–67) in the 1740s. Although Newbery's early books had much in common with the older traditions of chapbooks and ballads, he was also the publisher of one of the earliest children's science books, the pseudonymous Tom Telescope's *Newtonian System of Philosophy, adapted to the capacities of young gentlemen and ladies* (London, 1761). Many of the features which became typical of 'instructive and amusing' books were present in the *Newtonian System*, including moral lessons, the use of the conversational style, and recourse to examples from everyday life.

Alongside lectures, demonstrations and periodicals, books were one way in which the sciences were commercialised and made available to a wider audience in the eighteenth century. That the sciences could appear in children's books demonstrates their widespread social acceptance, since children are usually regarded as the most vulnerable members of society. Yet these books are more than merely indicative of the context in which they were written, for they were explicitly intended to mould the future of society. In addition to underlining the importance of the sciences and industry to the future development of the nation, authors included a moral framework intended to produce upstanding adult members of society.[3] The ubiquity of these moral lessons has led to the children's literature of the 1780s–1820s being referred to as 'moral didacticism', although the lessons varied depending whether the author placed more emphasis on the ideals of the Enlightenment or the serious

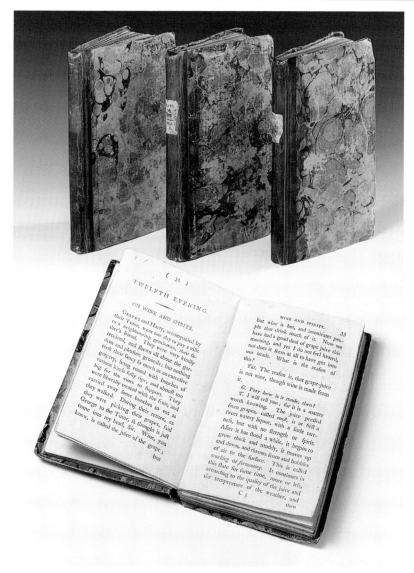

Figure 15.1 The first four volumes (of six) of the 2nd edition of *Evenings at Home* (London, 1794–8). Each volume cost 1s. 6d., was 15 cm tall, and contained about 156 pages. The volumes are in their original binding of marbled boards with paper title-labels on the spines. (By permission of the Syndics of Cambridge University Library)

religion associated with the evangelical revival. One group tended to produce moral systems grounded on reason and logic, while the other based conduct on evangelical religion.

The authors of 'rational' and 'religious' children's books competed in the same literary marketplace, and although the books frequently appealed to parents of similar backgrounds to the authors, there was no firm dividing line between the two audiences. However, with the increasing influence of evangelicalism, and the suspicion attached to Unitarians and political liberals after the French Revolution, 'rational' books came to be associated with radicalism and more overtly religious books benefited from the conservative reaction. The core of this chapter is an examination of two books which represent these 'religious' and 'rational' variants, demonstrating that the professional classes, despite their differences, shared a commitment to educating children in the sciences.

Why write books especially for children?

Literacy was the key to the adult library, but learning to read was seen as a stage to be passed through as swiftly as possible. The development of the new children's books did not greatly change this, as very few of them were written specifically for instruction in literacy.[4] The *Lessons for Children* (London, 1778), by Unitarian author Anna Barbauld (1743–1825), was one exception. Many children, especially those from the lower ranks of society, continued to learn to read from hornbooks, and latterly from the texts prepared for the Sunday and day school movements. Literacy skills could then be practised on the Bible, chapbooks, tracts, broadsides, or whatever printed material parents had available.

Children's books supplemented that material for those who could afford them, for although children's books were cheaper than books for adults, with a typical volume costing 2*s*. rather than 10*s*., they were far from universally accessible.[5] The first generation of children's books provided little more than amusing stories for practising reading skills, but by the 1780s, children's authors were aiming to educate as well as amuse. As the Quaker author Priscilla Wakefield (1751–1832) put it, 'Nonsense has given way to reason; and useful knowledge, under an agreeable form, has usurped the place of the Histories of Tom Thumb, and Woglog the Giant.'[6] The 'agreeable form' usually meant setting the instruction in a narrative framework, such as letters or conversations between an adult and some children. The child reader of such narratives was supposed to learn through identification with his fictional counterpart, and the characters' story provided some variety from the kernel of instructional material.

For about forty years, morally didactic works were seen as the ideal form for children's books. Both of my examples were classic works of this type, written by well-respected authors and frequently reprinted. *An Easy Introduction to the Knowledge of Nature and the Holy Scripture* (London [1780], 8th edition 1793) was written by Sarah Trimmer (1741–1810), the wife of a prosperous brick-maker from Ealing, who had twelve children and a deep concern for children's books and school provision. Trimmer was the daughter of a former royal tutor, and was received at the court of George III and Queen Charlotte on several occasions. Although she did not move in evangelical circles, Trimmer was a staunch member of the Church of England and took her religion seriously. Her book is

Figure 15.2 This image of Mother and Charlotte examining something Henry has found was used as the frontispiece for the first edition of Trimmer's *An Easy Introduction to the Knowledge of Nature and the Holy Scripture* (London, 1780, British Library shelfmark 12835.aa.115). Significantly, this takes place out of doors, as does most of the action in the narrative, and implicitly urges the readers to look at nature themselves. (By permission of The British Library)

written as a series of conversations between a mother and her children, Henry and Charlotte. As they walk together in the countryside around their home (Figure 15.2), mother points out animals, flowers, trees and other natural objects.

John Aikin (1747–1822) and Anna Barbauld's *Evenings at Home; or the Juvenile Budget opened* (London, 1792–6) was written by a Unitarian brother and sister whose father taught with the young chemist and minister Joseph Priestley at the Nonconformist Warrington Academy.[7] Aikin later taught there himself, and practised as a physician, while his sister lived at Warrington until her marriage to a Unitarian minister and schoolmaster. Their book contains a mixture of short stories, poems and conversations. Most of the scientific discussions take place between George, Harry and their tutor, or between Lucy and her father. As with *An Easy Introduction*, the conversation is typically set out of doors, and focuses on natural objects which the characters discover during their walks.

What do the books look like?

From the late nineteenth century onwards, children's books have tended to feature large print, lots of pictures and attractive bindings, and Figure 15.3 shows three editions of *Evenings at Home* which benefited from these innovations. The contrast with the original books enjoyed by Sneyd Edgeworth (Figure 15.1) is stark. Early children's books rarely had pictures, were printed in a small typeface to save paper and were bound in pasteboards with paper labels stuck on the spines. They also tended to be small duodecimo volumes, which were thought to be more 'child-sized' than octavos.

Pictures were expensive in the late eighteenth century. Although woodcuts were cheaper than engravings, and were used in some children's books, they still tended to be relatively crude. Engravings could show more detail, but were usually limited to a frontispiece (such as that used as Figure 15.2), or to plates of technical diagrams in works for older children or adults. By the time the books in Figure 15.3 were published in the late 1870s, better and cheaper methods of wood engraving had been introduced, metal engraving had become cheaper, and colour printing had been invented, all of which transformed the appearance of children's books.

The decorative (and colourful) bindings of the late-nineteenth-century books are particularly striking in comparison with their earlier counterparts. The development of cloth case-bindings and the partial mechanisation of the binding process in the 1830s and 1840s completely changed the outward appearance of books. Publishers' bindings became common, and those for children's books were soon decorated with elaborate stamped patterns, often in several colours, and embossed with gold, as seen in Figure 15.3. Previously, books had been issued bound in boards as a temporary measure until the purchaser arranged for a leather binding. In the case of children's books, the wear and tear to which they would be exposed may have meant that they never received expensive bindings, and the patterned boards seen in Figure 15.1 indicate an attempt to provide a more attractive binding.

The absence of most of these decorative arts in early children's books meant that the text was almost solely responsible for keeping the child's attention. As Maria Edgeworth (1767–1849) and her

Figure 15.3 Three late-nineteenth-century editions of *Evenings at Home*. At the front is a small book from Warne & Co. (1879, 13cm); back left is a reward book from Routledge & Co. (1875, 17cm); back right is a reward book from Ward, Lock & Co. (1879). Note the elaborate bindings, and the use of illustrations. (By permission of the Syndics of Cambridge University Library)

father Richard noted in their *Essays on Practical Education* (London [1797], new edition 1815): 'To fix the attention of children, or, in other words, to interest them about those subjects to which we wish them to apply, must be our first object in the early cultivation of the understanding' (I, 73). Thus authors made the effort to be amusing in order to be more effectively instructive. The characters and events of the fictionalised story were intended to prevent boredom. *Evenings at Home* went one step further than most by using lots of short stories on different subjects and in different genres. The short length was adapted to the child's attention span, while the ever-changing subjects stimulated curiosity.

Why learn about the sciences?

Evenings at Home and *An Easy Introduction* were not unusual in their focus on the sciences, and the variety of authors from different political and religious backgrounds who wrote on the sciences highlights their importance. Unitarian works included Thomas Percival's *A Father's Instructions* (London, 1775–9) and Jeremiah Joyce's *Scientific Dialogues* (London, 1807), the latter of which claimed to be a sequel to *Evenings at Home*. Richard and Maria Edgeworth wrote a series of conversations on the sciences in *Harry and Lucy* (London, 1801), while the Quaker Priscilla Wakefield wrote epistolary works on botany and entomology, as well as her *Mental Improvement* (London, 1794–7). Jane Marcet, whose *Conversations on Chemistry* (London, 1806) were read by the young Michael Faraday, was a Swiss Protestant married to a Huguenot physician.

These authors justified their focus on the sciences in some or all of the following ways: as an aid to religion; as subjects of some practical utility; or as a source of rational recreation. Most authors made at least some mention of the importance of studying nature as God's creation. In the case of *Evenings at Home*, this was limited to a very few references to 'the Creator', while *An Easy Introduction* was imbued throughout with a sense of wonder at the Creation and its God. This is not because Aikin and Barbauld were less religious than Trimmer, but because they believed that religion was a private matter that should not be forced on other people. They agreed with the Edgeworths, who introduced *Practical Education* saying, 'We have no ambition to gain partizans, or to make proselytes, and . . . we do not address ourselves exclusively to any sect or to any party' (I, vi). As Nonconformists, Aikin and Barbauld were fully aware of the potential divisiveness of religion, and chose to avoid the issue.

Trimmer completely disagreed with such an approach, for she was convinced that religion could not possibly be omitted from the education of young children without endangering their chances of salvation. She reviewed *Evenings at Home* in the *Guardian of Education* in 1803, and was mercilessly critical of its failure to '*lead from Nature up to Nature's God*' (306). Trimmer's own *An Easy Introduction* was intended to be 'a kind of general survey of the works of Providence . . . as a means to open the mind by gradual steps to the knowledge of the SUPREME BEING' (vii). While most of the book urges an enthusiasm for nature – 'Look at those tulips! Examine those carnations! Observe that bed of ranunculas! And then admire that stage of auriculas!' (36) – towards the end there is an explicit statement of the argument from design: 'It is evident from the construction of every part of Nature, from the noblest to the most insignificant, that they are all admirably formed; they must all therefore have been the work of some wise, powerful BEING, infinitely our superior' (160). The work culminates with discussions of the soul, the character of God, and the Bible, and for Trimmer, this justified the inclusion of so much scientific material.[8] She made clear in her review that early childhood was 'the most important [time] for the purpose of *Religion* of any in the whole course of human life' (305).

Another way of justifying the inclusion of the sciences was to stress their practical applications. While Trimmer argued that demonstrating the existence and benevolence of God was a form of utility,[9] *Evenings at Home* was filled with the medical, culinary and industrial uses of natural objects. Parts of the oak tree, for example, were used in shipbuilding, textile dyeing, and tanning. This practical approach to the natural world was shared by many Nonconformists, especially those from the industrial north of England.[10] Even for those children who would not directly use the sciences in later life, learning about them could be useful in a more indirect way, since no one wanted to make 'foolish mistakes' when geography, astronomy or botany arose in conversation (I, 92).

The order and method of the sciences could also make them useful as logical training for the mind. *Evenings at Home* contains separate conversations on each of the major families of plants, minerals and metals, in which the characteristics and uses of each member of the group are dealt with in turn. This contrasts with *An Easy Introduction*, where a conversation on the animal kingdom can move from dogs to deer, to cats, to wild beasts, to elephants, to camels, and back to chickens in the course of a pre-breakfast walk (54–66). Trimmer's subject order fulfilled her aim of creating a sense of awe at the Creation, whereas Aikin and Barbauld needed a systematic presentation for training young minds in logical methods of ordering and thinking.

That the sciences provided rational recreation was implicit in most books. After all, the number of children who were going to become physicians or men of science was relatively small, but women and non-professional men would have leisure time, and sciences such as botany, entomology or astronomy, which could be practised with a minimum of expensive equipment or education, were one way of using that time. Many parents preferred to encourage their children to study the sciences rather than to take up gambling, drinking or novel-reading.[11] That the sciences often encouraged outdoor activity, yet at the same time were part of the morally virtuous study of the Creation, added to their attractions for both children and parents.

Are the sciences suitable for children?

In one of the dialogues in *Evenings at Home*, a mother tells her daughter, 'all things are not equally necessary to everyone; . . . some things that are very fit for one, are scarcely proper at all for others'. She continues, 'It is the purpose of all education to fit persons for the station in which they are hereafter to live' (I, 86). Other conversations in the book add 'profession' and 'gender' to the list of features to which education should be adapted. Girls might learn arithmetic so they could keep accounts, whereas boys might do so on the way to more advanced mathematics. Botany might be useful to medical practitioners or those keeping house in the country, but to most other readers, male or female, it was likely to be a form of recreation.

The profession of the young boys who were expected to be reading *Evenings at Home* was probably still in doubt, so they were to receive a broad and varied education in preparation for later specialisation. Similarly, although for 'the woman who intends to become a wife', professional options were restricted to 'the arts of housewifery', girls should also have a broad basic education.[12] The main effect of these different professional prospects in *Evenings at Home* concerns the method of learning. Girls are taught the same variety of subjects as boys, but since most of the material could never be used professionally, they learn informally from their father. In contrast, the boys learn from a tutor. The frontispiece from a Nonconformist encyclopaedia (Figure 15.4) depicts a similar attitude to gender differences: father and sons examine a book together, while mother and daughter carry out domestic tasks, but this all happens in the same space, and the family could be sharing a conversation. The gender division is more marked in *An Easy Introduction*, where Henry is promised the opportunity to learn more, but his sister is not to enter too deeply into subjects like astronomy. The complete absence of their father from the book emphasises the difference Trimmer saw between men's and women's roles.

How much science should be presented to young children was vigorously debated. In *Practical Education*, the Edgeworths wrote, 'We have found, from experience, that an early knowledge of the first principles of science may be given in conversation, and may be insensibly acquired from the usual incidents of life' (I, v). In contrast, Trimmer maintained in her preface to *An Easy Introduction* that 'delightful as these things are to children, if communicated in a way that is accommodated to their capacities, they can never be brought in their early years to attend to scientific accounts of causes and effects, or to enter far into each particular branch of knowledge' (vi). Thus, while Aikin, Barbauld and the Edgeworths went into enthusiastic detail, Trimmer explicitly limited her scientific

Figure 15.4 The gathered family, as depicted in the frontispiece by George Cruikshank to James Jennings' *Family Cyclopaedia* (London, 1821). Jennings estimated that a family with an income of £400 per year could spend £10 on schooling and a further £11 16s on 'pocket expenses for all the family' (vol. I, p. 449). Despite these financial restrictions, the family illustrated have managed to amass a good collection of books, as well as a globe and a wall-map. (By permission of the Syndics of Cambridge University Library)

content, telling Henry, 'till you are older you cannot understand much about the stars' (142).

If there was disagreement about how *much* science children could understand, *which* sciences they should be taught was also a matter for debate. Natural histories of animals and plants were the most common topics for children's science books.[13] Many authors recommended natural history as an ideal introduction to religion. It could also suggest opportunities for instructive conversations, since natural objects were often encountered in everyday life, and children were easily 'charmed with the frolicksome motions of animals, the fine forms and beautiful colours of vegetables, [and] the appearance of the sky and of the ocean'.[14]

Subjects other than natural history were more controversial. Trimmer believed astronomy was too complicated for young children, and felt that it could demonstrate only the magnificence of the heavens. In *Evenings at Home*, however, a father teaches his daughter about gravity and planetary motion, and Trimmer commented 'Too philosophical for children, who are supposed to be learning the principles of religion' (345). She also had qualms about geology. As Wakefield's *Mental Improvement* showed, geology and Scripture could be reconciled, as when the discovery of fossils on top of mountains is described as 'a convincing proof of the truth of the history of the deluge' (49). But such reconciliations had to be made explicitly, or Trimmer feared that geology would appear to contradict the Bible.

The role to be played by religion was the cause of the most significant differences between Trimmer and Aikin and Barbauld. Trimmer believed that religion ought to take first place in any educational scheme, and she also did not approve of the rational approach to religion taken by Unitarians. In *Evenings at Home*, Aikin and Barbauld were not trying to introduce children to God as much as to the sciences. Such an education would equip children to approach everything in life, including religion, rationally.

How do you read a children's book?

When first published, *Evenings at Home* and *An Easy Introduction* were expected to be read aloud by children of about seven to ten years, in the family circle with parents and other siblings. We now tend to think of reading aloud as a stage young children pass through before progressing to silent reading, and we forget the popularity of reading aloud socially in the eighteenth century. From his study of the diaries of Anna Larpent (1758–c. 1828), John Brewer described the role of reading aloud in the Larpent household. Not only was it part of Anna's lessons with her children, but it was central to her family life: 'In October 1792, for example, the Larpents were reading Joseph Priestley on *The Origin of Government*, "rather to lead to conversation & observation than as a followed reading". In a lighter vein, novels and plays were read and discussed "in the family circle" of children and servants.'[15] There were several reasons for recommending reading aloud for children, the most significant being its ability to facilitate discussion.

For Trimmer, discussion enabled parents to ensure that their children had understood the story and derived the correct moral. When reviewing *Evenings at Home*, she worried lest the absence of theological material might lead children to regard the sciences as dangerously atheistic. A 'judicious parent, or teacher' (353) would wish to control the interpretation of the text, adding a commentary when necessary and making sure their children did not pick up dangerous ideas.

Aikin and Barbauld, and their Warrington acquaintance, the physician and author Thomas Percival (1740–1804), agreed with Trimmer that parents should 'explain the terms, point out the analogies, and enforce the reflections'. However, once 'the words, the subject, and the moral are clearly understood, [the child's] curiosity concerning whatever may be connected with, or suggested by them, should be gratified and encouraged'.[16] Discussion was not just about ensuring understanding, but also encouraging curiosity, and in this way it offered opportunities for further education and self-instruction.

Like books, objects in the natural world could act as a focus for discussion, and the widespread use of conversations about nature in instructive works promoted this as a family activity. Thus, although the family in Figure 15.4 is portrayed discussing books indoors, the countryside through their window offers more opportunities for conversation – as the family in Figure 15.2 demonstrates. *Evenings at Home* is full of examples of children asking about something they can see, but do not recognise or understand. The mentor encourages them to think about what they already know, and relate that to the new circumstance. In the conversation 'On Earths and Stones' (V, 1–35), George and Harry are walking with their tutor:

Harry   I wonder what all this heap of stones is for.
George I can tell you – it is for the lime-kiln; don't you see it just by?

Their tutor begins to explain the uses of lime and the reactions of acids and bases, and refers to domestic examples. Later, he mentions that sea shells are chalky. George connects these facts and is enthused with a passion for experiment – until he remembers past experience:

George I will pour some vinegar upon an oyster shell as soon as I get home. But now I think of it, I have often done so in eating oysters, and I never observed it to hiss or bubble.

The tutor then explains that acids have different strengths, and that vinegar will not react with oyster shells. Thus, a conversation which started with curiosity about a heap of stones is turned into a lesson on acids and bases, in which George shows that he can connect facts, and make use of his own experiences. As Richard Edgeworth explained in *Practical Education*, while some children have 'learned only how to *talk* – we wish to teach our pupils to *think*' (II, 131).

As the catechisms discussed by Eugenia Roldán-Vera in chapter 18 of this volume also demonstrate, fictional conversations could take many forms. Trimmer's *An Easy Introduction* appears to consist of conversations, but there is no real dialogue. Henry and Charlotte are walking with their mother, when she asks:

> Do you not smell something very sweet? Look about in the hedges, Henry, and try if you can discover what it is. See, Charlotte, what a fine parcel of woodbines he has got; they are quite delightful: but notice the woodbine is very different from the oak. (22–3)

Superficially, this conversation uses a child's curiosity to proceed to a lesson on botanical identification, but neither Henry nor Charlotte has a voice, and their mother dictates what they should investigate, just as Trimmer herself wished to decide how her own children should interpret their reading. The form of the conversations about nature *in* the book reflects the expectations of discussion *about* the book.

And the moral is...?

The appeal of new ideals of childhood associated with Romanticism encouraged the development of imaginative fiction, and by the 1820s, it was beginning to replace 'instructive and amusing' works as the dominant genre of children's literature. By mid-century, evangelical writers had discovered how to turn fiction to religious ends, and began to write moral fiction rather than moral didacticism. Instruction on subjects like the sciences was then more likely to appear in books for the schoolroom rather than the home. The prominence of the sciences in 'instructive and amusing' children's books was thus a short-lived, though characteristic, phase.

Evenings at Home exemplifies familiar connections between Nonconformity, industry and the sciences.[17] But studies of commercialisation, and science in the public sphere make it clear that far more people than just liberal Nonconformists were reading and conversing about the sciences. The existence of a market for children's books on the sciences is a clear indication of this, as is the fact that such books were written by authors from varied backgrounds. Sarah Trimmer thought Unitarianism was 'so erroneous', but she still believed that young children should learn about the sciences.[18]

Beyond agreement that the sciences should be taught to young children, authors differed over what material was appropriate, and how it should be presented. The sciences could be part of a theology of nature, sources of conversation and rational amusement, practical preparation for a profession, logical training for both sexes, or any combination of these. The role to be played by religion was at the root of the differences between authors, for while Unitarians minimised references to a potentially divisive subject, and stressed instead the importance of reason and experience, evangelicals defended the centrality of religion. Their concern that books and natural objects should be read in the light of Scripture led evangelicals to recommend a much stronger parental control of discussion than is to be found in the Unitarian authors.

Children's authors saw their combination of the sciences with moral or religious lessons as an attempt to improve the future prospects of society. The sciences were a part of this programme for social improvement no matter whether it was based on a predominantly rational or religious foundation. These ambitions for the future were expressed explicitly in the epilogue to *Evenings at Home*, which hoped that Britain would, with the next generation:

> Commence with them a *better* age!
> An age of light and joy, which we,
> Alas! In promise only see. (VI, 152)

Notes

I would like to thank Jim Secord, Jon Topham and the Cambridge Historiography Group for all our instructive conversations about children's books.

1 M. Edgeworth to a friend, 1796, quoted in F. V. Barry, *A Century of Children's Books* (London, 1922), pp. 150–1.
2 Quoted in A. L. Le Breton (ed.), *Memoir of Mrs. Barbauld, including letters and notices of her family and friends* (London, 1874), p. 115.
3 See S. F. Pickering, *Moral Instruction and Fiction for Children, 1749–1820* (Athens, Georgia, 1993).
4 D. Vincent, *Literacy and Popular Culture: England 1750–1914* (Cambridge, 1989).
5 Like other books, children's books became more widely and cheaply available in the late nineteenth century, particularly due to the growth of the trade in reprints from the 1840s.
6 P. Wakefield, *An Introduction to the Natural History and Classification of Insects, in a Series of Familiar Letters* (London, 1816), p. iii.
7 On Nonconformist circles, see D. L. Wykes, 'The contribution of the Dissenting academy to the emergence of Rational Dissent', in K. Haakonssen (ed.), *Enlightenment and Religion: Rational Dissent in Eighteenth-Century Britain* (Cambridge, 1996), pp. 99–139; A. Thackray, 'Natural knowledge in cultural context: the Manchester model', *American Historical Review*, 79 (1974): 672–709; and R. E. Schofield, *The Lunar Society of Birmingham: A Social History of Provincial Science and Industry in Eighteenth-Century England* (Oxford, 1963).
8 On natural theology, see J. H. Brooke, *Science and Religion: Some Historical Perspectives* (Cambridge, 1991), ch. 6.
9 S. Trimmer, *An Easy Introduction to the Knowledge of Nature, and the Holy Scriptures* [1780], 11th edn (London, 1802), p. 5.
10 I. Kramnick, 'Children's literature and bourgeois ideology: observations on culture and industrial capitalism in the later eighteenth

century', in P. Zagorin (ed.), *Culture and Politics from Puritanism to the Enlightenment* (London, 1980), pp. 203–40.
11 Thackray, 'Natural knowledge', p. 692.
12 J. Aikin, *Letters from a Father to his Son, on Various Topics, Relative to Literature and the Conduct of Life*, 3rd edn., 2 vols. (London, 1796–1800), vol. I, p. 338.
13 See H. Ritvo, 'Learning from animals: natural history for children in the eighteenth and nineteenth centuries', *Children's Literature*, 13 (1985): 76–9.
14 Wakefield, *Insects*, p. iv.
15 J. Brewer, *The Pleasures of the Imagination: English Culture in the Eighteenth Century* (London, 1997), p. 196. Also p. 56.
16 T. Percival, *A Father's Instructions; Consisting of Moral Tales, Fables and Reflections* [1775], 4th edn. (London, 1779), pp. xiii–xiv.
17 Thackray, 'Natural knowledge'; Kramnick, 'Children's literature'.
18 S. Trimmer, *Some Account of the Life and Writings of Mrs. Trimmer* (London, 1814), p. 93.

Further reading

F. J. H. Darton, *Children's Books in England: Five Centuries of Social Life*, 3rd edn. (Cambridge, 1982)

L. Davidoff and C. Hall, *Family Fortunes: Men and Women of the English Middle Class, 1780–1850* (London, 1987)

M. V. Jackson, *Engines of Instruction, Mischief and Magic: Children's Literature in England from its Beginnings to 1839* (Aldershot, 1989)

G. Myers, 'Science for women and children: the dialogue of popular science in the nineteenth century', in J. Christie and S. Shuttleworth (eds.), *Nature Transfigured: Science and Literature, 1700–1900* (Manchester, 1989), pp. 171–200

M. Myers, 'Impeccable governesses, rational dames, and moral mothers: Mary Wollstonecraft and the female tradition in Georgian children's literature', *Children's Literature*, 14 (1986): 31–59

J. H. Plumb, 'The new world of children', in N. McKendrick, J. Brewer and J. H. Plumb (eds.), *The Birth of a Consumer Society: The Commercialization of Eighteenth-Century England* (Bloomington, Indiana, 1982), pp. 286–315

J. A. Secord, 'Newton in the nursery: Tom Telescope and the philosophy of tops and balls, 1761–1838', *History of Science*, 23 (1985): 127–51

A. Shteir, *Cultivating Women, Cultivating Science* (London, 1996)

D. Vincent, *Literacy and Popular Culture: England, 1750–1914* (Cambridge, 1989)

16 The physiology of reading

Reading is perhaps the one practice central to both virtually all intellectual history and virtually all our ways of uncovering that history. Yet for such a powerful practice, reading itself remains curiously elusive. Indeed, we easily fall into the assumption that there is really nothing to be understood. Once acquired, the skill of reading is today practised routinely, and in becoming routine it ceases to be wondrous. Even if that were not so, we would obviously find it hard to imagine investigating the very craft without which no investigation could take place. But in fact such inquiries have been mounted in the past. Readers have long asked how it is that characters on a page yield impressions in the mind, and what happens when they do. They have invoked theories of sensation, perception, imagination and reason to explain these processes. We ourselves are the sometimes unconscious inheritors of their efforts. To speak of 'impressions', for example, is itself to refer to a view of perception with a history stretching back at least to Aristotle. Indeed, it may even be impossible to consider these processes today without invoking the terms of some once-respected doctrine relating the world to the body and the mind. For all that, however, the nature of reading today remains almost as obscure as ever, and as potent.

This provokes two questions. They are rarely articulated explicitly, but their potential scope is very great indeed, and without answering them we leave unexamined a central element in much – perhaps all – historical interpretation. First, what is the long-term history of the reader's changing knowledge of the reading process? And second, what consequences has that changing knowledge had for the character and powers of reading itself? Historians have begun to ask these questions only in the last few years. They are the questions that frame this chapter.

Answers to these questions need themselves to be rigorously historical. That is, they should accept the premise that as knowledge of reading processes has altered over time, so the constitution of reading – not just the contexts of its pursuit, but its very nature – may also have altered. This clearly has implications for how we understand cultural change in general. It means considering the consequences of reading not just in terms of the powers we now ascribe to the practice, but in those of the possibilities and constraints that its historical practitioners themselves recognised. For example, could one still present a forceful argument about the Scientific Revolution, as Pierre Duhem once did, by publishing a three-volume work that was entitled *Leonardo da Vinci: Those Whom He Read and Those Who Read Him*, but that never paused to reflect in any way on what reading actually was for its heroes?[1] The force of the argument would today be regarded as vitiated because the practice placed at its very focus remained, in effect, outside history itself. Historians are now increasingly likely to argue that we cannot properly understand the dynamics of cultural change – how people of different eras acted on the basis of their reading, and why – without appreciating the opportunities and constraints they themselves acknowledged. Historians of science have not been notably fast to adopt this view, but their special need to marry issues of epistemology and practice (most evident in the study of experiment) should attune them to its implications. Since people have drawn repeatedly on physiological knowledge in their efforts to define the propensities of reading, we must give pride of place to this history – a history of processes as well as practices.

Yet a history of sober reasoning is not enough, since reading often appears to transcend the expectations of readers themselves. Indeed, its most remarkable property of all is its power to change the reader in ways both unpredictable and ungovernable. These are the moments of its greatest significance, when it can inspire conversion, or even true novelty – new artistic creations, for example, or new scientific insights. Abraham Cowley provides one example: the apologist for the early Royal Society, reading Spenser in his mother's parlour, claimed to be converted to a life of poetry 'as irremediably as a Child is made an Eunuch'. Johannes Kepler offers another. Reaching the end of Euclid's *Elements*, he pronounced himself 'seized by an unbelievable rapture' as he perceived the role of the regular polyhedra in the cosmos. (Thomas Hobbes, too, underwent a famous conversion experience on reading Euclid.) Then there are the many readers of Rousseau, who spoke of suffocation and ecstasy in the face of his work; and, finally, the nineteenth-century consumers of sensation novels, who felt drawn to them irresistibly by what they called 'a sort of magnetism'.[2] A history of the reading process must account not only for routine reading experiences, but for these moments of transformation, that seemed so extraordinary to their very protagonists.

The inference that such moments were merely miraculous must be resisted. Striking they may have been, but they were not beyond all description; nor are they beyond all historical accounting today. Cowley himself said that his profound delight at encountering Spenser produced 'Affections of Mind' leading him to abandon earlier ambitions and attend university, adopting thereby a contemporary terminology of the passions. Kepler's description likewise owed much to Neoplatonic rhapsodies on the ascent of Jacob's ladder and the contemplation of created forms. Rousseau's readers used the terms of sensibility, Wilkie Collins' those of mesmerism and reflex physiology. The language was in each case fairly specific. These readers attempted to understand and articulate the import of their experiences, and could only use concepts available when they were faced with making the attempt. The real problem lies in explaining how readers grappled with their experiences of radical transformation using the cultural resources to hand, and how their efforts to do so shaped the historical impact of those experiences. The physiology of reading was the common element in all their efforts.

To convey the extent of continuity and change in the physiology of reading, this chapter is framed around two readers hailing from very different periods and contexts. Each provided an account of a transformative experience brought about by reading. In both cases, that experience at least crystallised, and probably produced, representations of knowledge, the self and society. The first individual, George Starkey, was a seventeenth-century alchemist from colonial America. Starkey scraped a livelihood from the credibility of his persona as Eirenaeus Philalethes, most revered of adepts. The second, John Stuart Mill, was a rigorously rational adherent of political economy in early Victorian London, honed by his father's unrelenting educational drive into the formidable logician known today for his contributions to the philosophies of liberalism, emancipation and inductive science. There is no straightforward historical link between the two, and I very much doubt that anyone has ever juxtaposed Starkey and Mill before. Indeed, their viewpoints differed so enormously as to be utterly incommensurable. But this is in large part the point. Starkey and Mill represent extremes of period, context and *mentalité*; yet both tried repeatedly to understand their reading experiences in physiological terms, and those terms reveal striking continuities as well as contrasts. The very difference between them therefore serves to highlight the ubiquity of such attempts, and the unexpected similarities to emphasise lasting trends. In giving voice to the transformative power that they both felt reading to possess, each hinted at the most fundamental elements of literary culture in his time. It may be that it is their similarity as much as their difference that betrays the full importance of the history of reading.

The transmutation of a reader

It could be a tough life selling secrets in the 1640s, and alchemist and Harvard graduate George Starkey had clearly been working too hard at it. Poring over a borrowed copy of Lazarus Zetzner's huge, close-printed and prodigiously difficult collection of chymical texts, the *Theatrum chemicum*, he found himself dozing off. Finally he gave up the struggle and fell fast asleep. Yet his imagination did not flag. It immediately launched Starkey into a vivid dream, set in the very room where he had been trying so hard to concentrate on his reading. He was not alone now. An unseen spirit instructed him to rise. He found himself directed by the mysterious voice to the passage in Zetzner's book that dealt with 'the true Philosophical fire'. He obeyed, and 'turned to the very place'. Then, 'sleeping in the darke yet', he underlined the most salient part of the text, in which the profoundest mystery lay hidden. Waking with a start, he realised immediately that he had been the recipient of a divine hint. Starkey returned to work, convinced of the promise of his labours. He retained a conviction that this vision had helped him to penetrate the secrets of alchemy with success. Keeping the underlined page, years later he showed it to Samuel Hartlib, who recorded his tale and communicated it to his various correspondents. It confirmed the providential favour granted to Starkey in his alchemical quest.[3]

Chymists like Starkey referred commonly to moments of this kind, in which some 'genius' or divine messenger would vouchsafe privileged insights helpful in the quest for the alkahest. The practitioner might either claim the tale to derive from an actual experience, or, more often, relate it as part of some intricate allegory for alchemical processes, its veridical status left for the reader to decipher in ways vividly reconstructed in William Newman's recent book on Starkey. Either way, such stories clearly served a purpose. Alchemists routinely sought to legitimate their knowledge by telling of its being gifted by more or less direct divine infusion. For example, Jan Baptista Van Helmont, perhaps the most influential chymist of the century, collected a whole series of these 'intellectual visions'. The reading of a book was often the stimulus sparking such epistemic infusions. But their being conventional does not mean that Starkey was simply making up his stories, and since they recurred in his private notebooks as well as in his public statements their straightforward fabrication seems unlikely.[4]

Motifs of this kind could grow into remarkably rich and complex narratives. A revealing example arose during Starkey's own figurative description of the alchemical process of transmutation, which he published under the name of Eirenaeus Philalethes in *Ripley Reviv'd* (1678).[5] In the course of this elaborate allegory, Philalethes at one point described encountering a queen-like figure of Nature, who handed him a book called *Philosophy Restored to its Primitive Purity*. He ate the book, at which 'my Understanding [was] so enlightned, that I did fully apprehend all things which I saw and heard'. The allegory corresponded to Starkey's reported dream – and incorporated millennial themes too, the eating of the book recalling a similar passage in the Book of Revelation. 'Then I considered with my self', he continued, 'and behold the Book that I had devoured (like a Charm) had so commanded my Spirits, that I could think of nothing more than the enjoyment of this rare Beauty which I had beheld.' A voice then asked of Philalethes what it was that he desired, and after a conversation he experienced another vision of the Queen (Nature) accompanied by her King. Emboldened, he then advanced into a castle, finding that doors and gates opened to him as he approached. He wandered through its halls. His candle failed to light the thick darkness, since it was 'not homogenial' with the light of its flame; but instead the dark condensed into shadowy figures of birds, creatures and monsters. Eventually he chanced upon a crowd of men, who fled, dazzled and terrified, before his candle's light. They had been reading and commenting aloud on medieval alchemy books, lit by a much paler glow, 'as it were of Fox-fire, or rotten Wood'. Philalethes now put down his own candle, and, struck by sudden giddiness, fell asleep. On awakening he found that the darkness had apparently lifted, to reveal a labyrinth; the thread which he had brought with him to track his path was invisible in the new light. 'I took out of my Pocket a small Book to see if I could read in it', he recorded. It was Jean d'Espargnet's *Enchiridion physicae restitutae* – an authoritative work of chymistry. But he 'could not read one word in it'. Luckily, a decrepit old man appeared, who offered to help. He asked what the book was. On being told, he replied, 'It is a good book . . . He and *Sendivow* are the two best that ever wrote.' The old man peered at the book, and 'read out of it such strange things that I never heard of before'. He began to read aloud 'such Processes that I had never heard of'. Philalethes interrupted him: 'I do remember well the Authors, and what they wrote', he declared, 'but never to my remembrance did I find what you read in them.' The old man urged him to look again himself. With his approbation and assistance, Philalethes did so. Still he could read nothing. Then, thinking that this might be due to the different light, he turned to the abandoned copies of Geber and Rhasis; and suddenly they became clear. He could see that they were complete records of required processes, allowing only for blank spaces where there had been 'Truth . . . couched in few words'. The spiritual voice again returned, to explain that the fleeing readers had been 'such who wrot in *Alchymy* according to the Light of Fancy, and not of Nature'. They saw only

'Phantasticall' things, wrote mysteriously to seduce 'doters', and could not endure true light. Newly perceiving the meanings of these texts and the character of their interpretation, Philalethes could proceed to both a personal redemption and a practical transmutation.[6]

Bizarre though it may seem to modern eyes, Starkey's allegory was fairly routine by the standards of seventeenth-century alchemy. Reading was a pervasive element in such testimonies of inspiration, which were quite typical of alchemical or chymical reports. Sometimes it participated as a trigger to a vision, sometimes as a participant in one; either way, it underpinned the status and content of claims to transmutation. And in many cases it appeared as a peculiarly powerful stimulus to visionary experiences. All this alchemists understood. They comprehended the status and power of reading in terms of a sophisticated account of human nature and the created world. Accepting the central role of the faculty of imagination in their reading practices, they construed that role in a particular way. 'Experience' in general they took to be a form of *sympathy*. This sympathy operated between the imagination and the greater world, represented in images. Reason interpreted those images, but depended on them too. That was why characters from alchemical books sometimes reappeared to their readers in dreams: what they were experiencing was the power of the imagination to guide reason in the absence of stimuli from the material world. This had two notable consequences. First, it encouraged chymists to create a rich succession of illustrated books. Like the 'signatures' left by the Creator in nature, their engravings appealed to the imagination. The intricate iconography of alchemical works here served a distinct purpose, impressing on the imagination figures that it would later revive to facilitate the experience of inspirations (Figure 16.1). Second, moreover, images expressed a simultaneous multiplicity of meanings – something that chymists believed was very hard, if not impossible, to achieve in linear prose. They facilitated the same kind of epistemological cornucopia as existed in the books of Scripture and Nature themselves. The magnificent engravings to be found in the works of Robert Fludd, Michael Maier, Heinrich Khunrath and Starkey were not simply *illustrations*. They were an independent – and sometimes primary – means of embodying and reproducing 'experience' in this specific sense of the term (Figure 16.2). A similar desire to express multiple meanings at once led many to experiment with verse, and even music. Maier actually claimed that poetry had been invented to facilitate the communication of alchemical truths.

This had important implications both for the character of alchemical books and for the act of reading them. As Lauren Kassell argues in chapter 7 such books were not just 'vehicles of alchemical secrets', but 'testaments to the necessity of alchemical revelation'. They were deliberately esoteric, not just to conceal their knowledge from the vulgar, let alone to foster a cult of incomprehensibility for its own sake, but to accomplish a specific end. Their obscurity took a distinct form that allowed for them to be read in a characteristic way, according to a particular understanding of the process of reading itself. Chymists recognised that transmission of skills by texts – any texts – was profoundly problematic. 'From the writings of Philosophers', warned the Hague practitioner Johann Friedrich Helvetius, 'this Art of Arts is most rarely learned'. In fact, a reader could only glean the all-important practical skills from an alchemical book in two circumstances. First, he might be proficient already: the 'true writings of the Philosophers are only understood by the truly adept', as many agreed. Or, second, he might be in training with an existing adept. Apprenticeship was the only means to be introduced to true alchemical reading, for 'no lover of this Art, can find the Art of preparing this Mystery in his whole life, without the Communication of some true Adept Man'. If such a 'father' could be found, however, all might be well, since it was quite possible to grasp the sense of alchemical writ 'by the Manuduction [guidance] of some Adept Philosopher'. The process of apprenticeship consisted of a long series of practical procedures, readings and exercises in personal conduct, which together would culminate in the transformation into a new adept, skilled in the reading techniques of a philosopher as well as his chymical techniques. Robert Fludd typically remarked that it amounted to a 'spirituall *refinement*' – that is, a personal transmutation. This, then, was why truths were obscure in alchemical books. But it also explained the *nature* of their obscurity. In the arcane texts and intricate engravings of chymical volumes, their authors remarked, the truth was 'latent'. It was hidden in works like Sendivogius', 'even as our Tincture of Philosophers is both included, and retruded, in External Minerals, and Metallic Bodies'. But truths like these *must* be 'latent' in order to be extracted in the process of apprenticeship. The process, then, was one of transmutation, to be carried out alongside the personal transmutation of the practitioner (from neophyte to adept) and the material transmutation of the substance (from ore to philosophers' stone). The successful reading of a work like Sendivogius' – for which Starkey was clearly grasping – was the apprentice-adept's first true transmutation.[7]

The idea of reading embraced by Starkey was, then, grounded in a chymist's understanding of the microcosm and macrocosm, and had important consequences for the character of his knowledge and conduct. But in broad terms the dramatic experiences that Starkey and his like recorded were by no means restricted to alchemists.

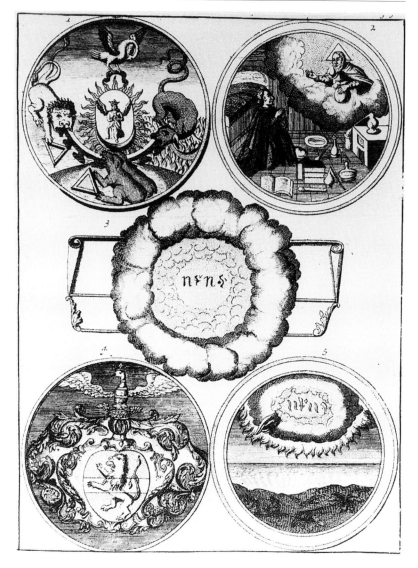

Figure 16.1 A representation of alchemy. The successive images here present a composite portrait of the practice. (1) is a symbol of the philosopher's stone, surrounded by emblems of gold (the lion), antimony (the wolf), and the philosophical mercury (the dragon); (2) portrays the alchemist receiving divine assurance in his work, through reading, fasting and labouring; (3) shows the chaos of Genesis, which is recreated at the heart of an alchemical process; (4) displays a heraldic representation of the philosopher's stone; and (5) shows the restoration of the elements to harmony. Although this illustration comes from a relatively late work, similar images had long appeared in alchemical books of the Renaissance and early modern period. J. C. Barchusen, *Elementa chemiae* (Leyden, 1718), p. 503.

René Descartes' well-known dreams replicated some of their more evident aspects, in particular the appearance of an illustrated book and its reading within the dream itself. Religious radicals, too, frequently appealed to visions experienced while, or shortly after, scouring the Scriptures – and sometimes alchemical works too. Their experiences typically involved a symbolic destruction of the self and its reconstitution (or, metaphorically, resurrection). The 'experimental faith' they professed as a result has been explored in depth by Nigel Smith and Pamela Mack. And the scourge of the enthusiasts himself, Henry More, conceded that he himself had experienced very similar 'visions' as a result of seeing the frontispiece to Ptolemy's *Geography* before nodding off. The sumptu-

Figure 16.2 An alchemist's representation of experience. In Maier's work such representations were accompanied by poetry, prose, and music. The motto for this one reads: 'Make Latona white and tear up your books.' Latona here is the mother of Diana and Apollo, or the Sun and Moon. These in turn stand for the philosophical mercury and sulphur, so Latona is also representing the *materia prima*. The instruction to destroy books was to be taken in at least two ways: first, the alchemist must cast book-learning aside at the crucial stage, because books themselves remain obscure without divine assistance; and second, he must realise that the portrayal here is a hint of the central role of dissolution in alchemical work. The distinct leaves of the book further indicate what was taken to be the layered texture of nature itself. M. Maier, *Atalanta Fugiens* (Oppenheim, 1618). For the interpretation, see H. M. E. de Jong, *Michael Maier's Atalanta Fugiens: Source of an Alchemical Book of Emblems* (Leiden, 1969), pp. 113–19, and S. K. de Rola, *The Golden Game: Alchemical Engravings of the Seventeenth Century* (London, 1988), p. 98.

ously engraved figure of the ancient astronomer, filtering through More's imagination, had reappeared as a celestial sign from God warning of the onset of civil war. Robert Boyle too, in what was a highly conventional lament, bemoaned the power of romances to captivate the mind and lead to 'raving'. Even if a Descartes, a More or a Boyle might not agree with the views of the alchemists, they could claim to understand them. The power of the page seemed to transcend faith, nation and philosophical conviction.[8]

This was because reading drew on sempiternal characteristics of human nature, and in particular on the passions. Passions were composite entities made up of emotion, physiology and morality. They appeared in human beings in response to stimuli that might come either from the outside world or from the imagination. Writers in the early-modern period produced elaborate taxonomies of the passions and their effects, with the aim of producing a workable knowledge of human nature. They dealt with the full range of motives, including curiosity – the ambiguous passion that motivated all inquiry and much sin.[9] In each case, a passion was generated by motions of the animal spirits in the body. It thereby produced identifiable symptoms, because the animal spirits were also the vehicles for communicating sensations and muscular actions. In moderation, those symptoms were merely revealing of the passions prevailing at a given moment. In excess, however, a passion could lead to serious physical, mental and moral impairment by misdirecting animal spirits. Unfortunately, one of the consequences of the expulsion from Eden had been a permanent disturbance in the previously harmonious correspondence between sense, imagination and nature. The result was that one's passions were all too often inappropriate, excessive or both. As Royal Physician Walter Charleton put it, 'it is from the *Imagination* alone that [the soul] takes all the representations of things, and the fundamental *ideas*, upon which she afterward builds up all her *Science*'. Passionate knowledge was therefore often not knowledge at all, but error. Many writers described the human condition as one of civil war between reason and these passionate states. Fortunately, individuals who failed to restrain their passions could have their conditions diagnosed by a skilled physician or minister, and remedies (both medical and moral) could be prescribed. The dominant mode of guidance was that available in conduct books: above all, they taught, the best self-policy was moderation in all things.[10]

This guidance proved useful to critics of chymistry, two of whom in particular deserve to be cited here. One was a sometime tutor to Charles I named Patrick Scott. Scott argued in his *Tillage of Light* (1623) that the figurative language inherited by the alchemists from the ancients – which, they insisted, veiled the profound mystery of transmutation – was in reality *solely* an artefact of the passions. He maintained that what the ancients had veiled was not alchemical profundity at all, but the very *lack* of obscurity of worthwhile moral knowledge. They had hinted at a real elixir only to provide weaker minds, tempted to vice, with some occupying labour to divert them from more harmful pursuits. In truth, the philosophers' stone was merely a 'secret direction of morality ordayned for the better exaltation of the inward man after the modest rules of wisdom'. Scott went on to interpret all the inferred practices of alchemy as internal disciplines for moderating the passions: 'the pounding and mixing of the matter', for example, was really only 'the beating down and qualifying of our affections in the mortar of a wise heart'. The unfortunate moderns who believed these to be allegories of practical transmutations were allowing their passions to run away with them.[11]

The second instance worthy of note here was the later critique of Henry More. More, Meric Casaubon and others used the physiology of the passions to argue that the inspirational epistemology of the alchemists allowed too much scope for religious enthusiasm. Crediting the stories of men like Starkey in a very different way from that intended, they treated their tales not as testimonies, but as symptoms. They represented the enslavement of the Reason to passionate Imagination gone astray. More and Casaubon then advanced the remedial recommendation that reading be made a 'habit' of hard labour, not a subjection to inspirational vision. This talk of habits, too, was a commonplace of the passions literature. It referred to the creation of customary responses to the passions so as to moderate and mitigate their effects. This seems to have been why Casaubon in particular found experimental philosophy so worrying. He believed that the virtuosi neglected the labours of reading, and assumed that collective witnessing, convincingly reported, would be enough to eliminate dangerous enthusiasm. In his eyes, such self-deception stood to undermine the achievements of Protestantism and the personal struggles of all readers against their roiling passions. Why must alchemy be rejected, then? Because it was the prime symptom of immoderate and immodest reading.[12]

Seventeenth-century reading produced all of the benefits of the passions, but all of their dangers too. It could be the basis for enlightenment and reason, but if practised immoderately it could also lead to vice, illness, insanity and even death. It led directly to representations of knowledge, but also to characterisations and condemnations of error. In particular, it was possible to interpret the reading of chymical books in terms of the passions, and both advocates and opponents did so. The accounts provided by Starkey and his contemporaries of their 'visions', whatever status they attributed to their messages, drew on widely distributed notions of 'experience', 'knowledge' and 'imagination'. These concepts were artefacts of the history of medicine, and in particular of physiology. But one did not need to be a scholar or physician to use them. Their consequences were broad indeed. Representations of reading in these terms underpinned rival accounts of knowledge, personal propriety and collective conduct.

Reading and regimen

George Starkey's accounts of transformative reading experiences drew on widely agreed representations of the passions. In attempting to express the inexpressible – and to make clear the grounds and purpose of its inexpressibility – he and other chymical writers employed concepts common to virtually all readers of the time. They revealed what the mid-seventeenth century took to be most self-evident about the act of reading and its effects. It is suggestive that opponents of 'enthusiasm' and of the magical practices that they elided with it also drew upon the same language of the passions in their diagnoses. The accounts of the world and of human nature advanced by the two camps were, of course, distinctly at odds, as were their prescriptions for practice; yet in terms of describing the processes of reading themselves they turned out to be remarkably congruent.

This longevity of *durée* is, it turns out, revealing. The history of the physiology of reading is not quite the history of physiology itself extended into a new realm. Advice traditionally founded in humoral physiology outlasted by far belief in the reality of humours themselves. This was because it was a matter, not of formal medical philosophy, but of the routine daily conduct recommended to individuals, and pursued by them on a quotidian basis. That is, it was an element in what was called *regimen*: the proper conduct of the self so as to enhance virtue and prolong life. Regimen included therapeutics, dietetics and moral conventions. It drew legitimacy from medical theories, but it was not dictated by those theories. Different physiologies and natural philosophies might come and go – Vesalian anatomy might replace Galenic, and Harveian circulation supplant traditional theories of the blood – but this practical advice remained remarkably consistent across countries and centuries. Based in an experiential knowledge of the passions as much as in a theoretical account of their sources, it embraced a few maxims of great simplicity and eminent good sense. A reasonable summary might run as follows. Shun physicians who recommend novel and heroic remedies. Instead, be your own physician; you know yourself better than anyone else does. Learn what foods agree with you, and the manners of living – of exercise, study and sleep – that suit you best, then stick to them. Ration your thinking-time – more than a few hours a day is bad for you. Eat soup (as Descartes told the sick Pascal). Do not eat too many spices, salty foods or purgatives. Drink wine sparingly, but do not abstain altogether. Consume as much as you relish, but no more – as animals do. Manage your passions: you cannot eliminate them, but you can work with them to optimise their effects. Distract the imagination from dwelling on illness, for example, let alone death.[13] In all, obey the central principle of *moderation*, or *balance*. A reasonable amount of wealth, emotion, red meat and wine is to be preferred to both an excess and a total abstinence. This will lead to longevity, virtue and learning. Avoid it at your peril: as Arditi has rendered it, 'the immoderate squanders his estate, makes a poor counselor, and endangers his friends'.[14]

Conduct books frequently elided this kind of advice about medical regimen with cautions about reading. Montaigne remarked that excessive reading enervated the body, and recommended walking around to ease the problem.[15] It was a lasting cliché that the learned read too much for their own good, and from Boyle and Francis Willughby well into the nineteenth century their sickly constitutions were blamed on this formative practice.[16] Francis Osborne thus urged his son not to fall victim to 'an overpassionate prosecution of learning', since 'a few books well studied and thoroughly digested nourish the understanding more than hundreds but gargled in the mouth'. In many cases, as here, advice specifically mirrored that offered about nourishment or respiration. Daniel Tuvill thus concurred, adding that students were often 'like the greedy glutton that swallows much good meat, incorporates none'. Osborne made the link explicit, remarking that 'it is an aphorism in physic that unwholesome airs, because perpetually sucked into the lungs, do distemper health more than coarser diet used but at set times. The same may be said of company, which if good is a better refiner of the spirits than ordinary books'. And he told his reader to 'spend no time in reading, much less writing[,] strong lines, which like tough meat ask more pains and time in the chewing than can be recompensed by all the nourishment they bring'. As part of the bodily system, it made sense that reading should be guided by such advice. In addition, well-tempered reading, as a form of exercise, strengthened the body and honed it for more reading in future. And Osborne in particular gave vent to some more specifically puritan opinions, such as the caution that it was unwise to read while still in bed after waking up in the morning, 'since the head during that laziness is commonly a cage for unclean thoughts'. In all cases, however, inappropriate or excessive reading was 'a symptom of intemperancy'.[17] As late as 1799, a German pedagogical tract incorporated reading into regimen as completely as had Montaigne and Descartes. You should never read while standing, it warned, nor after a meal. Wash your face in cold water and leave the house before starting to read; read only in the bosom of nature; do so aloud, so that the voice will facilitate the penetration of virtuous ideas.[18]

As conduct manuals and guides to gentility instructed, health was a sign of virtue. Being a good person was good for you. It was not that physiological knowledge was immaterial to such prescriptions. Rather, it was put to use in diverse ways, to support – and sometimes to question – this body of almost proverbial expertise. In the early decades of the eighteenth century, for example, John Locke and his successors put the common language of the passions to use to create both an authoritative philosophy of human reasoning and an educational regime based on that philosophy.[19] Lockean educational practice repudiated the visionary fervour of the enthusiasts and the chymists in favour of the inculcation of habits capable of moderating the passions. But they were still habits – that is, normative behaviours adopted to channel the animal spirits in the body and guide the passions – and the passions themselves continued to underpin virtually all attempts to understand reading and its power. As the great critic Dacier put it in explaining the continuing appeal of Aristotelian poetics, 'men are the same now, they were then, they have the same Passions'.[20]

The best-known model of experience and reasoning to provide

an Enlightenment physiology of reading based in the inherited discourse of the passions was David Hartley's *Observations on Man* of 1749. Hartley's work derived not only from Locke's philosophy but, more immediately, from conversations with the playwright John Gay on the powers of association and the fostering of virtue. It sprang directly from discussions of poetic power. Hartley elaborated an extensive account of how impressions (vibrations in the sensory system) combined with the resultant ideas ('vibratiuncles') during the process of education to create a knowledgeable human subject. Reading, he explained, came about by a habitual linking of vibratiuncles to the experience of seeing certain symbols on a page. As in previous cases, the practical strategy of disciplining the passions remained consistent with earlier conduct even while its physiological underpinnings changed. Hartley insisted that all true thinking, and in particular all moral thinking, depended on this process being inculcated correctly. He urged that language itself be purged of excesses on the basis of his theory, to the extent of proposing John Byrom's shorthand as a way of imbuing communication itself with the qualities of industry, candour, ease, certainty and thrift.[21] But the consequences of Hartley's association psychology for representations of reading were less innovatory than one might expect.

The identification of reading as an element in regimen gave rise to a series of recommendations for action. These recommendations were thoroughly conventional, and persisted through quite profound alterations in physiological knowledge, even though their proponents always claimed legitimation in terms of such knowledge. They set greatest store by moderation, above all other practical virtues. One must be sparing both in the extent and in the concentration of one's reading, lest the passions gain the upper hand and wreak havoc on both reason and health. Women, as particularly vulnerable to their passions, came to stand for a reading public collectively prone to whims of sentiment.[22] A lasting language of passions, spirits, imagination and reason pervaded all these representations. Hartley himself insisted on the same message at the climax of his *Observations*. He himself adopted the primacy of moderation, to the extent that even the search for truth should be subject to it. The consequences if it were neglected, Hartley warned, could be frightful, and were exemplified in all too many scientific thinkers:

the Study of Science, without a View to God and our Duty, and from a vain Desire of Applause, will get Possession of our Hearts, engross them wholly, and, by taking deeper Root than the Pursuit of vain Amusements, become in the End a much more dangerous and obstinate Evil . . . Nothing can easily exceed the Vain-glory, Self-conceit, Arrogance, Emulation, and Envy, that are found in the eminent Professors of the

Sciences, Mathematics, Natural Philosophy, and even Divinity itself. Temperance in these Studies is therefore evidently required, both in order to check the rise of such ill Passions, and to give room for the Cultivation of other essential Parts of our Nature. It is with these Pleasures as with the sensible ones; our Appetites must not be made the Measure of our Indulgences.

Reading and self-cultivation

By the end of the eighteenth century there were a hundred years of educational practice – and far more than that of moral advice – relating the passions to the avowal of moderation in reading practices. It was broadly accepted, not just that good citizens learned to read, but that they learned to read in particular ways. The inculcation of such habits was part and parcel of *being* good citizens. In the nineteenth century, as the enterprise of science came into being, this view would be reconstituted into a formal practice known as *ethology*, or 'self-cultivation'. Ethology was the science underwriting educational practice, and was touted by its proponents as the essential component in the making of a civil society. The prime mover in creating this new practice was John Stuart Mill. A transformative reading experience of his own propelled Mill to articulate and advocate his new science.

In the autumn of 1826, Mill found himself in what he later called 'a dull state of nerves'. Long convinced of the virtues of philosophic radicalism, he now found campaigns for political economy and utilitarianism to be enervating and 'insipid'. He likened his state of mind to that in which 'converts to Methodism' found themselves on being 'smitten by their first "conviction of sin"' – the descendant, that is, of the sense of annihilation that seventeenth-century religious radicals had experienced on perusing Scripture, and that was identified by themselves and their opponents with the pre-visionary state of the apprentice alchemist. Suddenly the vocation of his previous life seemed pointless. Even if the grandest aims of utilitarianism were achieved, they would bring Mill no happiness. 'I seemed to have nothing left to live for'. Mill tried all the standard remedies for melancholy – the very remedies recommended for centuries as part of sound regimen. He at first hoped that the spell would dissipate of its own accord, then, when that failed to occur, tried reading. But even his favourite books did no good: 'I read them now without feeling'. No conversation assisted, largely because Mill's father was incapable of understanding his 'mental state', and would have been an improper 'physician' for healing it in any case. It was clear that more dramatic remedies were called for.

Mill fell back on self-diagnosis. He believed implicitly in the association psychology developed by Hartley almost a century earlier, being convinced that 'all mental and moral feelings and qualities' were outcomes of 'association' between ideas generated originally out of sensations. Humans came to approve of some things and disapprove of others because education and experience combined to make pleasure and pain 'cling' to particular ideas. So the paramount aim of education had to be 'to form the strongest possible associations of the salutary class'. Mill now persuaded himself that in his own case, his teacher – that is, his father – had dedicated insufficient attention to the means by which those associations were cemented in place, trusting to traditional instruments of reward and punishment. These had produced only 'artificial and casual' associations, creating no 'natural' tie between object and emotion. To make proper associations of creativity and feeling, then, and *a fortiori* to make those associations durable, such tactics would have had to be pursued with intensity and perseverance. Only then should education have moved on to inculcate the 'habits' of analysis that Mill's father prized. Those habits had the great virtue that they separated emotion and prejudice from ideas, thereby allowing natural connections between things to become clearly perceived, and permitting the calculations of rational utilitarianism. But in destroying prejudicial and artificial connections between ideas, analysis also attacked the emotional bases by which reasoning itself must be actuated. The 'habit' of analysis would inevitably erode these actuating feelings, Mill now concluded, unless counter-habits had already been cultivated as 'correctives' to its desiccating effects. So analysis itself became 'a perpetual worm at the root both of the passions and of the virtues', with only 'purely physical and organic' pleasures proving resilient in the face of its corrosive action. This was what had happened to Mill himself. He was suffering from a human nature bereft of inspiring associations. The habit of analysis had taken over completely. He therefore went about his occupations 'mechanically', constructing arguments and delivering them without 'spirit'. Mill was convinced that his condition was irremediable.

Relief came from reading. In Marmontel's *Mémoires* Mill came across a passage relating the author's reaction to his father's death, and it proved an 'inspiration'. 'A vivid conception of the scene and its feelings came over me', he recalled later, 'and I was moved to tears'. The experience persuaded Mill that he was, after all, capable of associations of this kind. 'I was no longer hopeless: I was not a stock or a stone'. It proved a major turning point. Mill began to emerge from his ordeal. As he did so, he formed a 'theory of life' in explicit contrast to that of his earlier years. This new theory centred on a new attention to 'the internal culture of the individual'. The practice of analysis had to be supplemented, Mill now argued, by efforts to maintain a 'due balance among the faculties'. In effect, what he did was to resurrect the old idea of a regimen of temperance. He began to seek cultural experiences capable of fostering this 'balance', beginning with music – Weber's unwieldy *Oberon* was an early choice – but soon moving to concentrate on books. It was reading that proved most effective in attaining the requisite poise. Mill particularly relished encountering the poetry of Wordsworth, which he found 'a medicine for my state of mind'. Here he discovered a 'culture of the feelings', and a profound source of 'sympathetic and imaginative pleasure'. With a 'culture' of this sort, he declared, 'there was nothing to dread from the most confirmed habit of analysis'. Henceforth it would be the reading of Wordsworth that Mill would credit for curing his 'depression'.

This experience coloured the rest of Mill's life; it would eventually form the crux of his *Autobiography*. In the 1830s, it certainly inspired a series of public reflections on the character of reading, and on what impact the practice had on the constitution of both knowledge and society. Mill retained an association psychology based closely on Hartley's. The practice of imagination, for example, he defined as a 'mental habit' of making associations richly clothed in circumstantial details. It was an aspect of 'organic sensibility' possessed by artists and poets in greater degree than historians. Thinking itself was a process of association between psychological states, that might in time all be recognised as 'produced by states of body'.[23] In his commentary on his father's *Analysis of the Phenomena of the Human Mind* – itself a Hartleian analysis – Mill applied these notions to produce a long account of the processes of sensation and perception involved in reading. Could one properly be said to be conscious of a sensation too fleeting and unattended to have remained in the memory, he asked. Perhaps not. Reading a book provided an excellent example. The reader must in fact have 'perceived and recognized the visible letters and syllables', yet almost always would remember only the sense they conveyed, not the characters themselves. Many psychologists argued from thence that the impressions must have transited the mind without consciousness. But to Mill it seemed almost oxymoronic to speak of having feelings unconsciously. What one really meant was that one did not *remember* having them, and concluded from that that one had not attended to them when they were present. 'This attention for our feelings is what seems to be here meant by being unconscious of them'. In some cases this must have been so, such as in early reading experiences when syllables have to be enunciated explicitly. Each momentary confrontation was then to be considered 'a kind of electric shock', he continued, 'which is communicated to a certain portion, to a certain limited sphere, of intelligence: and the sum of all these circumstances is equal to so many shocks which, given at once at so many different points, produce a general agitation'. But through repetition, 'the process has become so rapid that we no longer remember having those visual sensations'.

> The usual impressions are made on our organs by the written characters, and are transmitted to the brain, but these organic states . . . pass away without having had time to excite the sensations corresponding to them, the chain of association being kept up by the organic states without need of the sensations. This was apparently the opinion of Hartley; and is distinctly that of Mr. Herbert Spencer.

Mill concluded with a remark about the speculative character of all such theorising. But he insisted that it was nonetheless of central importance to 'accounting for the course of human thought'.[24]

Reflecting on the power of reading entailed consequences for portrayals of knowledge, character and society. 'Since the discovery of printing', Mill announced as he emerged from his depression in 1831, 'books are the medium by which the ideas, the mental habits, and the feelings, of the most exalted and enlarged minds are propagated among the inert, unobserving, unmeditative masses'. The representation became something of a *Leitmotiv* in his writings during the 1830s, recurring in book reviews, polemical articles and debates. The press, he argued, underwrote a wise commonwealth. But there was nothing inevitable about the process; it occurred only when proper understandings of creativity and reception existed – understandings based in Hartleian principles and wedded to a sound regimen. Authorship was one resulting theme. Enlarged minds were a cardinal need for society, and one not always met. True opinions were not mere observations of external facts, and Lavoisier's chemistry or Newton's *Principia* could not have been 'evolved' by anyone possessed of the appropriate sensations. The Lavoisiers and Newtons of the age had to be forged. Mill resolved that this made it both appropriate and necessary for good writers to receive state subsidy – to be provided, he thought, in the form of university posts presently 'squandered upon idle monks'.

The character of reception, too, had to be be taken into account. Mill warned the French Saint-Simonians that anyone disposed to write for English readers must do so 'in the manner best calculated to make an impression upon their minds'. His compatriots had an 'extremely *practical* character', which disposed them to oppose whatever they saw as 'systems' of thought. At the same time, he warned that the invention of mass literature led to the 'disease' of subordinating creative individuality to the interests and consuming powers of the mass. 'This is a reading age', Mill remarked, 'and precisely because it is so reading an age, any book which is the result of profound meditation is, perhaps, less likely to be duly and profitably read than at any former period'. Such a book produced no more 'durable impression' in a reader's mind than an ephemeral newspaper article. 'And the public is in the predicament of an indolent man, who cannot bring himself to apply his mind vigorously to his own affairs, and over whom, therefore, not he who speaks most wisely, but he who speaks most frequently, obtains the influence'. As a direct consequence of steam printing, the public 'gorges itself with intellectual food, and in order to swallow the more, *bolts* it'.[25] The echo of traditional regimen advice was clear. And it would persevere. In the 1880s, controversies over public reading habits would revisit the same arguments, with even greater luridness. Appalled Victorians like Alfred Austin would declaim against the 'vice' of 'overfeeding' at the trough of W. H. Smith's as 'a vulgar, detrimental habit, like dram-drinking ... a softening, demoralising, relaxing practice, which, if persisted in, will end by enfeebling the minds of men and women, making flabby the fibre of their bodies, and undermining the vigour of nations'. The language of degeneration, demoralisation and nationalism was all relatively new, but the practical advice, as ever, stayed much the same. Indeed, it remained common to lament the 'dreaming or reverie' into which ill-disciplined reading could pitch the debauched. These were not, it should be explicitly stated, merely metaphorical notions. Improper reading *really did* deter sound physical exercise in the eyes of these writers, and it *really could* throw off the balance of one's mind and body, leading to the danger of thoroughgoing atavism.[26]

Mill's own conclusions extended from the personal to the grandly political. He prescribed reading courses for private correspondents, for example, instructing them to generate a 'habit' of reading every day. This involved 'working' over a set number of pages – 20, 50 or 100 – as early in the morning as possible, and then supplementing this with further reading that would not 'fatigu[e] the mind'. He warned one such student that his course of reading in formal texts 'cannot be done with safety to the health of those who eschew light literature'. Mill constructed a detailed and specific list of works to be read, divided into four 'courses', each of which was to be approached with different levels of intensity and at different times of day.[27] At the public level, he advocated the establishment of state-financed libraries as an essential component in 'national education', dedicated to imbuing not only the *ability* to read, but the *wish* – exactly the emotion that his crisis had taught him to be essential. Education for the proletariat must be 'mental cultivation', such as 'accustoms them to the use of their understandings'. The formation of trustworthy public opinion depended on this, as, eventually, would the chance of sustaining an emancipated populace in a stable state.[28]

Mill's transformation thus gave rise to far-reaching arguments about the making of knowledge and social order. But it found fruit above all in his renowned *System of Logic*. Begun virtually at the moment of his recovery, the *System* is known today for its treatment of induction; but for Mill himself this was perhaps not its major purpose. In fact, the work amounted to a long argument for the importance of what its author called 'self-culture', or 'ethology'. This was the body of knowledge abstracted from his own experience of transformation in the late 1820s. As 'the science which corresponds to the art of education', ethology was to be the *philosophia prima* corresponding to the art of self-formation in the future British Empire. It became Mill's candidate for an 'Exact Science of Human Nature'. From it, a comprehensive social science might also develop. Fittingly, the very last passage of the *System of Logic* returned quietly to Mill's reading experience at the depth of his depression. He had decided at the moment of reading Wordsworth that while the happiness of mankind, identified through a utilitarian calculus, might indeed be the distinctive end of good actions, it should not always be their *explicit* end. In many instances it was better to pursue personal standards of virtue, partly because they provided more inspiring associations, and partly because in fact the pursuit of such standards often led more effectively to the general good. The thought was eminently suited to traditional advice for moderation and balance in the passions. But the wording of the closing paragraphs of the *System of Logic* mirrored precisely that which Mill had used to describe his own transformative experience of reading at this, the moment of his lowest depression and his transformation into recovery. In its conclusion, then, Mill's 'system' reduced to the application of his own insight gained through reading to the fundamentals of knowledge.[29]

Conclusion

Here, then, are two examples of reading experiences, taken from the mid-seventeenth and mid-nineteenth centuries. Their protagonists would seem to be as different as possible. Between them lie 200 years, throughout which readers were being transformed by their books. But in Mill's time they were no longer being transmuted by them. The difference matters, and the significance of the terms in which readers described their experiences is not merely metaphorical. There is a physiology of reading that relates primarily to the mechanisms of perception and to the machinery of the mind. The terms used by these readers to understand their experiences were physiological in origin. In Starkey's day, chymists used the language of the passions and the doctrines of regimen to articulate the inarticulable. In Mill's, a personal life transformed by reading became the template for a new social science and a new educational practice at the moment when science itself was coming into being.

In the early modern period, attempts to account for the power of the page reached back into classical philosophy and medicine. They used the ideas of Aristotle, Galen and Hippocrates to create theories of interpretation as well as perception, based on rival concepts of extramission and intromission, and of species flitting through the air.[30] In the seventeenth century, changes in natural knowledge associated with Cartesian philosophy and new practices of anatomising contributed to reconceptualisations of the passions and hence of the practice of reading. In the Enlightenment, Rousseauvian readers reported being driven to bed, contracting colds, and almost going mad; some declined to read at all, fearing that death itself might result. In England, readers responded with almost equal outbursts to the epistolary novels issued by Hartley's printer, Samuel Richardson. Various physiologies were offered for the much-lauded (and much more lamented) dramatic power of the novel. Lord Kames, for example, spoke of an 'ideal presence' generated by reading, just as if the reader were creating a 'waking dream', or 'reverie', and proceeded to argue that factual history and 'fable' alike commanded one's passions in the same way, thus giving rise to 'sympathy' and the bonds of social order.[31] If we as historians take the sequence of such representations and 'turn the world upside down', what we get is a history of concepts of the body and its relation to the mind and the world. The language of passionate subjection and the practice of personal regimen were common to all of them.[32] As a result, history of science rests on the history of reading, of course, but the debt is reciprocal. The experience and impact of reading itself seems so constant because it has long rested on the pursuit of long life and sound reason.

Notes

I thank Steven Shapin, Andrew Scull, Alison Winter and the editors for helpful suggestions.

1 P. Duhem, *Études sur Léonard de Vinci: ceux qu'il a lus et ceux qui l'ont lu*, 3 vols. (Paris, 1909–13).
2 A. Cowley, *Works*, 11th edn, 3 vols. (London, 1710), vol. I, p. iv, vol. II, p. 782; pp. 143–4; M. Caspar (ed. O. Gingerich), *Kepler* (New York, 1993), p. 273; R. Darnton, *The Great Cat Massacre, and Other Episodes in French Cultural History* (New York, 1984), pp. 242–3; A. Winter, *Mesmerized: Powers of Mind in Victorian Britain* (Chicago, 1998), p. 324; J. Aubrey (ed. O. L. Dick), *Brief Lives* (Ann Arbor, 1957), p. 150.
3 S. Hartlib, 'Ephemerides', A_6–A-B1 (University of Sheffield Library); W. Newman, *Gehennical Fire: The Lives of George Starkey, an American Alchemist in the Scientific Revolution* (Cambridge, Mass., 1994), pp. 40, 64–5.

4 For alchemical epistemologies of infusion, see P. Smith, *The Business of Alchemy: Science and Culture in the Holy Roman Empire* (Princeton, 1994); N. Smith, *Perfection Proclaimed: Language and Literature in English Radical Religion, 1640–1660* (Oxford, 1989); O. Hannaway, *The Chemists and the Word: The Didactic Origins of Chemistry* (Baltimore, 1975).

5 'An Exposition upon the first Six Gates of Sir George Ripley's Compound of *Alchymie*', in E. Philalethes [i.e. G. Starkey], *Ripley Reviv'd* (London, 1678), pp. 103–4; Newman, *Gehennical Fire*, pp. 268–9.

6 Philalethes, *Ripley Reviv'd*, pp. 120–6; Newman, *Gehennical Fire*, pp. 122–3; Rev. 10:9–10.

7 L. Kassell, 'Reading for the philosophers' stone', q.v. ch. 7; J. F. Helvetius, *The Golden Calf* (London, 1670), pp. 67–69, 102–4, 117–18; M. Maier, *Atalanta Fugiens* (Oppenheim, 1618); B. T. Moran, *The Alchemical World of the German Court: Occult Philosophy and Chemical Medicine in the Circle of Moritz of Hessen (1572–1632)* (Stuttgart, 1991), pp. 102–14; Hannaway, *The Chemists and the Word*, p. 61. For Helvetius see also L. Principe, *The Aspiring Adept: Robert Boyle and his Alchemical Quest* (Princeton, 1998), pp. 93–5. There is an excellent reconstruction of a student adept's reading practice in Newman, *Gehennical Fire*, pp. 125–69.

8 S. Gaukroger, *Descartes: An Intellectual Biography* (Oxford, 1995), pp. 107–8; Smith, *Perfection Proclaimed*, pp. 132–3; P. Mack, *Visionary Women* (Berkeley, 1992), pp. 1, 7–10, 23–33, 57–8, 84–5; A. Johns, *The Nature of the Book: Print and Knowledge in the Making* (Chicago, 1998), pp. 419–28; L. Principe, 'Virtuous romance and romantic virtuoso: the shaping of Robert Boyle's literary style', *Journal of the History of Ideas*, 56 (1995): 377–97.

9 L. Daston and K. Park, *Wonders and the Order of Nature 1150–1750* (New York, 1998).

10 [W. Charleton], *The Natural History of the Passions* (London, 1674), sig. A4^v, 48–50, 54–9, 64.

11 P. Scott, *The Tillage of Light* (London, 1623); see also C. H. Josten, 'Truth's golden harrow: an unpublished alchemical treatise of Robert Fludd in the Bodleian Library', *Ambix*, 3 (1949): 91–150.

12 M. Heyd, '*Be Sober and Reasonable*': *The Critique of Enthusiasm in the Seventeenth and Early Eighteenth Centuries* (Leiden, 1995); Johns, *Nature of the Book*, pp. 426–8.

13 N. G. Siraisi, *The Clock and the Mirror: Girolamo Cardano and Renaissance Medicine* (Princeton, 1997), pp. 70–90; R. M. Bell, *How to Do it: Guides to Good Living for Renaissance Italians* (Chicago, 1999); S. Shapin, 'Descartes takes physick' (unpublished paper, 1999).

14 J. Arditi, *A Genealogy of Manners: Transformations of Social Relations in France and England from the Fourteenth to the Eighteenth Century* (Chicago, 1998), p. 175.

15 H.-J. Martin, *The History and Power of Writing* (Chicago, 1994), p. 363.

16 Johns, *Nature of the Book*, pp. 380–4.

17 L. B. Wright (ed.), *Advice to a Son: Precepts of Lord Burghley, Sir Walter*

Raleigh, and Francis Osborne (Ithaca, NY, 1962), pp. 43–6, 52–3; D. Tuvill (ed. J. L. Lievsay), *Essays Politic and Moral, and Essays Moral and Theological* (Charlottesville, VA, 1971), pp. 81, 84; A. Bryson, *From Courtesy to Civility: Changing Codes of Conduct in Early Modern England* (Oxford, 1998), pp. 151–92, esp. p. 179.

18 Cited in Darnton, *Great Cat Massacre*, p. 250.

19 [J. Locke], *Some Thoughts Concerning Education* (London, 1693); I. Green, *The Christian's ABC: Catechisms and Catechizing in England c. 1530–1740* (Oxford, 1996), pp. 233–43.

20 A. Dacier, 'The Preface', in *Aristotle's Art of Poetry* (London, 1705), sig. A7^r.

21 D. Hartley, *Various Conjectures on the Perception, Motion, and Generation of Ideas* (1746), trans. R. E. A. Palmer (Los Angeles, CA, 1959), pp. 53–5; Hartley, *Observations on Man, His Frame, His Duty, and his Expectations*, 2 vols. (London, 1749), vol. I, sig. a3^r, pp. 268–323; II, p. 255.

22 J. Brewer, *The Pleasures of the Imagination: English Culture in the Eighteenth Century* (New York, 1997), p. 193.

23 Mill, 'A system of logic' (1843), in J. M. Robson (general ed.), *Collected Works of John Stuart Mill*, 33 vols. (Toronto, 1981–91), vols. VII–VIII, esp. vol. VII, pp. 480–1, vol. VIII, pp. 849–52, 857–9.

24 Mill (ed.), 'Analysis of the phenomena of the human mind' (by James Mill): *Works*, vol. XXXI, pp. 93–253, esp. pp. 138–40, 192, 194–6; Mill, 'An examination of Sir William Hamilton's philosophy', in *Works*, vol. IX, pp. 257–8. See also Sir W. Hamilton, *Lectures on Metaphysics*, 2 vols. (London, 1870), vol. II, pp. 250–8; J.-J. S. de Cardaillac, *Etudes elémentaires de philosophie*, 2 vols. (Paris, 1830), vol. II, pp. 124–38.

25 J. S. Mill, 'Attack on literature' (1831), in *Works*, vol. XXII, pp. 318–27; Mill, 'Comparison of the tendencies of French and English intellect' (1833), in *Works*, vol. XXIII, pp. 442–7, esp. pp. 445–6; Mill, 'Civilization' (1836), in *Works*, vol. XVIII, pp. 117–47, esp. pp. 133–5; Mill, 'Austin's lectures on jurisprudence' (1832), in *Works*, vol. XXI, pp. 51–60, esp. pp. 54–5; Mill, 'The present state of literature' (1827), *Works*, vol. XXVI, pp. 409–17, esp. pp. 410–13.

26 K. J. Mays, 'The disease of reading and Victorian periodicals', in J. O. Jordan and R. L. Patten (eds.), *Literature in the Marketplace: Nineteenth-Century British Publishing and Reading Practices* (Cambridge, 1995), pp. 165–94.

27 Mill to Florence May, c. November 1868: *Works*, vol. XVI, pp. 1472–5.

28 Mill, 'The education bill (2)', and 'Election to the school boards (2)', in *Works*, vol. XXIX, pp. 391–6 and 398–401.

29 J. S. Mill, 'Autobiography' (1873), in *Works*, vol. I, pp. 3–290, esp. pp. 137–92; Mill, 'A system of logic', in *Works*, vols. VII–VIII, esp. vol. VIII, pp. 861–70, pp. 875–8, pp. 951–2; M. Poovey, *A History of the Modern Fact: Problems of Knowledge in the Sciences of Wealth and Society* (Chicago, 1998), pp. 323–5.

30 A. Manguel, *A History of Reading* (London, 1996), pp. 27–39.

31 Johns, *Nature of the Book*, ch. 6; R. Darnton, 'Readers respond to Rousseau: the fabrication of Romantic sensitivity', in Darnton, *Great Cat Massacre*, pp. 215–56, esp. pp. 242–3; G. J. Barker-Benfield, *The*

Culture of Sensibility: Sex and Society in Eighteenth-Century Britain (Chicago, 1992), ch. 2.
32 Henry Home (Lord Kames), 'Elements of criticism (1762)', in S. Elledge (ed.), *Eighteenth-Century Critical Essays*, 2 vols. (Ithaca, NY, 1961), pp. 838–47.

Further reading

G. J. Barker-Benfield, *The Culture of Sensibility: Sex and Society in Eighteenth-Century Britain* (Chicago, 1992)

J. Brewer, *The Pleasures of the Imagination: English Culture in the Eighteenth Century* (New York, 1997)

R. Darnton, 'History of reading', in P. Burke (ed.), *New Perspectives on Historical Writing* (Cambridge, 1991), pp. 140–67.

A. Johns, *The Nature of the Book: Print and Knowledge in the Making* (Chicago, 1998)

J. O. Jordan and R. Patten, *Literature in the Marketplace: Nineteenth-Century British Publishing and Reading Practices* (Cambridge, 1995)

A. Manguel, *A History of Reading* (London, 1997)

J. Raven, H. Small and N. Tadmor (eds.), *The Practice and Representation of Reading in England* (Cambridge, 1996)

III Publication in the age of science

The new magazine machine, from George Cruikshank, *The Comic Almanac*, 1846. (Courtesy of the Syndics of Cambridge University Library)

17 A textbook revolution

> The true faith will never flourish till a book has been published in *English*, in *octavo*, on the *plan* of Woodhouses's [Principles of] Anal.[ytical] Calcul.[ation (1803)], & in a **Compact** & **tangible** shape. Whoever does this will be the father of a new English School.
>
> Edward Bromhead to Charles Babbage (December? 1813)

Some eighteen months after its formation, the student members of the short-lived but ambitious Analytical Society thus came to the conclusion that their longed-for mathematical revolution would only be effected through the publication of an adequate textbook. The Society's members had arrived in a University of Cambridge where the pedagogical framework prescribed by the dominating Senate House examination embodied a geometrical, intuitive approach to mathematical studies, presented as standing in a Newtonian tradition. The university repudiated in partisan terms the analytical approach by which the great innovations of Continental, and particularly French, mathematicians of the later eighteenth century had been achieved. The Analytical Society had been established with an explicitly evangelical zeal to proselytise the 'true faith' of analysis. Yet, its only publication was a volume of *Memoirs* (1813) modelled on the publications of learned societies, and its direct impact on the studies of the university was insignificant. As historians have recognised, the introduction of a more analytical approach to mathematics in the university context owed more to the changes in pedagogy effected by former members of the Society and others in the later years of the decade.

Yet, as the former Analyticals who sought to effect such pedagogical reform soon discovered, there was more involved than met the eye. With this in mind, Andrew Warwick has recently examined some aspects of what the analytical revolution might have looked like 'from below', discovering that the process was much more protracted than previously thought. He has sought to unpick the threads of pedagogical practice, exploring the introduction of analytical approaches within the context of college lecturing and tutoring, private tutoring, college examinations, the Senate House

Figure 17.1 The 'reading' man. While early nineteenth-century Cambridge pedagogy depended heavily on public and private tutors, the close study of the standard pedagogical texts, including those by Newton and Vince, was also essential. '*Helluones librorum*' (Devourers of books), by Richard Corbould Chilton; aquatint by Francis Jukes, engraving by J. K. Baldrey. (Reproduced by kind permission of David McKitterick)

examination and the whole gamut of reading matter – both printed and manuscript – which a student might encounter (Figure 17.1). In a parallel way, this chapter relates the analytical revolution to contemporary publishing practices, exploring the way in which the commercial context of textbook publication impinged upon the would-be reformers. This case study elucidates the issues involved in the vastly important educational publishing market in nineteenth-century Britain, demonstrating their significance for the history of science in the period.

The pedagogical estate

You never hear of a book published by a Cambridge man. What do you mean by that? It is a convincing proof to me that you never were at Cambridge; you must have heard of some there; of course no one could expect you to hear of any elsewhere.

E. S. Appleyard, *Letters from Cambridge* (Cambridge, 1828), p. 129

The copyrights of standard works of instruction, with constant and predictable markets, and relatively stable texts, have long been recognised as a particularly valuable species of property. However, until the landmark 1774 House of Lords ruling against the London booksellers' claims to perpetual copyright, most such standard works rested in the hands of a relatively small group of London traders, who generally co-operated in financing the new editions which were regularly demanded. The sudden cancellation of these and other valuable copyrights sparked a tremendous upheaval in the book trade – resulting in a dramatic reduction in the prices of standard works, and an even more dramatic increase in the production of printed matter. Moreover, copy-holding booksellers attempted to create replacement property by financing new works, often carefully guarded as private ventures by individual entrepreneurs. In such an uncertain and competitive climate, and with a rapidly growing readership, standard educational works were particularly attractive. Indeed, a number of the more prominent copy-holding booksellers who successfully negotiated the financial uncertainty of this period by transforming themselves from small-scale family businesses into highly capitalised, non-retailing publishing houses, like Thomas Longman (1771–1842), did so to a large extent by investing in such books.

There was also abundant scope for newcomers with an eye to the main chance. Numerous publishers in the early nineteenth century built up extraordinarily valuable businesses within a short space of years chiefly on the basis of their instructional or elementary publications. Among the more striking were Richard Phillips (1767–1840) and William Pinnock (1782–1843), both of whom started as schoolmasters, and both of whom had by the second decade of the century obtained copyrights worth many thousands of pounds. Yet the volatility of the competitive market also brought dangers. Ironically, the bankruptcies of Pinnock and Phillips gave another of the new men, George Byrom Whittaker (1793–1847), a chance to build up his own stock of educational books. The son of a schoolmaster, Whittaker had been apprenticed to a prosperous wholesale bookseller and publisher in London's traditional book-selling quarter, and within a decade of taking over the business in 1814, had expanded it to such an extent that he was bracketed together with Longmans as one of the 'two greatest Publishers in England'.[1] Although Whittaker shortly afterwards suffered bankruptcy in the financial crisis of 1825–6, he recovered to become one of the leading trade publishers of the first half of the century. Like other later publishers – notably John William Parker (1792–1870) from 1836, George Bell (1814–90) from 1839 and Daniel Macmillan (1813–57) from 1843 – Whittaker's educational business relied heavily on his cultivation of both the authors and the readers of the universities of Oxford and Cambridge.

At the start of the nineteenth century the University of Cambridge had one of the leading presses in the country, used to print both scholarly and pedagogical publications, as well as university ephemera of various kinds. However, the only commercial publishing carried out by the university was that which exploited its ancient privilege (shared with the King's Printer and the Oxford University Printer) of printing the Bible and the Book of Common Prayer. There were certainly good reasons for the university to decline to speculate on any of the other books issuing from its presses at the start of the nineteenth century. As many of the new provincial printers and booksellers discovered in the years after the lapse of licensing in 1696, the London book trade was vastly favoured not only by the population distribution, but also by the transport and commercial infrastructure of Britain. Moreover, compared to the new urban centres which developed in the late eighteenth century, Cambridge was a relative backwater, with poor transport links and a sparse rural population in the surrounding district.

However, these difficulties were countervailed by the presence in the town of so many potential authors and readers. Of course, many of the authors were well supported by the increasingly valuable financial rewards of a college fellowship, and could afford to publish books at their own expense. Moreover, with so many interested readers on hand, other books were readily financed by subscription. Nevertheless, some books were issued at the risk of local booksellers, who were keen to exploit the limited but reliable local market, and to generate credit with London booksellers, which could be exchanged for additional retail goods (Figure 17.2). Moreover, this market increased towards the end of the eighteenth century, as the university began to grow in both size and prestige. After a century of marked decline, matriculations at the university showed an upward trend after 1770. In addition, there were significant changes in pedagogy, reflected in the increasingly demanding nature of the Senate House examination, which was extended from three days to four after 1779. These changes were reflected in a new generation of authors (most notably James Wood (1760–1839) and Samuel Vince (1749–1821)), who began in the 1780s to rewrite the textbooks that had been the staple of the university in the last half-century.

Into this expanding market came John Deighton (1748–1828), who arrived in Cambridge in 1778 as a thirty-year-old bookbinder from London, and had, by the turn of the century, established himself as the leading bookseller and publisher in the town. As for so many educational publishers, the timing of the expansion in the trade was important for Deighton, since it allowed him to build up a large and valuable property in copyrights fairly quickly. Particularly at first, however, many of the books bearing Deighton's name were

Figure 17.2 The flamboyant Cambridge bookseller John Nicholson (1730–96), gained the sobriquet 'Maps' from his habit of announcing his arrival at the doors of his customers with the words: 'Maps! Anything wanted today Sir?' Engraved by James Caldwell from a portrait painted by Phillip Reinagle (which now hangs in Cambridge University Library), and published at the request of the university in 1790. (Reproduced by permission of the Syndics of the Cambridge University Library)

published on commission (at the authors' expense), although even these were useful in creating credit for Deighton in London. In addition, titles in which Deighton owned a stake were at first usually published in concert with other booksellers in Cambridge and London (and more occasionally in Oxford). Such arrangements were requisite not only to raise the necessary capital, but also to ensure adequate wholesaling in London and the country.

These growing links between the Cambridge and London trade indicate the increasing market nationally for books by Cambridge authors. In particular, the market for educational and mathematical books expanded disproportionately in the late eighteenth century. Deighton capitalized on this, and over the years that followed, he became the first Cambridge publisher with claims to operate on a national scale. He cemented his close ties with London during a further period there between 1785 and 1794. On his return to Cambridge, he purchased the business of Joseph and John Merrill, the leading copy-holding booksellers in the town, and rapidly established himself as the leading point of contact with the London trade. By this stage his capital was very considerable indeed, and like others of the new breed of publisher he moved away from publishing in partnership with numerous other booksellers. Nevertheless, with the object of reaching a national and even an international market, he continued routinely to publish in partnership with at least one London publisher – after 1814 typically Whittaker.

However, despite the growing importance of Cambridge books in the national market, many (especially textbooks) were written with the local market almost exclusively in mind. Indeed, given the high cost of transporting such heavy merchandise, it made little sense to print and publish a book in Cambridge which was not aimed primarily at the local market, unless for the convenience of the author. From this perspective, it is easy to see why, unlike some of the new metropolitan publishers, it would have been inconceivable for Deighton to specialise in publishing, to the exclusion of a retail business. Deighton needed to sell a large proportion of the books he published to members of the university. His retail premises, situated prominently within a few hundred yards of Trinity and St John's (much the largest colleges), were thus inseparably linked with his publishing activity (Figure 17.3).

Deighton's shop rapidly became a central focus of Cambridge life – in the words of the *Cambridge University Calendar* for 1802, a 'public part of the university'. Its most obvious attraction was its retail stock. The late eighteenth century saw the introduction of the display of books on booksellers' counters, and it was not uncommon for members of the university to spend time 'lounging' at the bookseller's.[2] The stock on Deighton's counter was particularly worth perusing. Despite the war with France, he announced in 1802 that he had established 'a Continental Connexion' by which he was 'enabled to sell Foreign Classical Books on as Low Terms as any Person in the Kingdom, and to procure any Foreign Work with all possible expedition'. Such trade gave Deighton a great advantage in serving both the institutional and private needs of the university. The University Library relied very heavily on the Deighton business throughout the nineteenth century for the supply of foreign (as well as British) books (Figure 17.4).[3] In addition,

Figure 17.3 On his return to Cambridge from London, Deighton purchased the business of the leading copy-holding booksellers in the town, taking prominent premises on Trinity Street, not far from the main gate of Trinity College. In this photograph, dating from the 1870s, it is the first shop from the right, the frontage largely unchanged from fifty years before. (From the Cambridgeshire Collection, courtesy of Cambridge Public Library.)

numerous former members of the university and others relied on the catalogues of stock which Deighton began to issue regularly.

There were, however, other incentives for students to frequent Deighton's shop. This was the bookseller who published not only the standard guide to the university, but also the semi-official *University Calendar*, a broadside Cambridge almanac, and a dictionary of university slang, any of which a freshman might desire. Moreover, while for many years the Cambridge booksellers had taken students' names for the various professorial lecture-courses, this was a function which Deighton now increasingly took to himself (Figure 17.5). Most strikingly, Deighton had earned for himself the privilege of having posted in his shop the results of the disputations in the schools, giving the preliminary ranking of candidates in advance of the Senate House examination.

Deighton's increasing domination of the university trade, and his semi-official role within the university, clearly gave his growing list of publications particular authority with Cambridge readers. As he gradually increased his hold on the key mathematical textbooks of the university, he began to add to all his publications a list entitled:

Figure 17.4 The West Room and the Dome Room of Cambridge University Library, 1809. From a watercolour by Thomas Rowlandson. (Reproduced by permission of the Syndics of the Cambridge University Library)

'Mathematical books, by members of the University of Cambridge, sold by J. Deighton' (Figure 17.8). While the recommendations of tutors – both public and private – and of older students, were doubtless of the first importance, Deighton's acceptance of a new mathematical text clearly defined its chances of success to a considerable extent.

Moreover, of all Cambridge traders, Deighton had the closest link with the University Press. By 1802 he was so trusted within the university that, following the sudden death of the University Printer, he was appointed in a caretaking capacity until a suitable replacement could be found. He was also appointed as the Cambridge agent for the Press's Bibles and prayer books, regularly taking thousands of pounds worth of books off their hands. These close links became increasingly cemented as his publishing activity steadily grew, and the printing of his books soon constituted a large proportion of the commercial printing jobs carried out at the Press, a position which gave him considerable leverage.

It is clear that Deighton held considerable power over any author wishing to write for the Cambridge market. While publishers in London frequently received unsolicited manuscripts from unknown authors, the university was a closed world in which authors and

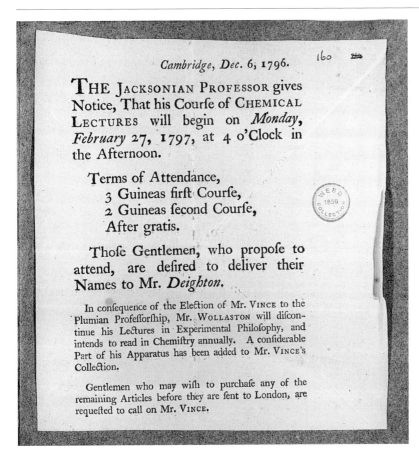

Figure 17.5 Metaphorically as well as literally, Deighton's shop was a central focus of university life. Lecture bills, as with this for the Jacksonian Professor's lectures in 1796, directed students there. Cambridge University Archives, University Papers, UP1 fol. 160. (Reproduced by permission of the Syndics of the Cambridge University Library)

readers alike were generally known. For a publisher like Deighton, the evaluation of authors' manuscripts was based on first-hand knowledge and face-to-face transactions. Only by successfully negotiating such transactions could an aspiring Cambridge author hope to secure Deighton's valuable services in reaching his intended audience.

The perils of scholarly publishing

> The publication of a Math^l. work, particularly if it goes one step beyond the comprehension of Elementary readers is a dead weight & a loss to its author.
> John Herschel to Bromhead, 19 November 1813

According to Charles Babbage (1791–1871), the scheme of the Analytical Society arose from his parody on one of the most active purveyors of print in the early nineteenth century – the British and Foreign Bible Society. At the time, Cambridge was in the midst of a controversy occasioned by the founding of an auxiliary branch of the Society. The key focus of debate was whether or not the Bible should be circulated without any commentary, or together with the prayer book, in order to enforce Anglican doctrine. Babbage reports that in May 1812, at the height of the controversy, he devised a satirical sketch of a 'society to be instituted for translating the small work of Lacroix on the Differential and Integral [Calculus]. It maintained', he later recalled, 'that the work of Lacroix was so perfect that any comment was unnecessary'.[4]

Babbage reported that the circulation of his spoof among student friends resulted in the formation of 'a society for the promotion of analysis', comprising fewer than a dozen members, who over the next year met in a rented room to read and discuss papers on analytical mathematics. Yet for all that the Society's members continued to refer to their enterprise in evangelistic language, their practice was anything but reminiscent of the British and Foreign Bible Society. Far from distributing the 'small work of Lacroix', or making similar attempts to proselytise, they pursued their own concerns with mathematical research in isolation from the studies of the university.[5]

When in the Michaelmas Term of 1812 the Society resolved to publish a volume of memoirs, it was thus the Royal Society, rather than the British and Foreign Bible Society, that constituted their explicit exemplar. However, it was not until 1821 that a learned society in Cambridge successfully issued transactions – and even then it was financed by the University Press syndicate, rather than being a purely commercial enterprise. With no obvious relevance to an identifiable market, and with no status or patronage to recommend them, the Analyticals could not expect to interest a publisher in taking on the *Memoirs of the Analytical Society* at his own risk.

Nor did they. From the start the Society expected to have to cover the expenses of composition, and of the paper and presswork for 100 copies. Moreover, at least some members anticipated that this would entail a loss. It was also agreed that the authors of the individual memoirs – in the event only Babbage and John Herschel (1792–1871) – should be allowed to pay for the paper and presswork of an additional 150 copies for private use. However, while the authors hoped for some financial recompense from the sale of these, they were also clearly intended for distribution among potential patrons. This Babbage and Herschel did with some effect.

The Society clearly made efforts to ensure a wide distribution for the *Memoirs*. The Analyticals had firmly rejected the idea of putting 'Cambridge' in the society's name, and viewed their primary audience as existing in London and beyond. Thus they arranged not only for Deighton, but also (somewhat untypically for a Cambridge

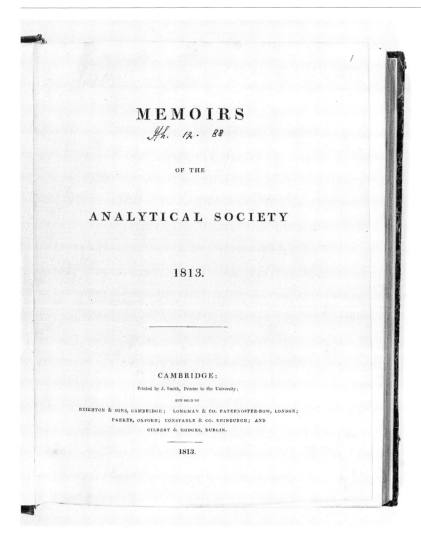

Figure 17.6 The *Memoirs of the Analytical Society* (Cambridge, 1813) were presented anonymously as the proceedings of a learned society. Unlike most mathematical works by Cambridge authors, they were wholesaled in Oxford, Edinburgh and Dublin, as well as London and Cambridge. (Reproduced by permission of the Syndics of the Cambridge University Library)

production) for Longmans in London and respectable houses in Oxford, Edinburgh and Dublin to act as wholesalers (Figure 17.6). Yet, those who wholesaled privately funded works rarely took trouble to advertise or otherwise promote them, and even in Cambridge the Deightons declined to put the *Memoirs* on their list of mathematical books.

With a London market primarily in mind, the Analyticals initially sought a printer there, only to fall back on the University Press, probably for the sake of convenience.[6] There were, however, more substantial reasons for having the work printed in Cambridge. Mathematical composition was a highly skilled activity, and the University Printer, John Smith, was probably justified in claiming for himself the distinction of training unusually skilled mathemat-

Figure 17.7 The mathematical lecturers at the Royal Military College, Sandhurst, declared of Babbage's memoir that 'they never saw its equal in typography'. The 'awful brackets' and 'small numerals' required were among the many difficulties in printing posed by the Analyticals' notational reform. *Memoirs of the Analytical Society* (Cambridge, 1813), p. 54. (Reproduced by permission of the Syndics of the Cambridge University Library)

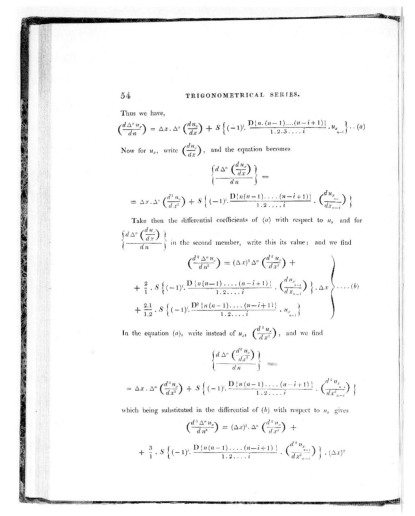

ical compositors, who could accomplish what many a London-trained compositor refused to attempt. Nevertheless, the notational novelty of the work caused considerable typographical problems (see Figure 17.7). As William Whewell later wrote: 'the extraordinary complexity and symmetry of the symbolical combinations sorely puzzled the yet undisciplined compositors of that day'.[7] Very soon after the job was begun, Smith was complaining of great difficulty in printing Babbage's material, claiming that he had 'never put together such crabbed stuff in his life'. He was soon reporting that 'awful brackets' were requisite for the expressions, and that he had none large enough, and had been forced to send to London for some more. Then it was 'the want of a particular kind of small numerals' that slowed the job down. Smith even had some types specially cast, but in other cases the authors were obliged to adapt their notation to the available types.

Having begun composition in February 1813, the 114 pages of the *Memoirs* were not all set until November. The authors were impatient, amusing themselves with puns at the expense of the printer's devils. Babbage wrote to Herschel: 'I most heartily join with you in condemnation of those heretical printers who in opposition to the true faith, and at the instigation of some geometer have presumed to mangle a paper in the sacred language'. Herschel contemplated the need to stand over the compositors with a cat o' nine tails, and thought that Smith had 'gone to sleep in the world' and needed 'to have his memory fly-flapped'. Yet, like many Cambridge authors (as Smith later complained) they provided their manuscript in small batches, and on occasion kept the compositors waiting and looking 'Devilish black'.

When the *Memoirs* finally appeared, prohibitively expensive at 15s., they signally failed to meet their authors' expectations. In Cambridge they were generally not understood. Babbage wrote to Edward Bromhead (1789–1855): 'Of course much nonsense is talked about them here; but I have not heard criticism yet venture beyond the second line of the first Memoir: of which men ask "is it to be found in Jemmy Wood" [i.e. James Wood's *Elements of Algebra*] and if not they divide by x and are lost in the clouds of ψs which follow'. Much more disturbing, however, was the failure of the *Memoirs* to make any impact in the national press. This was quite typical of Cambridge publications of the period, but Babbage was distrait. 'What shall we do?' he wailed to Herschel when the *Memoirs* failed to attract any notice in the *Edinburgh Review*. 'Surely our impudence might have deserved *some castigation*. But to die unnoticed – This, this is *misery*.' Babbage even contemplated anonymously sending a hostile review of the *Memoirs* to a monthly magazine, just to arouse some interest.

Moreover, the *Memoirs* were a costly learning experience, literally as well as metaphorically. Smith's bill came as a shock. They had been warned that the most expensive hot-pressed superfine writing paper would be required to obtain a good impression of the small numerals used in the algebraic work. However, the paper cost twice Smith's estimate, and the bill amounted to £132 9s. for only 250 copies. Even after the authors had paid for their copies, the members each had to find £9 6s. 4d. to cover the cost. Moreover, when Herschel wound up the accounts in 1822, he considered it better to regard the proceeds of the sale of the *Memoirs* as '£0 0 0 – lest if any Enquiries be made of the booksellers they should bring in a bill for Commission & Warehouse-room'.

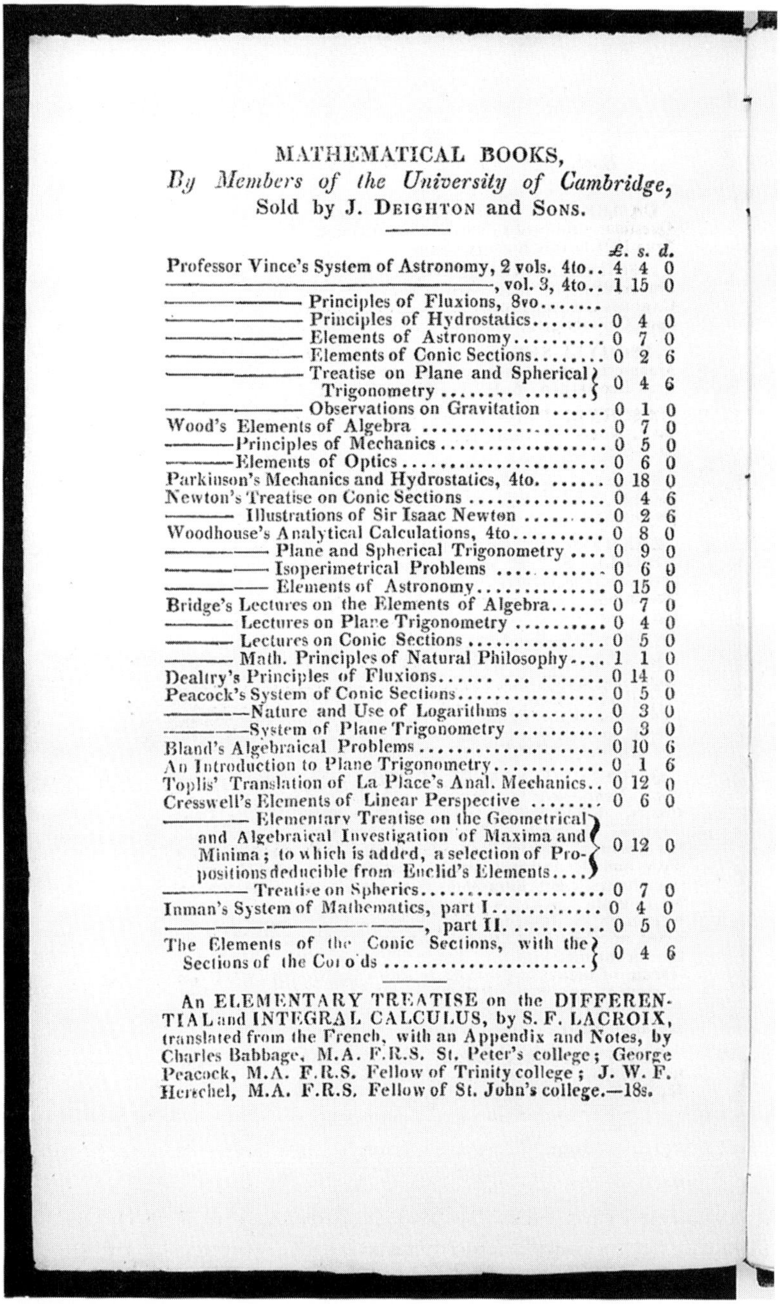

Figure 17.8 The translation of Lacroix's *Elementary Treatise on the Differential and Integral Calculus* (1816) received particular notice in the *Cambridge University Calendar*, being advertised both on the paper cover and in Deighton's list of mathematical books. *Cambridge University Calendar* (1818), p. [330]. (Reproduced by permission of the Syndics of the Cambridge University Library)

MATHEMATICAL BOOKS,
By Members of the University of Cambridge,
Sold by J. Deighton and Sons.

	£.	s.	d.
Professor Vince's System of Astronomy, 2 vols. 4to..	4	4	0
———————————, vol. 3, 4to..	1	15	0
——————— Principles of Fluxions, 8vo.......			
——————— Principles of Hydrostatics........	0	4	0
——————— Elements of Astronomy..........	0	7	0
——————— Elements of Conic Sections........	0	2	6
——————— Treatise on Plane and Spherical Trigonometry	0	4	6
——————— Observations on Gravitation	0	1	0
Wood's Elements of Algebra	0	7	0
——————Principles of Mechanics	0	5	0
——————Elements of Optics	0	6	0
Parkinson's Mechanics and Hydrostatics, 4to.	0	18	0
Newton's Treatise on Conic Sections	0	4	6
——————— Illustrations of Sir Isaac Newton	0	2	6
Woodhouse's Analytical Calculations, 4to...........	0	8	0
——————— Plane and Spherical Trigonometry	0	9	0
——————— Isoperimetrical Problems	0	6	0
——————— Elements of Astronomy..............	0	15	0
Bridge's Lectures on the Elements of Algebra......	0	7	0
——————— Lectures on Plane Trigonometry	0	4	0
——————— Lectures on Conic Sections	0	5	0
——————— Math. Principles of Natural Philosophy....	1	1	0
Dealtry's Principles of Fluxions.....................	0	14	0
Peacock's System of Conic Sections.................	0	5	0
——————Nature and Use of Logarithms	0	3	0
——————System of Plane Trigonometry	0	3	0
Bland's Algebraical Problems	0	10	6
An Introduction to Plane Trigonometry............	0	1	6
Toplis' Translation of La Place's Anal. Mechanics..	0	12	0
Cresswell's Elements of Linear Perspective	0	6	0
——————— Elementary Treatise on the Geometrical and Algebraical Investigation of Maxima and Minima; to which is added, a selection of Propositions deducible from Euclid's Elements....	0	12	0
——————— Treatise on Spherics..................	0	7	0
Inman's System of Mathematics, part I............	0	4	0
———————————, part II.	0	5	0
The Elements of the Conic Sections, with the Sections of the Cor o'ds	0	4	6

An ELEMENTARY TREATISE on the DIFFERENTIAL and INTEGRAL CALCULUS, by S. F. LACROIX, translated from the French, with an Appendix and Notes, by Charles Babbage, M.A. F.R.S. St. Peter's college; George Peacock, M.A. F.R.S. Fellow of Trinity college; J. W. F. Herschel, M.A. F.R.S. Fellow of St. John's college.—18s.

The pleasures of pedagogical publishing

I have no doubt of the ultimate success of the true faith but I have many as to the question whether its propagators will derive any profit from its establishment.

Babbage to Herschel, 20 July 1816

I think we may fairly calculate that this work [*A Collection of Examples* (1820)] will constitute a little annuity to us, of at least 25£ per annum for each of us: a very considerable sum & when we consider the present state of the country, it will be as good as a farm which pays us rent.

Peacock to Herschel, 18 February 1822

The publication of the *Memoirs of the Analytical Society* was an 'expiring effort'. Even before the volume's appearance in December 1813 the Society's activities were in decline, and there were apparently no further meetings.[8] Various former members repeatedly urged the revival of the Society over succeeding years, and Herschel in particular was keen to produce another volume of *Memoirs*. Indeed, with the passage of time it became clear that the work had made some limited inroads into a wider mathematical readership, and had secured them some credit with potential patrons. Nevertheless, the former members now directed their analytical memoirs (with financial impunity) to such learned journals as the *Philosophical Transactions of the Royal Society*. Moreover, as several of them became involved in pedagogic practice in Cambridge, they radically reassessed the kind of publication which was likely to effect the longed-for introduction of analysis in England.

In 1813 Herschel and fellow Analytical George Peacock (1791–1858) shared the honours as first and second Wranglers and Smith's Prizemen, respectively. Herschel was elected a fellow of St John's College the same year, and in 1814 Peacock was elected a fellow of Trinity. The following year, Peacock became a mathematical lecturer at Trinity, and Herschel, having flirted with the idea of a legal career, was appointed a sublector at St John's. By contrast, Babbage, who was a year younger, had taken only an ordinary degree, after seriously offending the university's Anglican orthodoxy in one of his 'Acts' (formal disputations), and had removed to London, where, supported by his father's wealth, he pursued his mathematical interests in more congenial surroundings.[9]

Confronted with the reality of Cambridge pedagogy, Herschel and Peacock quickly appreciated that appropriate pedagogical texts were essential for the introduction of analytical mathematics into the studies of the place. All the existing textbooks embodied the geometrical approach followed in the Senate House examination and in college and private tuition. The only work of pure analysis in circulation, Robert Woodhouse's *Principles of Analytic Calculation*

(1803), was designed for accomplished mathematicians, and made no concessions to the demands of pedagogy. Yet it was an unspoken rule that questions on a new subject should not be set in the Senate House examination 'unless it had been discussed in some treatise suitable and available for Cambridge students'.[10]

It was in this context that the idea of translating Lacroix's *Elementary Treatise on the Differential and Integral Calculus*, on which the Analytical Society had been founded, was revived. Concluding that 'it was hopeless for any young and unknown author to attempt to introduce the notation of Leibnitz into an elementary work', Babbage had commenced to translate the work while still an undergraduate. However, it was not until late in 1815, when Peacock and Herschel came to the conclusion that notational reform would not be accomplished 'until some foreign work of eminence be translated into English', that the three of them endeavoured to prepare a translation for publication.[11]

The authors were keen to make the work appropriate to the Cambridge market, reshaping it significantly in the process of translation.[12] Drafting an appendix to replace Lacroix's own, Herschel in particular sought to make it useful to 'men cramming for degrees' who might thus get 'a large tinge of the true faith'. Such endeavours were conceived within the competitive conditions of the local market for books. In particular, Herschel hoped the appendix might prevent the reprinting and new publication of several geometrical texts, and that it might 'sell the book'.

Of course, there was a financial as well as an ideological incentive for wanting to sell the book. Deighton was not the only one for whom pedagogic publications were an important and reliable source of income. The expansion of the textbook trade at the end of the eighteenth century had gradually increased the income of successful authors, though to a lesser extent in Cambridge than in London. For some without college fellowships, like Samuel Vince, authorship combined with private tutoring to provide the main means of subsistence. As junior fellows Peacock and Herschel received in the order of £100 per annum, which they supplemented by tutoring. However, the rewards of pedagogic authorship were nonetheless attractive. Indeed, Herschel wanted to 'get money and *reputation at the sa<me> time*', and was thus also keen to write for the scientific journals and encyclopaedias then providing increasingly important sources of both for men of science. Moreover, by the time the Lacroix appeared, Herschel was busy at work on an analytical textbook of algebra, and Peacock was planning a work on the application of algebra to geometry.

With a local market firmly in mind, Peacock and Herschel approached Deighton with the translation. Deighton was happy to publish it at his own risk, in partnership with Law & Whittaker in London, offering £60 for the first edition, and adding that he would 'make more considerable allowance' if the edition sold within two years. Although the return was small when divided three ways, the offer was not ungenerous, especially as Deighton had raised the possibility of a future edition. Indeed, the translators were unable to obtain a better offer in London, highlighting Deighton's advantageous position in commanding a readership for the work. Deighton's decision to publish was not surprising. To begin with, the Lacroix was a work which already circulated to some extent among Cambridge undergraduates in the French, and it is likely that he had received requests for it.[13] Moreover, the structural position of two of the proposed translators clearly indicated that it had good prospects of being incorporated in the studies of the university.

When the translation appeared, a 700-page volume priced at a lordly 18s. in boards, it sold as rapidly as its projectors could have hoped. Before publication, Deighton had inserted announcements in the *Cambridge University Calendar* announcing that the work was in preparation, and on its publication in December 1816, he followed his usual practice of sending a man around the university with extra copies of the title-page to put up as posters. By then, the book had been so often asked for that Deighton had almost a quarter of the standard edition of 1,000 copies bound in boards immediately it came off the press. The rather optimistic assessment of the translators was that nearly 200 copies had been 'disposed of' in Cambridge in the first fortnight, although Deighton did not in fact require further copies from the warehouse for a further fifteen months. Deighton's stock was exhausted by 1823, which, while not so rapid a sale as with some standard textbooks, was a very respectable sale for so expensive a book. Moreover, half the edition had been sold through Whittaker, supplying the metropolitan and national trade – a far higher proportion than was common with Cambridge textbooks.[14]

The book caused no little stir in Cambridge. Peacock had heard of the work occasioning 'several sudden conversions', but, as Babbage readily conceded, its impact on the practice of pedagogy was strictly limited. Nevertheless, coinciding with Peacock's introduction of differential notation into the Senate House examination, the work was perceived by many as an assault on the traditional studies of the university, and even provoked a 'furious pamphlet' on this account, published at the expense of the university.

However, the authors' next move was to produce a 'supplement' which related analysis yet more firmly to the pedagogical practices of the university. Even before the Lacroix was published, Babbage had suggested the desirability of publishing a volume of examples of the differential and integral calculus which would have pragmatic appeal even for those tutors who disapproved of the change to the differential notation.[15] Peacock and Herschel leapt at the idea, and without further consultation Peacock announced in the

'Advertisement' to the translation that the supplement was in preparation and would be ready shortly. In Peacock's hands especially, the work was adapted to the current Cambridge practice of teaching by exemplification. Its 'principal object', he told Herschel, was to provide 'such examples as are likely to be most useful to a *student* only'.

Like Peacock, Deighton saw the relevance of the proposed supplement to the Cambridge market, and was 'very anxious to treat for it'. Flushed with the success of the Lacroix, Deighton had even offered to publish a translation of another of Lacroix's works, planned by William Whewell and Hugh James Rose. However, Peacock thought Deighton had treated them 'very shabbily' in his payment for their translation, arguing: 'if we let him have the other, he shall pay for it as well'. It did not take mathematicians of the Analyticals' ability to determine that less than half of the proceeds of the sale of the Lacroix had covered the manufacturing cost. The remainder – something in the order of £500 – had been divided between the publishers and booksellers. Judging that the new work was 'quite certain to sell', the authors decided – like not a few Cambridge textbook authors at this period – to publish at their own expense, paying Deighton and Whittaker commission (at 25 and 35 per cent respectively) to act as wholesalers.

They thus found themselves once again relying to a large extent on the advice of Smith, the University Printer, to guide them in the mechanics of putting together a book. With the Lacroix, Deighton had used his own stocks of paper, but this time they had to order their own, through Smith. Moreover, unlike Deighton, they had little leverage over Smith, who was inclined to let work on the *Examples* lapse when there were more pressing demands. Though similar in length, the book was in the press for three years compared to eight months for the Lacroix, and while the authors (especially Peacock) were certainly largely to blame, Smith was far from innocent. Had they not already bought the paper, they would have changed printers for one in London, and Herschel became fearful that a 'caesarian section' would be required to deliver the press of this 'portentous birth'.

There was also advertising to consider. Herschel thought that Deighton had 'greatly neglected' the advertisement of the Lacroix in the commercial scientific journals which were becoming an increasingly important part of the contemporary book trade. With their eyes on a wider market, the authors hoped to reach a national readership by advertising the *Examples* not only in the Cambridge newspapers, but also in the national newspapers, magazines and reviews. Likewise, while Whittaker was apparently the main London wholesaler, the authors also approached two other London publishers – J. Mawman and Longman & Co. – to act as wholesalers. Of course, Cambridge continued to be a critically important market for the *Examples*, and Deighton advertised the work as forthcoming in the *University Calendar*. Even the form of the book, which was ultimately published in two volumes at 30s., was determined to a large extent by the consideration of what it would look like 'on Deighton's table'.

On its publication in October 1820, the sale of the *Examples* was so great that the binders could not supply the local demand, let alone the London booksellers. At the end of sixteen months, 340 copies had been sold, a number far beyond the authors' expectations. Moreover, more than three-quarters of the costs of production had been recovered, and the authors were expecting a 'decent recompense' for their labour. Thereafter, sales were slower, and Deighton continued to advertise stock of the work as late as 1829. Nevertheless, the success of the *Examples* indicates the extent to which the former Analyticals had bent their necks to the realities of Cambridge pedagogical practice in making analysis relevant to the practices of college lecturing, tutoring, and college and university examinations. In so doing, they had also aligned themselves with the practices of the Cambridge book trade, and obtained a share in its valuable property.

Conclusion

That the reform of the Cambridge mathematical curriculum was necessarily achieved in part through the introduction of new textbooks has long been clear. However, such a transformation, involving the financial concerns of both publishers and authors, required careful navigation through a complex print culture. In the volatile publishing market of early-nineteenth-century Britain, the commercial value of standard educational works gave them a particular importance. Moreover, the ever greater expansion of education at all levels throughout the century served only to increase this. The opening in 1828 of the University of London – many of whose professors were Cambridge men – created an important potential market, particularly for Cambridge mathematical textbooks. A further significant innovation occurred in the 1850s, when the introduction of public examinations for middle-class schools, organised by the universities, also increased the market for books by university authors. The textbook market was again radically increased by the provision for mass elementary education in the Education Act of 1870, and it was only thereafter that the University of Cambridge finally became a major publisher of educational works in its own right. It was in these diverse contexts of commercial speculation that the transformation of the sciences in the nineteenth century took place, and the changing publishing practices of the period are thus of great consequence in constructing the history of the sciences.

Notes

The research for this chapter was conducted during my tenure of the Munby Fellowship in Bibliography, Cambridge University Library (1992–3) and of a Leverhulme Special Research Fellowship, Department of History and Philosophy of Science, Cambridge (1997–8). The chapter is based on a paper presented at a conference of former Munby Fellows held at Darwin College, Cambridge in 1994. I am grateful for helpful comments made by the contributors to that conference, and by Jim Secord and the editors of this volume.

1. On Whittaker see C. Matthew (ed.), *New Dictionary of National Biography* (Oxford, forthcoming).
2. D. McKitterick, *A History of Cambridge University Press*, vol. II, *Scholarship and Commerce, 1698–1872* (Cambridge, 1998).
3. D. McKitterick, *Cambridge University Library: A History*, vol. II, *The Eighteenth and Nineteenth Centuries* (Cambridge, 1986).
4. C. Babbage, *Passages from the Life of a Philosopher* (London, 1864), p. 28. Although there are clearly elements of myth-making in this account, written as it was some half a century after the events it describes, many of its essential features also appear in the 'History of the origin and progress of the calculus of functions during the years 1809 1810 [. . .] 1817' (Oxford University, History of Science Museum, Buxton Ms. 13), which was written in September 1817. The incorrect suggestion of some historians that the Society was actually founded in the autumn of 1811 is based on letters, one of which was misdated by its author, and the other by a manuscript librarian.
5. See P. C. Enros, 'The Analytical Society (1812–13): precursor of the renewal of Cambridge mathematics', *Historia Mathematica*, 10 (1983): 24–47.
6. This account is based on the extensive extant correspondence between Babbage, Herschel, Peacock, Bromhead, and other members of the Society. The correspondence consulted is in the Herschel Papers, St John's College, Cambridge, the Herschel Papers, Royal Society, London, the Babbage Papers, British Library, and the Bromhead Papers as quoted in Enros, 'The Analytical Society'.
7. [W. Whewell], 'Cambridge Transactions – Science of the English Universities', *British Critic*, 9 (1831): 71–90, at p. 85. On Smith see McKitterick, *Cambridge University Press*, pp. 220–1 and 314.
8. The suggestion that the Society continued in existence until December 1817 is based on a too literal reading of a whimsical letter from Bromhead to Babbage, 20 December 1817 (British Library, Add. Ms. 37182, ff. 91–2).
9. H. W. Becher, 'Radicals, Whigs and conservatives: the middle and lower classes in the analytical revolution at Cambridge in the age of aristocracy', *British Journal for the History of Science*, 28 (1995): 405–26, at pp. 403–4.
10. W. W. Rouse Ball, *A History of the Study of Mathematics at Cambridge* (Cambridge, 1889), p. 128.
11. Babbage, *Passages*, p. 38.
12. J. L. Richards, 'The art and science of British algebra: a study in the

perception of mathematical truth', *Historia Mathematica*, 8 (1981): 23–45.
13 Alexander D'Arblay to his father, 7 July 1814, quoted in E. A. and L. D. Bloom (eds.), *The Journals and Letters of Fanny Burney (Madame D'Arblay)*, vol. VII, *1812–1814* (Oxford: Clarendon Press, 1978), p. 435n.
14 Authors' Books Delivered (CUP 29/5), Cambridge University Archives. Most of the 226 copies sent out on the day of publication went to a Cambridge bookbinder named Underwood, not to the London bookseller Thomas Underwood as David McKitterick supposes. See McKitterick, *History of Cambridge University Press*, p. 302. The twenty-five wasted copies were probably damaged stock. The continuing sale of the work, taken with the free warehousing enjoyed at the University Press by members of the university, would hardly have justified their destruction otherwise.
15 Babbage, *Passages*, p. 40.

Further reading

H. W. Becher, 'Radicals, Whigs and conservatives: the middle and lower classes in the analytical revolution at Cambridge in the age of aristocracy', *British Journal for the History of Science*, 28 (1995): 405–26
P. C. Enros, 'The Analytical Society (1812–13): precursor of the renewal of Cambridge mathematics', *Historia Mathematica*, 10 (1983): 24–47
J. Feather, *The Provincial Book Trade in Eighteenth Century England* (Cambridge, 1985)
J. Gascoigne, *Cambridge in the Age of the Enlightenment: Science, Religion, and Politics from the Restoration to the French Revolution* (Cambridge, 1989)
L. Howsam, *Cheap Bibles: Nineteenth-Century Publishing and the British and Foreign Bible Society* (Cambridge, 1991)
D. McKitterick, *A History of Cambridge University Press*, vol. II, *Scholarship and Commerce, 1698–1872* (Cambridge, 1998)
J. Raven, *Judging New Wealth: Popular Publishing and Responses to Commerce in England, 1740–1800* (Oxford, 1992)
J. R. Topham, 'Two centuries of Cambridge publishing and bookselling: a brief history of Deighton, Bell and Co., 1778–1998, with a checklist of the archive', *Transactions of the Cambridge Bibliographical Society*, 11 (1998): 350–403
  'Scientific publishing and the reading of science in nineteenth-century Britain: an historiographical survey and guide to sources', *Studies in History and Philosophy of Science*, 31A (2000)
A. Warwick, *Masters of Theory: A Pedagogical History of Mathematical Physics in Cambridge from the Enlightenment to World War I* (Cambridge, forthcoming)

18 Useful knowledge for export

> Peoples, history, geography, religion, morals, politics, all that was already written [in my mind] as in an index; I lacked however the book which treated them, I felt alone in the world . . . But there must be books, I said to myself, especially about those matters, for children; and, understanding them well, one can learn with no need of teachers; and I launched myself in search of those books, and in that remote province, at that hour of my resolution, I found what I was looking for, just as I had conceived it, prepared by patriots who loved America and who, from London, had provided for that South American need for education, answering my clamor: *Ackermann's catechisms* . . . I have found them!, I could shout like Archimedes, because I had forseen them, invented them, looked for such catechisms.
> Domingo F. Sarmiento in 1826[1] (President of Argentina, 1868–74)

This is a story about writing, publishing and constructing identities through books of a simple form. The story of how certain knowledge about nature became printed in a series of textbooks: how it was thought, negotiated, paid for and transformed by a variety of people involved in a remarkable publishing enterprise. It is a story about books of no identifiable single author, and about how this lack of singularity in writing affected the 'messages' that the books conveyed. What the transatlantic readers of these books made out of such messages is part of a different story – but it is in part a result of this one too.

Books for Spanish America

The effects of the revolution that transformed British print culture in the first decades of the nineteenth century were not confined to the island. Thanks to the development of communications and the increased knowledge of the English language in continental Europe, there was, from the last years of the eighteenth century, an increasing demand for English books abroad. Moreover, cheaper editions of English books began to be reprinted in continental Europe and the United States for local consumption and, in some cases, for reintroduction into the English market – one of the motivations for the further establishment of agreements of international copyright.[2]

In the 1820s, English books began to reach a hitherto inaccessible market: the recently independent Spanish American countries. Commercial relations between Britain and Latin America had begun only in the last years of the eighteenth century: first through smuggling to evade Spain's (and Portugal's) monopoly of the trade with their colonies, and more openly after 1808, when the French invasion of the Iberian peninsula disrupted colonial relations, triggered independence movements and stimulated Britain's search for alternative markets to the blockaded continental Europe. Once the independence of these countries was consolidated and internationally recognised in the 1820s, the region became a magnet for British traders and investors in mining companies and the debt bonds of the first Latin American loans. Investment decreased significantly after the financial crash of 1825, but for the rest of the century Latin America remained an important market for the manufactures of Britain's industrialised economy.

In 1823, Rudolph Ackermann, by then a well-established London art publisher, decided to venture into the book market for the Spanish-speaking countries in America. Within five years, he published nearly 100 titles in that language, comprising four magazines, twenty-six textbooks in all subjects of the arts and sciences – known as 'Ackermann's catechisms', because of their question-and-answer style – one series of giftbooks and a number of books of general interest, from novels to political treatises.[3] These works were distributed widely in urban centres across Mexico, Guatemala, Gran Colombia (the temporary union of Venezuela, Colombia and Ecuador), Peru, Chile and Argentina. Many were reprinted locally, especially the textbooks, of which dozens of editions were published throughout the century not only in the Spanish-speaking countries themselves but also – on an even larger scale – in France, for export to them. Besides Ackermann, during the 1820s there was a handful of French publishing houses – some of which had branches in London – exporting cheap books in Spanish to that region, whilst a group of Spanish American exiles in England were publishing individual works and short-lived 'instructive' magazines for their compatriots. Meanwhile, the British and Foreign Bible Society (BFBS) was committed to the production and distribution of cheap New Testaments and Bibles for those 'virgin' lands. Yet none of those publishing enterprises rivalled Ackermann's in the scale and the success of their products over the years.

Ackermann's catechisms were a novelty in independent Spanish America because they came from England (during the colonial era most of the educational material had come from Spain), because of the encyclopaedic range of their content, and because of the use of the interrogative style – a style used almost exclusively by the religious catechism, widespread throughout the Spanish empire since the sixteenth century. Produced in a period when the notion of *textbook* as a school object that was both book and teacher – a text especially designed for education in the classroom and not an ordinary book used as an aid to teaching – was almost non-existent in Spanish America (and incipient in England), Ackermann's catechisms were aimed at – and read by – a wider market constituted by schoolchildren and adults alike.[4] In the following years these books were not only reprinted several times, but they influenced, in form and content, the first textbooks produced locally. Overall, in a decade of faith in the power of education to emancipate the recently independent peoples from ignorance and superstition and direct them towards economic and moral progress, these texts became a symbol of enlightenment.

Authors and authority

Ackermann's catechisms played an important role in the shaping of identities of nations which were at the beginning of a process of political and cultural self-definition. It is natural for us, then, to ask: who wrote them? This question was not so important for their contemporary readers. Half of the catechisms were published without an author's name, and people used to refer to them rather by the name of their publisher: *'los catecismos de Ackermann'*.[5] In the words of the young Sarmiento, quoted above, these books were prepared merely by 'patriots who loved America'. For the future Argentine leader, the catechisms seemed to exist already somewhere, they only needed to be *discovered*: they were not *made*, they were out there like natural objects ready to be *used*.

Indeed, in a period of individualisation of the notion of authorship, the emerging genre of textbooks did not fully fit within the romantic ideology of the author as creator, or as the 'genius' from whom the authority of the printed object emanated. This genre imposed a series of conventions upon their publisher and writers which, on the whole, tended rather to the effacement of the figure of the author. A catechism is the most self-contained form of textbook, where both the questions and the answers are provided by the text (Figure 18.1). The writer has to use an impersonal voice and the knowledge about nature is presented not as a matter of individual opinion, but as something that has an independent existence. The printed questions are supposedly asked by the readers in search of knowledge, while the answers are provided by Nature herself in her clear, undisputed voice. In a sense, then, the authority

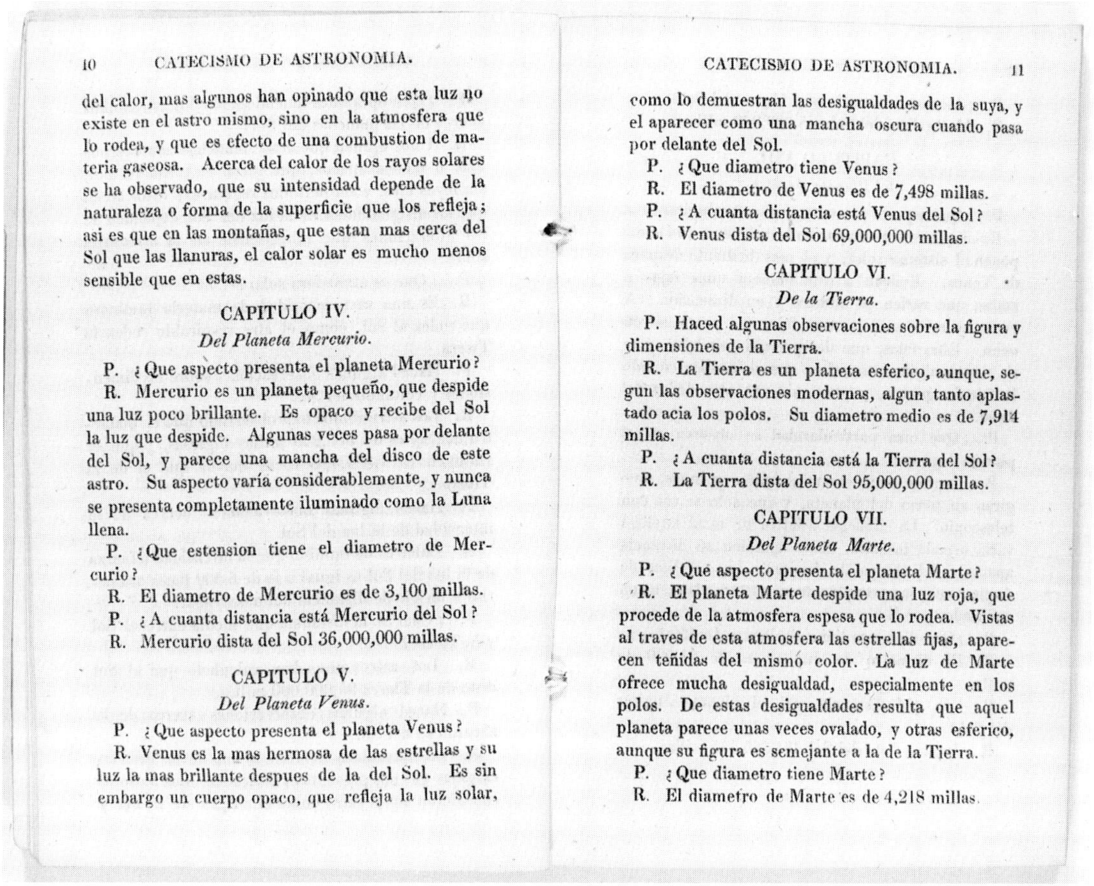

10 CATECISMO DE ASTRONOMIA.

del calor, mas algunos han opinado que esta luz no existe en el astro mismo, sino en la atmosfera que lo rodea, y que es efecto de una combustion de materia gaseosa. Acerca del calor de los rayos solares se ha observado, que su intensidad depende de la naturaleza o forma de la superficie que los refleja; asi es que en las montañas, que estan mas cerca del Sol que las llanuras, el calor solar es mucho menos sensible que en estas.

CAPITULO IV.

Del Planeta Mercurio.

P. ¿Que aspecto presenta el planeta Mercurio?

R. Mercurio es un planeta pequeño, que despide una luz poco brillante. Es opaco y recibe del Sol la luz que despide. Algunas veces pasa por delante del Sol, y parece una mancha del disco de este astro. Su aspecto varía considerablemente, y nunca se presenta completamente iluminado como la Luna llena.

P. ¿Que estension tiene el diametro de Mercurio?

R. El diametro de Mercurio es de 3,100 millas.

P. ¿A cuanta distancia está Mercurio del Sol?

R. Mercurio dista del Sol 36,000,000 millas.

CAPITULO V.

Del Planeta Venus.

P. ¿Que aspecto presenta el planeta Venus?

R. Venus es la mas hermosa de las estrellas y su luz la mas brillante despues de la del Sol. Es sin embargo un cuerpo opaco, que refleja la luz solar,

CATECISMO DE ASTRONOMIA. 11

como lo demuestran las desigualdades de la suya, y el aparecer como una mancha oscura cuando pasa por delante del Sol.

P. ¿Que diametro tiene Venus?

R. El diametro de Venus es de 7,498 millas.

P. ¿A cuanta distancia está Venus del Sol?

R. Venus dista del Sol 69,000,000 millas.

CAPITULO VI.

De la Tierra.

P. Haced algunas observaciones sobre la figura y dimensiones de la Tierra.

R. La Tierra es un planeta esferico, aunque, segun las observaciones modernas, algun tanto aplastado acia los polos. Su diametro medio es de 7,914 millas.

P. ¿A cuanta distancia está la Tierra del Sol?

R. La Tierra dista del Sol 95,000,000 millas.

CAPITULO VII.

Del Planeta Marte.

P. ¿Que aspecto presenta el planeta Marte?

R. El planeta Marte despide una luz roja, que procede de la atmosfera espesa que lo rodea. Vistas al traves de esta atmosfera las estrellas fijas, aparecen teñidas del mismó color. La luz de Marte ofrece mucha desigualdad, especialmente en los polos. De estas desigualdades resulta que aquel planeta parece unas veces ovalado, y otras esferico, aunque su figura es semejante a la de la Tierra.

P. ¿Que diametro tiene Marte?

R. El diametro de Marte es de 4,218 millas.

Figure 18.1 Two pages of Ackermann's *Catecismo de astronomía* (Londres, R. Ackermann, 1825). In a catechism, the information is systematically organised and differentiated in short and progressive sections, with the function of fixing the attention and facilitating memorisation. Each question constitutes a unit in itself, which favours its memorisation and repetition in different order. (Courtesy of the Syndics of Cambridge University Library – shelfmark: 1825.5.39)

of Ackermann's catechisms derived from Nature herself, while their legitimacy was provided by the reputation that the publisher had created of himself as 'friend' of the education of the Americans. The authors were just instruments of this transmission of knowledge.[6]

However, if we insist on asking who wrote these texts, the answer is elusive. When a name was mentioned on their title-pages, this actually referred not to the author but the translator. In fact, most of these catechisms were adapted translations of English educational works, and in the process of their publication so many factors were involved that the changes cannot be attributed solely to the role of the translators either. In Ackermann's enterprise as a whole, the selection, writing, translation and editing of what was published was not the decision of a single will but the result of a series of activities and negotiations among the different participating agents. In Ackermann's catechisms the author is not a single or unified figure, but rather a diffused one, that it is nonetheless possible to reconstruct to some extent in historical terms.

The translators of the catechisms – José Joaquín de Mora, Joaquín Lorenzo de Villanueva, José de Urcullu and José Núñez de Arenas – were members of the community of Spaniards exiled in London in the 1820s, who up to then had had little to do with Spanish America. None of them had been there and, with the exception of Villanueva, author of an important political treatise on the Bourbon regime in the late eighteenth century, none of them had produced any literary work that had reached those lands.[7] And whilst Mora had some experience in political and literary journalism in Spain, Núñez and Urcullu had hardly published anything before they took part in this educational series. Yet after their participation in Ackermann's enterprise, most of these writers continued producing literary or educational works in the different countries to which they emigrated.

The Spanish exiles had links with a number of Spanish American diplomats in England in the same period; members of both communities shared political ideals and some Spaniards sympathised with the independence of the former colonies. The latter were liberals who had fought for the establishment of the constitutional monarchy in Spain and who had fled to England after the restoration of the absolutist regime in 1823; deprived of their basic means of subsistence, they had to find alternative ways to earn their livings, and the craft of writing became the most viable option for some of them. The Spanish Americans were there with the mission of obtaining Britain's official recognition of their independence, negotiating loans, or organising companies for the exploitation of mines with British capital; they were much concerned with the moral improvement of their new countries, and some of them decided to sponsor the publication of instructive materials for export.[8]

These two groups converged in the business of Rudolph Ackermann, known for his philanthropic recruitment of European exiles in his workshop, and who also had a variety of entrepreneurial interests in Latin America: he invested largely in mining companies and debt bonds from different countries, and even printed the bond certificates of the first Spanish American loan, for Colombia, in 1822.[9]

Ackermann had the initiative to publish a literary magazine in Spanish; he employed the Spanish theological writer Joseph Blanco White (settled in England long before the 1823 wave of Spanish emigration) as the editor of *Variedades o Mensagero de Londres* (1823–25). Later, two agents of the Mexican government, Manuel Borja Migoni and Vicente Rocafuerte, as well as the Argentine envoy, Bernardino Rivadavia (future president of his country), sponsored the publication of a variety of works from the house of Ackermann, whilst the above-mentioned group of Spanish exiles were employed as hack writers or translators. It was Rocafuerte who, in the middle of 1824, suggested the publication of the series of catechisms to Ackermann. A liberal known for his strong sense of the cultural and political unity of all the Spanish American countries, Rocafuerte's idea of providing some basic education for Americans through these elementary texts was motivated by his conviction that 'intelligence and virtue are the true elements of freedom, and that the people who lack some kind of knowledge which has already been generalised among the popular masses of Europe, cannot be free'.[10] It is not known whether he sponsored the publication of the catechisms as he did others of Ackermann's works, but he certainly contributed to their promotion and distribution in the Spanish American countries.[11] This was one of Rocafuerte's multiple activities related to the publishing of educational works for Spanish America, which included subsidising a literary and instructive periodical published by a different group of Spanish exiles, and collaborating (as did many other Spanish Americans in London) with the works of the British and Foreign Bible Society and the British and Foreign School Society in that area.

The variety of people involved in Ackermann's enterprise implied that there was a variety of interests at play which were not always compatible and had to be negotiated. On the one hand, the Latin American sponsors were interested in publications of an educational character; this explains why Ackermann published very few works in Spanish on art, which formed the core of his English productions – lavishly decorated and inevitably expensive. On the other, the Spanish exiles, translating by commission – with a fee of around £10 per catechism[12] – rather than writing at will, still managed to imprint in their texts some of their own political ideals, relating to their sympathy for the independence of Spanish America and their own need to construct an identity as Spaniards in exile. Moreover, the publisher, although genuinely interested in the education of the Spanish American peoples, also aimed at producing commercially successful works, and therefore tried to ensure that his books and magazines were relatively cheap, attractive to a general audience and innocuous to the religious and political beliefs of their readers.

An illustration of the kind of negotiations going on within the publishing house may be observed in the relationship between Ackermann and Blanco White as editor of *Variedades*. The publisher's project of a miscellaneous magazine that would promote good taste among the Spanish American middle classes – something similar to his successful *Repository of Arts* for the English public – had to be adjusted to the strong convictions of Blanco White, who was more concerned with theological and political affairs. Reluctant to write about fashions and furniture, which he considered superficial issues that degraded him into 'a literary gallantee show man', White required Ackermann to employ someone else for the explanation of the plates; in return, the editor had to compromise and agree not to write anything that could offend the religious beliefs of the readers in order not to endanger the acceptance and circulation of the magazine. White repeatedly wrote in his journals that he felt 'harassed' by Ackermann, who wanted him to include in *Variedades* pieces of writing with which he disagreed. White took on the services of Pablo de Mendíbil to assist him in the magazine, first with indexes and translations, but later with more substantial pieces of writing; by 1825 White was paying Mendíbil one-third of the £75 he received from Ackermann for every quarterly number. Eventually he quitted this well-paid job to engage in more personally fulfilling writings.[13]

This was an early sign of the negotiating criteria that defined the character of the rest of Ackermann's Spanish publications. Behind almost every work published by Ackermann lies a story of sponsors, publishers and writers (necessarily more than one in the case of a translation) negotiating its content before it went into print. In the case of the catechisms, a close comparison between the Spanish and the English editions reveals the different concerns that were involved in the process of their translation.

Natural sciences translated

Ackermann's catechisms covered the following subjects: ancient and modern history, 'domestic and rural industry', Spanish and Latin grammar, literature, morality, music, mythology, political economy, rhetoric, algebra, arithmetic, geometry, agriculture, astronomy, geography, natural history and chemistry. The series was inspired by similar series of textbooks or books of an encyclopaedic range produced in Britain for a working-class audience, such as the series of William Pinnock's *Catechisms* (first published in the 1820s), Irving's *Catechisms* (1810s–1820s), later to be followed by those produced by the Society for the Diffusion of Useful Knowledge (from 1827) and by the Society for the Promotion of Christian Knowledge (1840s). Ackermann also published ten manuals in a non-interrogative form for school use, on such subjects as gymnastics, drawing and the English language. Of the catechisms of natural sciences, those of geography, astronomy, natural history and agriculture were adapted translations of Pinnock's *Catechisms*.[14] The manual on chemistry was an abridgement of Samuel Parkes' *Chemical Catechism* (1st edn. 1806), and the catechism of geography for the use of the terrestrial and celestial globes is closely related to John Greig's *An Introduction for the Use of the*

Globes, for Youth of Both Sexes (1816).[15] None of the Spanish catechisms gives credit to these works, but there is a striking similarity in the format and title-pages of Ackermann's and Pinnock's texts in their 1820s editions (Figures 18. 2 and 18.3).[16]

Of the catechisms of natural sciences, it is known that the geography text was translated by Mora, the one on natural history by José de Urcullu, and the one on geography for the use of the globes by Núñez de Arenas, although not all these names appeared in the first editions. All the catechisms were printed in small 18°, and they had no illustrations apart from the title-page and an occasional plate attached at the beginning of the book. The texts were sold at 2s. in London, and at 1 peso in America, or at a reduced price if bought wholesale.[17]

Although very similar in appearance to their English counterparts, Ackermann's catechisms were aimed at very different audiences. The English catechisms – in particular Pinnock's – were addressed to the subjects of the British crown, they were directed mainly to a school audience, they set out to simplify information as much as possible, they were suitable for people of different religious denominations, and they had a moralising content that reflected the conservative values produced in Britain by the reaction to the French Revolution. Ackermann's catechisms were addressed to the citizens of newly independent and republican countries, they were not only intended for school readers, they were aimed at a deeply Catholic audience, they attempted to convey as much information as possible in every text, they praised knowledge as the key to economic progress and intellectual liberation, and they aimed to contribute to the creation of a Spanish American identity.

Attending to the different audiences addressed, oriented by the needs of the publisher and sponsors, and following their own convictions, the translators omitted, added to and modified parts of the original texts. This can be illustrated by the study of the treatment of two main topics: the values conveyed through natural knowledge, and the image of America.

Useful values

Both the English and the Spanish texts stated that they wanted to convey 'useful' knowledge, but this evidently meant something different in the two series: the utility of science changed from the English to the Spanish texts. Whereas Pinnock's catechism of natural history asserted that the 'utility' of this discipline was that its study, 'rationally applied, is the basis of true religion', in Ackermann's catechism such utility lay in the practical benefits that mankind could derive from knowing the properties of each being in

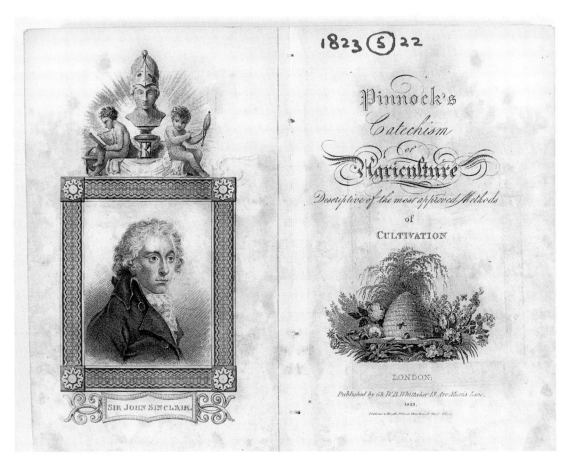

Figures 18.2 and 18.3
Title-pages of Pinnock's *Catechism of Agriculture* (London, G. & W. B. Whittaker, 1823); and Ackermann's *Catecismo de agricultura* (1824). The plates are steel engravings by Perkins and Heath. Pinnock's catechisms had as a frontispiece the portrait of a remarkable personality from the subject treated; Ackermann's texts had no frontispiece, but in some cases a folded plate showing scientific instruments or diagrams was attached opposite the title-page.

the three kingdoms of nature: animals, plants and minerals.[18] Even though the translators of the texts were deeply Catholic, and though Ackermann did not want to offend the religious beliefs of their readers in order not to endanger the circulation of his publications, these texts are more secular by far than the English ones. The *Catecismo de química* eliminated almost all of the providential remarks of Parkes' *Chemical Catechism*, and the translator of the *Catecismo de astronomía* had no compunction in mentioning that the Church condemned Galileo for heresy – a confrontation between science and religion absent from Pinnock's text. Moreover, in the *Catecismo de geografía* there is an emphasis on the different religions practised around the world and on the tolerance for different cults in many of the countries described, something not to be found in Pinnock's text. With strategies like these, the texts implicitly criticised without directly contesting the constitutional regulation that prohibited the practice of any non-Roman Catholic creed in the Spanish American countries.

The emphasis on religious tolerance is indeed surprising, and

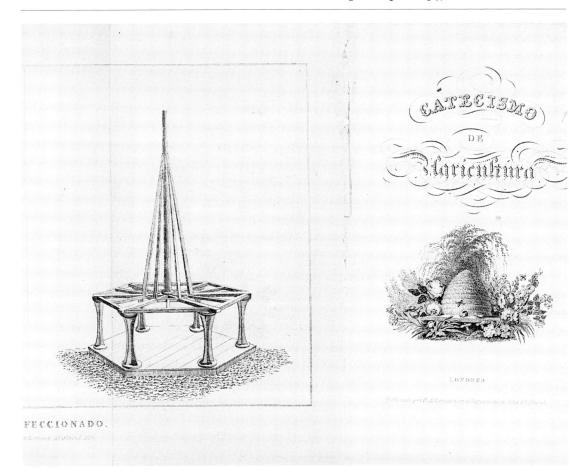

indicates a transformation of the political views of the Spanish by their experience of exile in a foreign country, where their religious beliefs made them a minority. It is even possible that the hand of Blanco White or Rocafuerte in the edition of the texts was responsible for these changes, since both were active promoters of religious tolerance in Spain and Spanish America. Without denying the importance of religious education, both Blanco White and Rocafuerte – like many Spanish Americans of this period – aimed at a morality deprived of the fanatical excesses of a 'baroque' religiosity and wanted to reduce the political and social power of the Catholic Church; this explains their collaboration with the BFBS, whose Bibles and New Testaments promoted a more intimate and less mediated approach to the word of God. On the whole, the Spanish texts were not, like the English ones, fighting against a materialist and potentially atheist view of nature; rather, the utility of scientific knowledge lay in the way it could contribute to the reduction of superstition and fanaticism, and thus to greater individual freedom of the citizens.

This made Ackermann's texts less circumscribed to the English context and at the same time gave science a more independent, universal status. (Courtesy of the Syndics of Cambridge University Library – shelfmark: 1823.5.22; and by permission of the British Library – shelfmark: 7077.b.30, respectively).

Science was also useful to convey moral and social values. The catechisms of geography and natural history offered examples of good and bad qualities of peoples and animals, which also differed from the English to the Spanish texts. The case of the dog is revealing: both texts praise the loyalty of the animal to its master, but Pinnock's catechism emphasises more the dog's unconditional obedience, even its servitude:

Pinnock's *Catechism of Natural History*	Ackermann's *Catecismo de historia natural*
The dog is the most intelligent of all quadrupeds, and on no one can we depend for equal fidelity and attachment. He is obedient and attentive, and performs whatever we wish him to do with alacrity. Satisfied with the most ordinary food, he watches his master's property day and night, and will risk his life in defending it. The ill-usage he may receive he seems to forget, and will lick the hand that chastises him. (28)	[The dog] is the true friend of man; he is attached to man, he caresses, amuses, and defends him, and never leaves him. Even if he is badly fed and maltreated, he is always loyal to his master. He guards the house, chases away the burglars, defends his master, saves his life, and if he cannot come to help him in time, dies at his feet, full of pain. (27)

The subtle difference in describing the dog as 'always loyal to his master' instead of a creature that 'will lick the hand that chastises him', may indicate the difference between a culture aspiring to reinforce the values of submission and obedience to authority and the culture of a community trying to keep loyal to their fatherland while living in exile and working in a different environment. For the latter, the value of 'giving oneself' for others was more important than blind obedience and social servitude.

An explicit 'republican' value is preached with the example of the lion. In Pinnock's text the lion has the attributes of royalty, namely power and magnanimity; in Ackermann's, the lion is not even called the 'king of beasts' and instead his inferiority to man is pointed out. This is a lion for a society with no monarch, constituted not by submissive subjects but by free citizens who are expected to be, as the dog, loyal to their government. Similarly, Pinnock's text calls the eagle 'the king of the feathered race' (48), whereas the Spanish one does not mention its 'royal' status but only says that, like the lion, 'it is used to decorate the coats of arms of the empires *and* the republics' (46).

Defining America

Ackermann's catechisms reflect a consistent effort to introduce information about the American continent which does not appear in the English originals. In the *Catecismo de geografía* America is given a long and detailed section at the end of the book. Animals and plants from America constantly augment the scope of the English text of natural history; and ancient Mexicans merit a space in the history of astronomy in the catechism on the subject. The text on agriculture adds a chapter on maize, the subsistence crop of Mexico and Central America.

However, the image of America presented in the texts illustrates some of the tensions that the translators were experiencing in their own exile. For example, the rich and varied nature of the American continent exerted a fascination on the translator of the *Catecismo de geografía*, who described it in the Humboldtian terms of extravagance and productivity that had so much shaped the European imagination of this period:

Q. What are the climate, territory and physical aspect of America?
A. In such an immense country there must be an extraordinary variety of these circumstances; so all the climates of the globe are found in America. The soil is in some parts incredibly fertile; the physical aspect, in general, is grandiose and beautiful, and America has the most prominent features of the physiognomy of the globe, such as the Niagara fall, the mountains of the Andes and the Amazon river.
Q. What are the main products of America?
A. All the richest that Nature provides in terms of precious stones, metals, timber, medicines, drugs, fruits and cereals.[19]

Yet, as in Humboldt's depiction, the image of an overwhelming nature sometimes seems to efface its human inhabitants, especially with regard to the indigenous population.[20] In their description of the human races, the Spanish texts display a more Eurocentric view than the English ones. The scale of races presented by *Pinnock's Catechism of natural history* gives American Indians a second place, after the inhabitants of Europe, whereas in the Spanish counterpart they appear third, after the Tartars; Negroes also occupy a higher place in Pinnock's text.[21] Both texts accept that all mankind belong to the same species and derive from 'one common parent', and both declare that the white race (Europeans and the peoples around the Mediterranean) is culturally superior to all others (Tartars, Americans, Malayans, Negroes and the inhabitants of the polar regions). But while the English version attributes the superiority of the complexion of the white race to the fact that Europeans inhabit the most temperate climate, the Spanish goes further by saying that the white skin is 'more beautiful', and more 'advantageous' in moral terms; this is because 'all the different passions, all the expressions of happiness or sadness can be easily shown in the cheeks, and the different colour they get is an expressive language of what the soul says', whereas the face of a coloured man shows very little alterations, 'even if the individual has a reason to blush, or even if he is ill, angry or desperate' (13–14). In this sense, being white meant being a more honest and reliable person. The Spanish translator seemed to have forgotten that whites only constituted a minority in the population of the countries where the texts would be read.[22]

Although *living* American Indians were ignored, *ancient* American Indians were not completely forgotten in the Spanish catechisms. The text on astronomy gives an important place to the discoveries of the 'ancient Mexicans' in the history of this science. The Aztecs are praised for their precision in defining the astronomical year, a discovery whose importance is placed alongside the astronomical contributions of the 'ancient' Chinese, Mesopotamians and Indians.[23] This was a way of giving America a place in the development of the sciences: Americans were linked to other peoples who already had a reputation as great civilisations, if only in ancient times.

If it was important to introduce America as a continent within the global world, the translators were also careful in differentiating the countries that constituted it. The geography text has a chapter for each of the American nations, in which their physical aspect, products, population, form of government, main cities and brief account of their history are described. Yet all the historical sections, with the exception of that of Peru, begin with the Spanish or Portuguese conquest. Though there is no defence or reproach of the European domination, the mere omission of the pre-Colombian past reveals the importance given by the translators to the role of their own country in the formation of the American nations. With examples like this, and despite the effort to give Spanish America a place in the world, these countries appear at times as a mere extension of Spain.

Yet America is not only defined in relation to Spain but also to Britain. In several texts Spain appears as a country in a politically backward state with which the Spanish Americans should break, whereas Britain is presented as the model of technological improvement and moral progress that they should emulate.

It has been argued that most of the European literature on Latin America in this period (much of it stemming from Humboldt) was shaped by a European ideology of economic expansionism that attempted to depict America as 'empty' of humans and abundant in natural resources ready to be exploited. It has also been suggested that the neglecting of the Indian, black and mixed population in the Latin American writings of the first decades of independence was related to the American Creole project of founding a decolonised society with European values and white supremacy.[24] These factors, even if they can be attributed to the interests of the English publisher or of the Spanish American sponsors, would not suffice on their own to explain the contradictions in the texts. Only if we look at the process of their production more closely can we appreciate the role of the translators, who projected onto the texts their own European awareness and their own ignorance about the American reality, as well as their feelings for their own country and for the country that had received them in exile.

Conclusion

What Ackermann's catechisms conveyed to the peoples of Spanish America was not the result of a single will, and rather expressed the variety of interests at play among the participants in the publishing enterprise. Here I have shown how in the process of production the content of the catechisms was 'customised' to the educational, moral and political needs of the Spanish American republics as the various participants perceived them. Understanding this transformation is of interest both in its own right and as a starting point for the study of the reception and use of the works.

It is an interesting paradox that books of this genre, whose authority derives from the idea that knowledge is something fixed, given by nature and not subject to opinion, were written not by one individual, but they were a composite, a manufactured product. In the case of textbooks, this lack of individual authorship enables us to gain a better understanding of the social mechanisms involved in a process of production and transmission of knowledge.

Notes

Grateful thanks to Nick Jardine, William St Clair, Jim Secord, Chris Stray and Jon Topham for advice and encouragement.

1 *Recuerdos de provincia* (Buenos Aires, 1979), pp. 157–8 (my translation).
2 See J. Feather, *Publishing, Piracy and Politics* (London, 1994), ch. 6.
3 For a list of Ackermann's publications in Spanish, see J. Ford, 'Rudolph Ackermann: publisher to Latin America', in *Bello y Londres. Segundo Congreso del Bicentenario*, 2 vols. (Caracas, 1980), vol. I, pp. 197–224.
4 For the origin and development of the textbook in the nineteenth century, see C. Stray, 'Paradigms regained: towards a historical sociology of the textbook', *Curriculum Studies*, 26/1 (1994): 1–29.
5 Later editions of some of the originally anonymous catechisms (published in France, Spain or the Spanish American countries) appeared with an author's name. Some of these names, like Urcullu or Núñez de Arenas, became authoritative for their texts on arithmetic throughout the 1860s and 1870s, several decades after the writers' deaths.
6 For a discussion on genre and the kind of conventions it imposes on publishers, writers and readers, see E. D. Hirsch, *Validity in*

Interpretation (New Haven and London, 1967), pp. 68–126, and R. Hodge and G. Kress, *Social Semiotics* (Cambridge and Oxford, 1988), p. 7. On the catechetical genre and its uses in a particular system of education, see E. Roldán Vera, 'Reading in questions and answers: the catechetical genre for the teaching of science in early-independent Latin America' (forthcoming).

7 In 1827 José Joaquín de Mora emigrated to South America and soon acquired a reputation for his political and educational activities in Argentina, Chile, Peru and Bolivia.

8 On the Spanish exiles in England in the 1820s, see V. Llorens, *Liberales y románticos: una emigración española en Inglaterra (1823–1834)*, 2nd edn. (Madrid, 1979). On the Spanish American community in London, see M. T. Berruezo León, *La lucha de Hispanoamérica por su independencia en Inglaterra 1800–1830* (Madrid, 1989), and K. Racine, 'Imagining independence: London's Spanish-American community, 1790–1829', PhD dissertation, Tulane University (1996).

9 See J. Ford, 'Rudolph Ackermann: culture and commerce in Latin America, 1822–1828', in J. Lynch (ed.), *Andrés Bello: the London Years* (Surrey, 1982), pp. 137–52.

10 Vicente Rocafuerte, *A la nación* (Lima, 1845), cited in *Vicente Rocafuerte, un americano libre*, prol. and notes by José Antonio Fernández de Castro (México, 1947), p. 49.

11 Indeed, I have found no evidence of any Spanish American sponsoring the publication of the catechisms themselves; they were only involved in the idea of the series and the contents of the books.

12 This is an estimate derived from the lists of payments to his translators from Ackermann's accounts with Coutts Bank. I am grateful to John Ford, former director of Ackermann & Co., for giving me access to them.

13 See Blanco White's 'Student's journals' for the years 1822 (Blanco White Papers, BW 3/i, Harris Manchester College, Oxford), 1823, 1824 and 1825 (Blanco White Papers, BW III/63–5, Sidney Jones Library, University of Liverpool). Also Blanco White's 'Book of accounts' (Blanco White Papers, BW III/59, Sidney Jones Library, University of Liverpool), and *The Life of the Rev. Joseph Blanco White, Written by Himself; with Portions of his Correspondence*, ed. John Hamilton Thom, 3 vols. (London, 1845).

14 Pinnock's catechisms covered sixty-four different subjects and reached several dozens of editions throughout the first half of the nineteenth century. They were published by William Pinnock in Newbury from *c.* 1810 to 1815; by the association of Pinnock and Samuel Maunder in London from 1815 to 1820, and by William Whittaker from 1821 to the 1840s.

15 For Parkes' text I consulted the 10th edition (1822). Both Parkes' and Greig's manuals were published in London by Baldwin, Cradock and Joy.

16 It is possible that Ackermann settled an agreement with G. B. Whittaker, the publisher of Pinnock's catechisms in the 1820s, for the translation of some of the catechisms. A meeting between Ackermann, G. B. Whittaker, William Blackwood and Blanco White took place in

London on 6 July 1824 (mentioned in Blanco White's 'Student's Journal' for 1824, p. 64). What happened in that meeting is not known.
17 By contrast with Pinnock's *Catechisms*, which were sold at 9d. Ackermann's texts, apart from being longer, obviously involved higher production and distribution costs, a reason why they soon began to be pirated in the Spanish American countries and France.
18 *Pinnock's Catechism of Natural History*, 6th edn. (London, Pinnock and Maunder, c. 1820), p. 2; José de Urcullu, *Catecismo de historia natural* (Londres, R. Ackermann [1826]), p. 1.
19 *Catecismo de geografía*, p. 48.
20 For a discussion of Humboldt's and Humboldtian accounts on the effacement of men by nature in America, and on how this served to legitimate a European ideology of expansion, see M. L. Pratt, *Imperial Eyes: Travel Writing and Transculturation* (London and New York, 1992).
21 *Pinnock's Catechism of Natural History*, p. 7; *Catecismo de historia natural*, p. 12.
22 By the 1820s, the total population of Spanish America was about 17 million, of which 3.2 million (18 per cent) were whites; around 60 per cent were Indians and the rest were *mestizo* or mixed blood.
23 *Catecismo de astronomía* (Londres, R. Ackermann [1825]), p. 3.
24 See Pratt, *Imperial Eyes*, chs. 6–8.

Further reading

J. Ford, *Ackermann, 1783–1983: The Business of Art* (London, 1983)

T. Halperin, *The Contemporary History of Latin America*, 14th edn., ed. and trans. John Charles Chasteen (London, 1993)

M. L. Pratt, *Imperial Eyes: Travel Writing and Transculturation* (London and New York, 1992)

J. Rodríguez, *The Emergence of Spanish America: Vicente Rocafuerte and Spanish Americanism, 1808–1832* (Berkeley, 1975)

E. Roldán Vera, 'Book export and the transposition of knowledge from Britain to Early Independent Spanish America', Ph.D. dissertation, University of Cambridge (forthcoming).

M. Rose, *Authors and Owners: The Invention of Copyright* (Cambridge, Mass. and London, 1993)

C. Stray, 'Paradigms regained: towards a historical sociology of the textbook', *Curriculum Studies*, 26, 1 (1994): 1–29

J. Topham, 'Science and popular education in the 1830s: the role of the Bridgewater Treatises', *British Journal for the History of Science*, 25 (1992): 397–430

M. Woodmansee and P. Jaszi (eds.), *The Construction of Authorship: Textual Appropriation in Law and Literature* (Durham and London, 1994)

19 Editing a hero of modern science

Mention the name of Francis Bacon to an early modernist and their response is guaranteed: 'Father of Modern Science', they will add, thereby distinguishing the great Verulam from his twentieth-century namesake. Yet it is curiously difficult in practice to justify that soubriquet. Bacon's inductive method is in large measure the retrospective construction of twentieth-century philosophers of science; his natural histories remained incomplete at his death, his proposed scientific investigations still unexecuted schematic proposals. To understand how Bacon comes to stand magisterially at the head of British history of science we have, in fact, to turn to two of his nineteenth-century admirers, and to a defining moment in the history of the book.

An unofficial intellect: the forming of James Spedding

Today the name of Francis Bacon is linked with one name above all others – that of the independently wealthy private scholar James Spedding. Spedding's life work – a seven-volume edition of Bacon's work followed by a seven-volume biographical edition of his *Letters and Life* – remains the standard edition for both scholars and biographers.[1] His love-affair with Bacon and his works developed in his youth, and was sustained throughout his life – on several occasions he turned down potentially prestigious and lucrative employment because it would have interfered with his ability to concentrate all his intellectual energies on his passionate pursuit of the 'real' Bacon. That passion was fuelled, and ultimately shaped, by a steady chorus of disapproval for Bacon, voiced in quality periodicals like the *Edinburgh Review*, and in historical publications on the political life of the Elizabethan and Jacobean eras in England, to which Spedding was bent on responding. His lifelong efforts to rehabilitate what he regarded as Bacon's unjustly tarnished reputation were consistently aimed at providing a convincingly unified picture of Bacon's 'life and thought' and targeted against a version of the great man which gave him a mean and nasty personality, and a flawed spirit. Ironically, it was Spedding's obsessive concern with what he regarded as the injustice of the received tradition which enshrined the memory of the 'bad' Bacon conclusively for posterity.

But then, James Spedding was the kind of man who clung doggedly to any undertaking once he had begun it – the product of a generation which invested heavily in its heroes, with a fervour which incorporated nationalistic pride and an idealistic, neo-classical regard for pure learning of a particularly nineteenth-century kind. Raised in the shadow of romanticism and European revolution, he took on the detractors of his idol as if his triumph over them would hold back a latent philistinism and pragmatism which threatened to overwhelm the civilised world. When we handle the sober volumes which are Spedding's lasting monument to his hero we need to remember that their editor saw his enterprise as offering a bulwark against the grimy opportunism of modernity.

Spedding made his 'first acquaintance with Bacon' as a schoolboy through *The British Nepos*, William Mavor's moralising textbook which aimed to instil virtue into its youthful readers through its uplifting biographies of great British men.[2] Mavor's Bacon exemplified a classic paradox: he combined greatness in intellectual production with baseness in his personal conduct, most dramatically in his apparent betrayal of his friend and patron, the Earl of Essex, and in his fall in 1621 on charges of accepting bribes as Lord Chancellor – Mavor followed Alexander Pope's memorable dictum that Bacon was the 'brightest, wisest, meanest of mankind'. Spedding later recalled how this standard version of Bacon captured his imagination:

There he was duly described as the 'brightest, wisest, meanest', &c. The greatness of his intellect was duly, his services to philosophy more than duly, celebrated. His ingratitude to Essex, his ambition, his servility, his duplicity, his hollowness of heart, his corruptions and his disgrace, were all set forth . . . I took it all in as freely as the Gospel. It woke in me neither suspicion nor wonder. I was a schoolboy, and of course thought as everybody thought. There was the outline of his face, in which I used to look for traits of the craft and cunning of the long-headed old villain. I read the Essays, and thought them very wise; but without the slightest wish to think the writer good. Not a doubt on the subject entered my head until after I had reached the years of discretion.[3]

By the time Spedding went up to Trinity College, Cambridge in 1827 (where he became a key member of the select circle of high-minded and cultivated undergraduates known as the Cambridge Apostles, alongside the young Tennyson), his life-long high-minded interest in Francis Bacon (another Trinity alumnus) was firmly in place.

Spedding graduated from Trinity in 1831. Thereafter his substantial private means created a curious problem for him in deciding on a career. In spite of the high regard in which he was held by his peers, he failed twice in his attempts to gain Cambridge fellowships which would have allowed him to stay on and pursue his scholarly activities. He did not need to work for a living; but he was not happy to be seen by his contemporaries merely as a 'private scholar'. In the end Spedding somewhat reluctantly accepted an administrative job at the Colonial Office, worth £150 per annum, which he was offered by Sir Henry Taylor, to whose notice he had come when he made a speech to a Cambridge debating club while an undergraduate. Sir Henry later wrote that Spedding 'was in a difficulty at the time about the choice of a profession; and, feeling that a life without business or occupation of some kind was dangerous, was glad to accept this employment as one which might answer the purpose well enough, if he proved suited to it, and, if not, might be relinquished without difficulty and exchanged for some other'.[4]

Spedding quickly regretted his decision to go to the Colonial Office, but could not make up his mind to leave. His friends noticed his unhappiness. J. W. Blakesley wrote to Thompson: 'He laments that he did not go to the bar. He is convinced that Nature meant him to be an unsuccessful barrister. He would begin to write a book in order to be able conscientiously to resign the Colonial Department, but that he cannot trust himself to doing anything else than lie in bed and smoke cigars'.[5]

Fortunately, Spedding was about to find himself an appropriately idealistic and elevated career, and a project which would occupy him for an entire lifetime. The Cambridge Apostles, with whom Spedding kept up his intellectual contact, turned their attention to Francis Bacon. One Apostle, Spedding's schoolfriend Edward FitzGerald reported to W. B. Donne in March 1834 that 'We were all talking the other night of Basil Montagu's new Life of Bacon – have you read it? It is said to be very elaborate and tedious. A good life of Bacon is much wanted. But perhaps it is as difficult to find a proper historian for him as for anyone that ever lived'.[6]

Spedding resolved to undertake that 'good life of Bacon' himself. Over the next five years he devoted increasing amounts of time to researching his hero's life, to the point when his employment at the Colonial Office became a positive encumbrance (he was never one to work long hours). Thomas Carlyle reported in September 1841 that 'Spedding is an Ex-Secretary; out of that task, and all connexion with the Colonial Office; by his own wish and resolution, it seems. He is not to quit London; but to live as an unofficial Intellect there'.[7] According to Sir Henry Taylor, Spedding 'took the opportunity of the Whig Government going out in 1841 to give up his employment. He then applied himself to edit the works and vindicate the fame of Lord Bacon'.[8] By 1847, Spedding was entirely absorbed in his Bacon project. When Sir James Stephen, with whom Spedding had worked in the Colonial Office, retired in 1847, the post of Under Secretary of State which he had held was offered to Spedding, at a salary of £2,000. Spedding turned down this very considerable career opportunity. According to Leslie Stephen, he 'could not be persuaded to abandon Bacon'.[9]

A very clever man: Macaulay writes on Bacon

It is not hard, then, to imagine how offensive Spedding found a characteristically rebarbative essay on Bacon which the eminent historian Thomas Babington, Lord Macaulay published in the *Edinburgh Review* in 1837, under the excuse of reviewing the *Life of Lord Bacon* by Bacon's early nineteenth-century editor, Basil Montagu, which had come out the previous year.

Macaulay was precisely the kind of high-profile essayist – cavalier when it came to facts – calculated to arouse James Spedding's indignation at the best of times. His public prominence meant that his pronouncements on any subject to which he turned his critical attention were (as Spedding had occasion later to bemoan) bound to gain wide dissemination. His urbanity and wit made it unlikely that anyone challenging him in print would get the better of the argument.

Macaulay's attitude to Francis Bacon was a complicated one, a combination of admiration and studied disdain. Writing to his sisters in 1815, Macaulay praises Bacon as 'the man who knew the human mind better than all others who ever lived';[10] he appears in Macaulay's roll-call of 'the Writers of modern times' – 'Milton, Pope, Spenser, and Racine, Bacon, Johnson, and Bossuet'.[11] But Bacon also stood for a particular type of fundamentally flawed character – a man who could colourfully be represented as standing for the dilemma of romanticised genius, whose brilliance coexisted uneasily with tragic imperfections of conduct and temperament. In 1842, Macaulay wrote a concerned letter to his sister Frances about his nephew Charley in which he invoked Bacon as a recognisable type of 'selfish intellectual voluptuary':

> I cannot help thinking that [Charley] is a good deal like what I was at his age. I am very anxious about the way in which he passes the next few years, and very anxious too that his affections may be developed as well as his understanding. I have sometimes feared that, in consequence of the peculiar circumstances in which he is placed, he may turn out a mere selfish intellectual voluptuary, learned, polished, curious, skilled in criticism and in all things appertaining to art, but loving nothing. Goethe's character is the sort of thing that I mean: and Bacon's has some resemblance to it . . . the tendency of such a life is not the best.[12]

Macaulay produced this version of Bacon's temperament again in discussions in which he participated in 1856, when Trinity College was engaged in heated debate as to whose statue should next be placed in the college chapel as the gift of Lord Lansdowne (like Bacon and Spedding, Macaulay was a Trinity man). The philosopher William Whewell had already presented a statue of Bacon (on whose philosophy he too had pronounced professionally); Macaulay now suggested Milton; Whewell suggested Herbert, Dryden, John Pearson, Isaac Barrow, or Bentley. In the course of the deliberations, Whewell wrote to Lansdowne pointing out that it might after all be less appropriate to erect a permanent memorial to those cultural figures who 'have a moral blemish, as Bentley or Dryden'. Macaulay retorted:

But surely you, to whom we owe that fine monument of Bacon, will, on reflection, admit that the faults of Bentley were not such as ought to be punished by permanent exclusion from public honours. Dryden was immoral as a poet, Bacon as a Judge, Bentley as Master of a College. I therefore would not set up any monument to Dryden in his character of poet, to Bacon in his character of Judge, or to Bentley in his character of Master of a College. But Dryden has no claim to a monument except as a poet. His licentiousness taints those very works on which alone his fame depends; and it is impossible to do honor to the writer without doing honor to the libertine. With Bacon . . . the case is quite different. You testified your respect for the great philosopher, although you knew that he had been a servile politician and a corrupt Chancellor.[13]

Macaulay's ability here to see both 'the great philosopher' and the servile, corrupt man, was central to his understanding of Bacon.

In January 1836 Macaulay informed the editor of *The Edinburgh Review*, Macvey Napier, that he was 'writing a review of Basil Montagu's Life of Lord Bacon. It will be immeasurably long, I fear, and very superficial in the philosophical part. But I rather think that it will be liked'.[14] The article was published in issue 132 of *The Edinburgh Review* in July 1837; running to 104 pages, it was prefaced with an apology for the 'unusual length of this article' which may 'startle some of our readers', but Napier believed 'we cannot bring ourselves to think it possible that there is any intelligent scholar, who, on perusal, could wish it shorter'.[15] Six years later, the piece received wider dissemination when included in a collection of Macaulay's *Critical and Historical Essays* from *The Edinburgh Review*, designed to combat an invasion of imported American editions of his work.[16]

Macaulay's article was controversial from the day it appeared. Sir David Brewster cautioned that 'the Reviewer has taken an extreme view of Bacon's conduct'.[17] Henry Brougham thought it 'very striking, and no doubt . . . the work of an extremely clever man', but pinpointed 'two grievous defects'. Firstly, 'a redundancy, an overcrowding of every one thing that is touched upon, that almost turns one's head; for it is out of one digression into another, and each thought in each is illustrated by twenty different cases and anecdotes, all of which follow from the first without any effort. This is a sad defect in Macaulay, and it really seems to get worse instead of better. I need not say that it is the defect of a very clever person – it is indeed exuberance. But it is a defect also that *old age* is liable to . . .'[18] Secondly, Brougham attacked Macaulay's style of writing, which mistook 'garrulity for copiousness'. But it was Macaulay's misunderstanding of Bacon's philosophy of induction that drew Brougham's strongest disapproval:

Greater blunder never was committed than the one Macaulay has made on the Induction Philosophy. He is quite ignorant of the subject. He may garnish his pages as he pleases with references: it only shows he has read Bacon for the *flowers* and not the *fruit*, and this is indeed the fact. He has no science at all, and cannot reason. His contemporaries at Cambridge always said he had not the conception of what an argument was; and surely it was not right for a person who never had heard of Gilbert's treatise, to discuss Bacon's originality, nay, to descant on Bacon at all, who seems never to have read the *Sylva Sylvarum* (for see p. 83 about ointments for broken bones); and who goes through the whole of his speculation (or whatever you choose to term it) without making any allusion to Bacon's notorious failure when he came to put his own rules in practice, and without seeming to be at all aware that Sir I. Newton was an experimental philosopher.[19]

Macaulay responded that 'Lord Brougham's objections arise from an utter misconception of my whole argument, and every part of it'.[20] In time, he was to have fiercer critics.

Countering Macaulay: Spedding's *Evenings with a Reviewer*

It took Spedding eleven years to put together a worthy and incontrovertible answer to Macaulay's 'libel' on his beloved Bacon: his *Evenings with a Reviewer; or A Free and Particular Examination of Mr Macaulay's Article on Lord Bacon, in a series of dialogues* (London, 1848). Even then he published the two-volume dialogue privately, in a tiny edition intended only for his friends. A proper edition was not published until after Spedding's death, omitting the first section, which was largely an *ad hominem* attack on Macaulay.

In this now neglected first section, Spedding (in the persona of 'B') responds to an intellectual *agent provocateur* who has read, and is inclined to respond favourably to, Macaulay's essay. Interlocutor A starts with the obvious case: 'Well, how does the cause of Bacon prosper? when is Macaulay to be overthrown? For at present, with us who know nothing of the matter but what he tells us, he has it all his own way'.[21] 'B' puts the problem of responding precisely:

To counteract impressions on a subject not generally understood, taken from a writer like Macaulay, whom everybody reads and nobody reviews, is not so easy a matter. He, you know, had no such difficulty to overcome. The impression he was to make was already there; it was easy to print it deeper. He had little to add, nothing to alter, still less to erase. He had merely to retouch the old plate, and put in shadows and background. It was only 'the wisest, brightest, meanest of mankind,' executed in the most approved style of modern art.[22]

The only way to counteract Macaulay's case, says B, is to assemble the detailed evidence to the contrary, and to lay it out in such minute detail, without colouring argument or gloss, that the material can speak for itself:

B
Don't you see that as things now stand such a narrative would make no impression at all; because nobody would believe it to be true? It would be necessary to discuss and argue at every turn; and where would be the use of arguing to an ignorant auditory?
A
You say you have evidence. Could you not produce it?
B
Evidence in abundance: but it lies in volumes not in pages; – evidence which cannot be truly estimated without patience, capacity, a suspended judgement, and an open heart; – evidence to be read
'– as the Scholiast of mankind
Should ever read their acts – conjunctively –
Interpreting the several by the whole.'
A few passages selected here and there will do nothing; because, when men's minds are preoccupied, as in this case, by a radical misconception of the character in question, each separate word and act will be misinterpreted according to that original misconception.[23]

With touching frankness, Spedding (or rather, B) confides in A the fact that his commitment to accuracy and completeness is such that there is no hope of his making the case against Macaulay so as to convince contemporary readers – it will be left to posterity to judge if he has succeeded in finally vindicating Bacon from his detractors. A asks him whether he intends simply to assemble Bacon's letters in a fully chronological order, thus allowing the reader to follow the course of his professional life for themselves. B confesses that he has an altogether more ambitious project in mind:

A
Then you propose to correct the text from the MSS. and former editions, and to print the letters all in their proper order. Is that all? That cannot take much time.
B
I propose to do more than that; though that of itself would be well worth doing. But if you look into the letters, you will find that a good deal more

should be done. I want not only to print the series, but to present it in such a shape that a modern reader, having no more than the average knowledge of a modern reader, may follow and understand it all, without getting out of his easy-chair. Therefore I must begin by (as well as I can) understanding it all myself; which if you knew the variety of subjects the correspondence touches upon, you would allow a good while for. Then having mastered it myself, I must consider how to present it to the reader in such a shape, that by the aid of connecting narrative, commentary, and (what is of no small consequence) a lucid typographical arrangement, he may take in the true meaning without any trouble of reflection.[24]

Not only must B (Spedding) master the confusion of Bacon's correspondence; the entire surviving corpus of Bacon's works must be reconstructed from scratch:

Besides, I include in my design not the letters only, but speeches, charges, tracts on matters of the day, and in general all writings which may be described as *occasional*; that is to say, everything which is addressed to the immediate business of the time, and not to posterity. And this is not all. This was the intention with which, concluding that there was not much *new* matter to be collected, I began. But on looking into it, I soon found that neither in correcting the old matter, nor in looking out for new, has any ordinary diligence been used for the last hundred years nearly. Not even the general indexes in the British Museum appear to have been carefully consulted. Still less has any effectual search been made in places where neither the general index nor the particular catalogues can be trusted as guides. I have already found a good deal of important matter which is quite new, and (except to MS. hunters) inaccessible. And the further I explore, the more I am convinced that there is hardly any collection of MSS. in which there is not a *chance* of finding something or other which will help me. And as I think it is of more consequence that the whole should be done effectually than that something should be done soon, I am resolved not to hurry, and you must be content to wait an indefinite time. I confidently expect, that if I live to do it, and do it tolerably well, it will have the effect (not in this generation but in the next, *i.e.* in the generation born after 1850) of turning the current of opinion on this subject; for I mean it to be the book which every youth of that generation who has any curiosity about Bacon will betake himself as the most readable book on the subject.

A is aghast – we are, after all, somewhere between 1837 and 1848 at the time of writing. What is the reader to do in the meantime?

A
Born after 1850! Then what are we to do, who were born before 1800, and have not so long perhaps to live after 1850?
B
You must continue to read Macaulay, and go uninstructed to your graves, unless you like to buy a copy of Bacon's works (which you can get for two or three guineas) and hunt in it for the facts on which Macaulay founds his fictions. It will give you some trouble; for as he gives no references, it will be difficult to find the particular passages. And when you have found

them, it will be difficult to recognize them as those you want, so little resemblance do they bear to his description. But if you have patience for it, it is very well worth doing.

There is something at once hopelessly idealistic and astonishingly uncompromising about B's response, which we would argue captures the essence of Spedding's scholarly consciousness. There are to be no half measures. An entire lifetime must be sacrificed, and every surviving scrap of documentation perused in order to vindicate Verulam from the affronts of his critics. Anything short of this is not worth undertaking. Anyone who wishes to know the truth in advance of the project's being accomplished must conduct the entire exploration themselves, or live with the current untruths.

Life and works: Spedding edits Bacon

For the rest of his life, Spedding devoted himself singlemindedly to accumulating the evidence in support of his passionate defence of Bacon. His work on his edition of Bacon's works, and his definitive *Letters and Life*, was only once briefly interrupted when he served on the Civil Service Commission in 1855. The first volume of Spedding, Ellis and Heath's *Works* was on the shelves by late January 1857 – 'some seven hundred pages; and the Reviews already begin to think it over-commentaried', as his friend Edward FitzGerald commented – and was promptly succeeded by further volumes, culminating in volume VII in January 1859.[25] This was followed by seven volumes of *Letters and Life*: in fact a collection of 'All his Occasional Works, Namely Letters, Speeches, Tracts, State Papers, Memorials, Devices and All Authentic Writings Not Already Printed among His Philosophical, Literary, or Professional Works, Newly Collected and Set Forth in Chronological Order', which strengthened a 'Commentary Biographical and Historical' which was Spedding's main concern (indeed, in 1878 he produced from these materials a two-volume *Account* of Bacon's life, with many of the 'occasional works' omitted). His stated objective was to provide

an explanatory narrative running between, in which the reader will be supplied to the best of my skill and knowledge with all the information necessary to the right understanding of them. [Such particulars] when finished will in fact contain a complete biography of the man, – a biography the most copious, the most minute, and by the very necessity of the case the fairest, than I can produce; for any material misinterpretation in the commentary will be at once confronted and corrrected by the text. [In this innovative method] the new matter which I shall be able to produce is neither little nor unimportant; but more important than the new matter is the new aspect which (if I may judge of other minds by my own) will be imparted to the old matter by this matter of setting it forth.[26]

While Spedding's motives are quite blatant in the *Letters and Life*, they are perhaps less clear in the *Works*. But, as he makes explicit in *Evenings with a Reviewer*, his *entire* editorial project is devoted to vindicating Bacon's name. The division of his writings into 'Works' and 'Letters' indicates that Spedding was dividing Bacon's oeuvre to produce 'Letters' to explain the 'Life', and 'Works' which would be read independent of, and untarnished by, the 'Life'. (The new Oxford edition of Bacon's writings significantly recasts many of Spedding's 'Letters' as 'Works'.) Spedding explained his rationale in the 'History and Plan of this Edition':

Bacon's works were all published separately, and never collected into a body by himself; and though he had determined, not long before his death, to distribute them into consecutive volumes, the order in which they were to succeed each other was confessedly irregular; a volume of moral and political writings being introduced between the first and second parts of the Instauratio Magna, quite out of place, merely because he had it ready at the time. In arranging the collected works therefore, every editor must use his own judgement.[27]

So Blackbourne in 1730 had attempted to order an *Instauratio magna*; Birch, producing in 1763 what had been the 'trade edition' until Spedding's, followed Blackbourne. About Montagu (1825–34), thought Spedding, the less said the better. Bouillet's 1834 French edition was cited as the best to date, although overly systematic. Spedding's answer was to put Bacon's work in context, 'arranging his works with reference – not to subject, size, language, or form – but to the different classes of readers whose requirements he had in view when he composed'. The works then fell 'naturally' into three categories: first, 'his works in philosophy and general literature; addressed to mankind at large, and meant to be intelligible to educated men of all generations'; second, 'his works on legal subjects; addressed to lawyers, and presuming in the reader such knowledge as belongs to the profession'; third, 'letters, speeches, charges, tracts, state-papers, and other writings of business; relating to subjects so various as to defy classification, but agreeing in this – they were all addressed to particular persons or bodies, had reference to particular occasions, assumed in the persons addressed a knowledge of the circumstances of the time, and cannot be rightly understood except in relation to those circumstances'.[28]

From this, the rationale for Spedding's edition is plain. The third category ('letters, speeches . . .') becomes the *Letters and Life*, with massive contextualisation (or apology) needed; the second category becomes the 'Legal and Professional' writings, of interest to a specialist audience of lawyers; and only the first category, on 'philosophy and general literature', becomes the Bacon 'intelligible to educated men of all generations'. By separating out Bacon in this way, Spedding put obstacles in the way of anyone trying to make connections between, say, his scientific and legal practices, his natural philosophy and legal theory, and, especially, his science and the circumstances of his life. It is a division put in place at one highly charged moment in the history of the book, but one which has had ramifications for the history of science as we know it.

The initial response to Spedding's first two volumes of the *Letters and Life* (published late in 1861) was not enthusiastic. Edward FitzGerald reported in January 1864 that 'Spedding's Bacon seems to hang fire; they say he is disheartened at the little Interest, and less Conviction, that his two first volumes carried'.[29] In private, FitzGerald doubted 'if [the edition] will be half as good as the *Evenings* where Spedding was in the *Passion* which is wanted to fill his Sail for any longer Voyage'.[30] Worse, Spedding's work signally failed to convince even his friends that Bacon's name was now cleared. W. H. Thompson felt that the papers Spedding adduced to clear Bacon in the Essex affair 'rather go against him'; Edward FitzGerald declared that his opinion 'is not the least altered of the Case: and (as I anticipated) Spedding has brooded over his Egg so long he has rather addled it'. Spedding himself was reportedly 'mortified' at the apathy with which his life's work was received, although more because 'he has awakened so little Interest for his Hero. You know his Mortification would not be on *his own* score'. To FitzGerald in late 1862, he confided that he 'could scarce lift up his Pen to go on', and nothing was yet written of volumes III and IV.[31] He complained that his eyes and memory were not what they used to be, and that research and composition had become irksome.

In 1867, Spedding published, at his own expense, two papers together entitled *Authors and Publishers*, denouncing the publishing business for destroying his work. These had been turned down by the periodicals because 'they would offend the Powers upon whom the sale of books depends, and might materially damage the value (as property) of the publication which admitted them'.[32] FitzGerald found the special pleading beside the point: 'The truth is that a solemnly-inaugurated new Edition of all Bacon was not wanted. The Philosophy is surely superseded; not a Wilderness of Speddings can give men a new interest in the Politics and Letters'.[33]

Spedding's final volume appeared in August 1874. His sponsor Sir Henry Taylor wrote in its support, praising its 'impartiality'.[34] Edward FitzGerald, rather more perceptively, thought the final summing up 'simple, noble, deeply pathetic', but 'rather on Spedding's own Account than his Hero's, for whose Vindication so little has been done by the sacrifice of forty years of such a Life as

Spedding's'. Indeed, he concluded, 'Positively, nearly all the new matter which S. has produced makes against, rather than for, Bacon: and I do think the case would have stood better if Spedding had only argued from the old materials and summed up his Vindication in one small Volume some thirty-five years ago'. The conclusion was little more than 'an Appeal "ad misericordiam", citing Witnesses of Character, which do not amount to much, I think'.[35] Thomas Carlyle was more sympathetic, detecting 'a grim strength in Spedding, quietly, very quietly, invincible, which I did not quite know of till this Book; and in all ways I could congratulate the indefatigably patient, placidly invincible and victorious Spedding'.[36]

Spedding himself was only too happy to concede the narrowness of his historical vision. When Prime Minister Gladstone offered him the Chair of Modern History at Cambridge in June 1869, Spedding declined it, claiming that 'the truth is that I am learned only in one chapter (or section of chapter) of English history. Even in that I have still much to learn. And it is too late in the evening to begin a new education'.[37] The only honour Spedding ever accepted was, tellingly enough, an Honorary Fellowship at Trinity College, Cambridge.

In March 1881, Spedding was knocked down by a cab at the bottom of Hay Hill, and taken to St George's Hospital. Typically, his main concern was to assure everyone that the driver of the cab was not to blame. It is perhaps a comment on his appearance by that date that the cab's occupant had Spedding sent to the hospital, rather than returning him (as befitted a gentleman of means) to his private address, to be cared for professionally there. He never returned home – like his hero Francis Bacon he died tragically, unnecessarily, in an incident that bordered on the absurd.

Conclusion

Spedding's Bacon is a magisterial and imposing figurehead, an appropriate 'father' for modern empiricism. We can no longer think of the works of Francis Bacon in any other way than that lovingly produced by James Spedding (loyally supported by his collaborators, Ellis and Heath). The voice of Bacon the philosopher is permanently enshrined in the elevated and high-minded language chosen by Spedding as the English in which to render Bacon's largely Latin oeuvre. What we have forgotten – as Spedding's justificatory *Evenings with a Reviewer* have been buried by the passage of time – is that that tone was self-consciously designed to counter Macaulay's denunciation of Bacon as ambitious, self-seeking and cowardly. The imposing volumes of Spedding's *Works* are, we have shown here, only one constituent part in a turbulent and passionate episode in the History of the Book. The *Works* are a formidable achievement as a work of scholarly reconstruction and retrieval. But as a record of Francis Bacon's contribution to knowledge they need to be handled with a certain amount of critical care.

Notes

Alan Stewart would like to thank the School of English and Humanities and the College Research Fund at Birkbeck, University of London, for their generous support.

Lisa Jardine wishes to thank the School of English and Drama, Queen Mary, University of London, for unfailingly supporting and facilitating her research activities.

1 But see now F. Bacon, *Works*, general editors G. Rees and L. Jardine, 20 vols. (Oxford, 1995–); L. Jardine and A. Stewart, *Hostage to Fortune: The Troubled Life of Francis Bacon 1561–1626* (London, 1998).
2 W. Mavor, *The British Nepos* (London, 1798).
3 J. S.[pedding], *Evenings with a Reviewer; or A Free and Particular Examination of Mr. Macaulay's Article on Lord Bacon, in a series of dialogues*, 2 vols. (London, 1848), vol. I, p. 7.
4 Sir H. Taylor, quoted in G. S. Venables, 'Preface' to J. Spedding, *Evenings with a Reviewer, or Macaulay and Bacon*, 2 vols. (London, 1881) (posthumously published edition), vol. I, ix.
5 J. W. Blakesley to Thompson, 20 September 1837. P. Allen, *The Cambridge Apostles: The Early Years* (Cambridge, 1978), p. 169.
6 FitzGerald to W. B. Donne, [21] March 1834. Edward Fitzgerald, *Letters and Literary Remains of Edward FitzGerald*, ed. W. Aldis Wright, 7 vols. (London, 1902–3), vol. I, p. 45.
7 T. Carlyle, September 1841. T. Carlyle, *Letters*.
8 Sir H. Taylor, quoted in Venables, 'Preface', p. x.
9 L. Stephen, 'Spedding, James (1808–1881)', *Dictionary of National Biography*, vol. LIII, pp. 315–16.
10 Macaulay to Selina and Jane Macaulay, 23 October 1815. T. B. Macaulay, *The Letters of Thomas Babington Macaulay*, ed. T. Pinney (London, 1974–), vol. I, p. 68.
11 Macaulay to Hannah More, 11 November 1816. Macaulay, *Letters*, vol. I, p. 83.
12 Macaulay to Frances Macaulay, 17 May 1842. Macaulay, *Letters*, vol. IV, p. 35.
13 Macaulay to William Whewell, 1 December 1856. Macaulay, *Letters*, vol. VI, p. 69.
14 Macaulay to Macvey Napier, 1 January [1836]. British Library, London, Additional MS 34, 617 fo. 293b; Macaulay, *Letters*, vol. III, p. 163.
15 Macaulay, *Edinburgh Review*, 132 (July 1837): 1–104 at 1 n*.
16 Macaulay, *Critical and Historical Essays, contributed to the Edinburgh Review*, 3 vols. (London, 1843).

17 Sir David Brewster to Macvey Napier, 27 July 1837. Macvey Napier, *Selection from the Correspondence of the late Macvey Napier, Esq.*, ed. Macvey Napier, Jr. (London, 1879), p. 194.
18 Henry Brougham to Napier, 28 July 1837. Napier, *Selection*, p. 196.
19 Brougham to Napier, 28 July 1837. Napier, *Selection*, p. 197.
20 Macaulay to Napier, 14 June 1838. Napier, *Selection*, p. 256; Macaulay, *Letters*, vol. III, p. 243.
21 Spedding, *Evenings*, vol. I, p. 1.
22 Spedding, *Evenings*, vol. I, p. 2.
23 Spedding, *Evenings*, vol. I, p. 4.
24 Spedding, *Evenings*, vol. I, p. 15.
25 FitzGerald to E. B. Cowell, 22 January 1857. FitzGerald, *Letters*, vol. II, p. 52.
26 Spedding, 'Preface' to F. Bacon, *Letters and Life*, ed. Spedding, 7 vols. (London, 1861–74), vol. I.
27 *Works*, vol. I, p. iii.
28 Spedding, 'Preface', to Bacon, *Works*, vol. I, p. iv.
29 FitzGerald to E. B. Cowell, 31 January 1864; FitzGerald, *Letters*, vol. II, p. 178.
30 FitzGerald to W. H. Thompson, 15 July 1861; FitzGerald, *Letters*, vol. II, p. 126.
31 FitzGerald to E. B. Cowell, 5 August 1863; FitzGerald, *Letters*, vol. II, p. 164.
32 J. Spedding, *Publishers and Authors* (London, 1867), p. vi.
33 FitzGerald to Frederic Tennyson, 29 January 1867. FitzGerald, *Letters*, vol. II, p. 226.
34 Sir H. Taylor quoted in Venables, 'Preface', pp. xv–xvi.
35 FitzGerald to Fanny Kemble, 24 August [1874] and FitzGerald to T. Carlyle, 3 November 1874; FitzGerald, *Letters*, vol. III, pp. 128 and 135.
36 T. Carlyle to FitzGerald, 6 November 1874; FitzGerald, *Letters*, vol. III, pp. 137–8.
37 Spedding to W. E. Gladstone, 29 June 1869. BL Add. MS 44, 421 fo. 71.

Further reading

F. Bacon, *Letters and Life*, ed. J. Spedding, 7 vols. (London, 1861–74)
  Works, ed. J. Spedding, R. L. Ellis and D. D. Heath, 7 vols. (London, 1856–9)
  Works, general editors G. Rees and L. Jardine, 20 vols. (Oxford, 1995–)
  The New Organon, ed. L. Jardine and H. Silverthorne (Cambridge, 2000)
R. Iliffe, 'Author-mongering: the "editor" between producer and consumer', in A. Bermingham and J. Brewer (eds.), *The Consumption of Culture 1600–1800. Image, Object, Text* (London and New York, 1995), pp. 166–92
L. Jardine and A. Stewart, *Hostage to Fortune: The Troubled Life of Francis Bacon (1561–1626)* (London, 1998)

T. B. Macaulay, review of F. Bacon, *Works*, ed. B. Montagu, in *Edinburgh Review*, 132 (July 1837): 1–104

M. Shortland and R. Yeo (eds.), *Telling Lives in Science. Essays on Scientific Biography* (Cambridge, 1996)

J. Smith, *Fact and Feeling: Baconian Science and the Nineteenth-Century Literary Imagination* (Madison, WI, 1994)

J. S.[pedding], *Evenings with a Reviewer; or A Free and Particular Examination of Mr. Macaulay's Article on Lord Bacon, in a series of dialogues*, 2 vols. (London, 1848)

    Evenings with a Reviewer: or, Macaulay and Bacon with a prefatory notice by G. S. Venables (London, 1881)

J. H. Wiener (ed.), *Innovators and Preachers: The Role of the Editor in Victorian England* (Westport, CT, 1985)

20 Progress in print

Travellers to Edinburgh in the early nineteenth century were drawn to 'the most busy scene in the Bibliopolic world of the North', Archibald Constable's high street bookshop. Many celebrated works issued from the low, dimly lit chamber, which echoed to 'the incessant cackle of the young Whigs'; among the most notable was the *Encyclopaedia Britannica*'s six-volume 'Supplement', published in 1824.[1] This featured extensive 'preliminary dissertations' which surveyed each science's framework and methods, introduced its history and defined its boundaries. The encyclopaedic enterprise had abandoned its unifying aims. Older forms from the Enlightenment, particularly the philosophic treatise in the mould of Adam Smith's *Wealth of Nations* (1776) and Isaac Newton's *Principia* (1687), became the template for new genres of reflective works on the sciences.

Systematic reflections and dissertations on the sciences blossomed throughout Europe after the French Revolution and the Napoleonic Wars. These were neither textbooks nor simplified popularisations, but books addressing a wide readership that included specialists. In Britain their greatest efflorescence was during the decade around the parliamentary reform year, 1832. These works have usually been discussed in connection with specific disciplines: induction in philosophy, uniformity in geology, the problem of 'systems' in natural history. But contemporaries read them as contributing to debates about a vision of nature appropriate to the emerging order of the machine. These books belonged to new genres of literary production, driven simultaneously by publishing finances and by campaigns for authority over an emerging mass readership. This chapter, then, analyses the production of scientific systems in a publishing climate engendered by reform agitation and industrialisation.

The industrial revolution in communication

The traditional picture of an industrial revolution focused on the later years of the eighteenth century has been questioned in recent years. At least until the 1840s, factory production characterised only textiles, iron and a few other sectors, and artisanal labour remained important even after parts of a process were mechanised. Industrialisation is now seen as a long-term process focused on the restructuring of finance, labour practices and patterns of consumption, extending back into the late seventeenth century. Yet the meaning of industrial revolution was as much in perception as in output. People who lived through the first half of the nineteenth century believed that they were witnessing unprecedented change. The consciousness of living in unique, historicised moments, which had come into being in response to the epochal events of the French Revolution, extended during the 1830s and 1840s into a sense that the material basis of human life was at a pivotal junction: the steam engine loomed over the intellectual landscape as the chief symbol of a new age. The machine dominated public debate partly because communication and transport were in the vanguard of the economic sectors undergoing industrialisation. New technologies, from steam presses to railway bookstalls, postage stamps and telegraph lines, were altering not only the terms of debate, but its fundamental forms.

The most visible and widely discussed change involved the movement of goods, people and information on the railways. The railways were the latest in a series of improvements in transportation which extended back into the later years of the eighteenth century, from the macadamed roads and canals to the high-speed horse carriage systems of the 1810s and 1820s. The first steam railway linked Manchester and Liverpool in 1830, and the network grew from about 500 miles in 1838, to 2,000 miles in 1844, to 7,500 miles in 1852. The geographical space of the British Isles shrank to between one-fifth and one-third of its former scale, at a cost which made travel available to a much wider proportion of the population. Establishment of the network forever altered the relation between different parts of the country. Most important of all, railways were progressive symbols of a technological age.

From the perspective of intellectual and literary life, the most significant changes involved printing and publishing.[2] Innovations ranging from paper-making machines to machine-stamped bindings, from improved mechanisms of distribution to sales at railway bookstalls, transformed processes that had altered little since the sixteenth century. In terms of output and innovation, the industry was small compared with textiles, iron or the railways. But print culture, reporting on itself, occupied a central place in public awareness of industrial revolution.

As in most economic sectors, the introduction of new printing technologies and forms of organisation was gradual, so that books and newspapers continued to involve a combination of hand and machine work. One of the early innovations had been the hand-

Figure 20.1 'The man wots got the whip hand of 'em all'. Cheered on by a printer's devil, the Stanhope iron-frame press drives away obscurantism and privilege. The press sports a French liberty cap, and sulfurous fumes rise towards a print of the Duke of Wellington. Hand-coloured engraving by William Heath, 1829. (By permission of the British Museum, Department of Prints and Drawings.)

operated iron-frame press, introduced at the start of the nineteenth century to replace the older wooden designs which had persisted almost unchanged since the introduction of printing. Although output was not much improved, the new presses were exceptionally durable and could print small type with great delicacy. Costing about three times as much as the wooden models, iron-frame hand presses remained cheap enough to be bought by small proprietors, yet sturdy enough to withstand larger print runs than had been possible before. Artisan-controlled, the hand press became a key symbol of popular rights (Figure 20.1). The freethinker Richard Carlile (1790–1843), in his 1821 *Address to Men of Science*, proclaimed that

Figure 20.2 An Applegarth and Cowper rotary 'Printing-Machine', powered by steam and engaged in printing the *Penny Magazine*, the useful-knowledge periodical in which this wood-engraving appeared. From 'The commercial history of a penny magazine', *Penny Magazine* (31 Dec. 1833), p. 509. (By permission of the Syndics of Cambridge University Library)

the press 'has come like a true Messiah to emancipate the great family of mankind' from 'kingly and priestly influence'.[3]

The steam printing machine – large, fast, expensive – was a different matter (Figure 20.2). First introduced to break the power of the pressmen at *The Times* newspaper in 1814, it came into widespread use in the 1830s, after important patents had expired. (Even then most specialised works continued to be printed by hand, although perhaps the first book printed by machine was the second

edition of John Elliotson's translation of Johann Friedrich Blumenbach's *Institutions of Physiology* (London, 1817).)[4] Unemployment in the print trades became endemic. Moreover, machine-printing made publication possible on a massive scale, at a price far beyond those without substantial capital. The process was ideal for entrepreneurs with large-scale financial backing, who could produce 'cheap, amusing and instructive' publications for a penny. Working-class publishers like Carlile were constantly in danger of having their wares drowned in the sea of useful knowledge pouring from the factories.

Industrialisation thus proved far more effective in combating radicals than censorship or swingeing taxes on paper and advertising. Campaigners against these 'taxes on knowledge' argued for 'a cheap press in the hands of men of good moral character, of respectability, and of capital'.[5] Publishing – though not reading – became dominated by middle-class entrepreneurs. As a leading entrepreneur said in 1835, 'Nothing, in our opinion, within the compass of British manufacturing industry, presents so stupendous a spectacle of moral power, working through inert mechanism, as that which is exhibited by the action of the steam-press'.[6]

The millennium of useful knowledge

The transformation of communication was implicated in a crisis of representation: an age of reflection, inward examination, national self-awareness. 'What, for example,' the essayist Thomas Carlyle (1795–1881) commented in the *Edinburgh Review*, 'is all this that we hear, for the last generation or two, about the Improvement of the Age, the Spirit of the Age, Destruction of Prejudice, Progress of the Species, and the March of Intellect, but an unhealthy state of self-sentience, self-survey; the precursor and prognostic of still worse health?'[7] From William Hazlitt to Robert Southey, from Henry Brougham to John Stuart Mill, authors diagnosed the 'condition of England' and debated how to characterise the kaleidoscopic changes. These books were symptomatic of the novel idea that an 'Era' might have 'Characteristics', that the 'Age' might have a 'Spirit'.

The second quarter of the nineteenth century witnessed an outpouring of works on the public meaning of science and technology, as the controversies of the immediate post-revolutionary period were transmuted into a national debate about the coming of an industrial society. As progressive forms of knowledge, political economy, geology, phrenology and nebular astronomy were at the heart of the response to the changes that were sweeping away the old order. The principles of the natural sciences could define the meaning of progress. The great debate was how to do this: and in line with contemporary political disputes, the problem was forging a viable programme of scientific reform in the wake of the French Revolution, especially as so many analytical methods and institutional forms were continental imports.

The aim was to survey the principles of a science. As the philosopher Dugald Stewart (1753–1828) said in introducing *Britannica*'s 'Supplement', the task of definition 'properly devolves upon those whose province it belongs, in the progress of the work'.[8] The concept of a general treatise prefacing a more detailed account had been prominent in France during the late eighteenth and early nineteenth centuries. The key works included the mathematician Pierre Simon Laplace on the evolution of the solar system (1796), the naturalist Jean-Baptiste Lamarck on species transformation

(1809), and the comparative anatomist Georges Cuvier on the revolutions of the earth (1812).

New notions of authorship accompanied the new genres of reflective science. Original work in science depended on the credibility of the author in a way that fiction and poetry did not. Discovery was not a democratic process, available to all through skill or practice; insights into nature came suddenly, and to select individuals in a moments of profound insight, those whom the chemist Sir Humphry Davy (1778–1829) had called 'the sons of genius'. Great men embodied in themselves the age's scientific spirit.[9] Publishers, more prosaically, encouraged the emergence of a heroic role for the man of knowledge because familiar names sold books. While it was possible in principle for any gifted individual, whatever their background, to contribute to science, the apotheosis of the author tended to imply that the production of knowledge should be kept in the hands of a few 'manly intellects' who could devote themselves to science as a vocation.

According to contemporary stereotypes women were unsuited to scientific authorship.[10] A lithograph (Figure 20.3) shows just how unfeminine 'a lady of scientific habits' was seen to be. A small *Craniology* volume tops her head, which is broadly based on a (French) *Encyclopédie*; her dress is *Pantologia*, a reference punning male trousers and *Pantologia* (1808–13), a twelve-volume encyclopaedia (there is a possible further swipe at 'pantisocracy', the communitarian utopia projected by Samuel Taylor Coleridge and some friends during the radical 1790s). Her feet are bound as Walker's *Tracts*. This woman is an author, as she holds a scroll under one arm ('Armstrong *On Slavery*') and quills under the other ('Handle, *Army Notes*'). Her garb of useful knowledge renders her deeply unfashionable; moreover, made of books, she has no body at all, and can give birth only to more books. The associations of knowledge with the masculine, the foreign, the rational and the controversial could scarcely be clearer.

Women authors were caricatured, not because they were being edged out, but because their participation in many aspects of cultural life was pervasive. The notion that science should be dominated by a male clerisy developed as a strategic response to the increasing heterogeneity of the reading audience. Without some select body of interpreters, Coleridge (1772–1834) had feared the '*plebification*' of knowledge and the collapse of society. Books needed to be appropriate for a public no longer limited to the genteel consumers of natural philosophical systems during the eighteenth century. The wealthy London astronomer John Herschel (1792–1871) agreed: without suitable reading matter the artisan population would be 'dull boys', and as he recognised, 'a

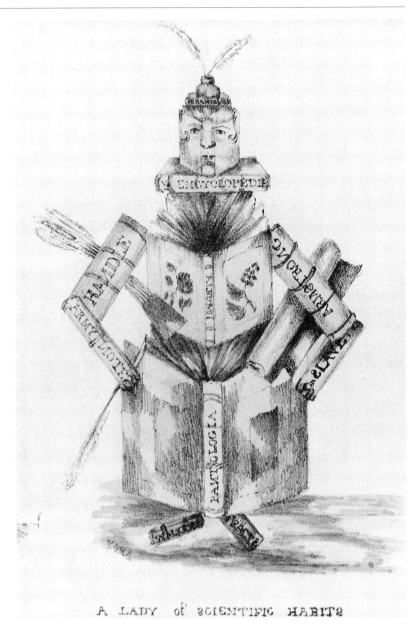

Figure 20.3 'A Lady of Scientific Habits', hand-coloured lithograph of the early nineteenth century. The tradition of constructing scholars from their books goes back to Giuseppe Arcimboldo (1537–93), painter to the court of Rudolf II at Prague. (Author's collection)

community of "dull boys" in this sense, is only another word for a society of ignorant, headlong and ferocious men'.[11]

An army of Lilliputians

The readership for reflective surveys of knowledge expanded rapidly. As the London bibliophile Thomas Frognall Dibdin (1776–1847) wrote in 1832, sales of ponderous quartos and classical folios were in the doldrums, but 'there were "brisk doings" below stairs', as smaller formats dominated the market.[12] The possibilities were canvassed during the early 1820s, most loudly in *The Edinburgh Review* and in Henry Brougham's *Practical Observations upon the Education of the People*, which ran through twenty editions in 1825. Brougham's pamphlet sketched a utopian vision in which inexpensive readings on science, issued in parts at a few shillings, became available in even the poorest cottage (Figure 20.4). In the same year, Archibald Constable (1774–1827) announced ambitious plans for cheap editions of history, biography and science. A few months later 'the Napoleon of the realms of print' was bankrupt and living in squalor, a victim of financial mistakes and the general depression of trade. The 1825–6 crash is often seen as a watershed, although few firms went under and the number of titles published soon recovered.[13]

Yet the financial crisis did bring forward trends important for the sciences. The paucity of well-known novelists between Sir Walter Scott (1771–1832) and Charles Dickens (1812–70) is only partly a consequence of modern literary obsessions with the canon; it also reflects the difficulty of selling untried titles in the decade after 1826. Publishers became more cautious, less willing to take risks on fiction and poetry. New novels were published as three-deckers priced at an extortionate thirty-one shillings and sixpence. Literary publishers focused on reprints of fiction and poetry. Their key author was Scott, whose hugely successful 'Magnum Opus' used a five shilling small octavo format – about the size of a modern paperback novel – which became standard for inexpensive book production.

The most striking development was the rise of cheap non-fiction series, in which many of the new scientific works appeared. Constable's utopian vision was revived on a reduced scale as a Miscellany, which began publication in 1827. The Miscellany, while not reaching so vast a readership as Constable had projected, did reveal an untapped market for non-fiction series publishing. The Society for the Diffusion of Useful Knowledge (SDUK, f. 1826) forwarded Brougham's scheme to enlighten a wider range of readers. The aim was, in part, to undermine political radicalism with rational information – as some working-class readers reportedly complained,

Figure 20.4 'A Box of Useful Knowledge'. The head of Lord Brougham (wearing his Lord Chancellor's wig) opens to reveal an inkpot and a pile of small books of the kind published by the Society for the Diffusion of Useful Knowledge. Hand-coloured wood engraving, c. 1832. Similar images are described in M. D. George, *Catalogue of Political and Personal Satires*, 11 vols. (London, 1954), vol. XI, pp. 648–9. (Courtesy Whipple Museum of the History of Science, Cambridge)

'to stop our mouths with *kangaroos*'.[14] A sixpenny number of the SDUK's 'Library of Useful Knowledge' appeared every two weeks, in double columns in small type, with the aim of covering all the main sciences. Brougham's captivating introduction had sold 33,100 copies by the end of 1829, proclaiming that science 'elevates the faculties above low pursuits, purifies and refines the passions, and helps our reason to assuage their violence'.[15]

SDUK campaigns were noisy, public and programmatic. The propaganda reflected a belief that the spread of useful knowledge among the working-class poor required paternalistic support analogous to that available through the Society for Promoting Christian Knowledge (f. 1698) and the British and Foreign Bible Society (f. 1804). Outside the realm of religious publishing, though, reliance on charitable donations and local committees proved a dead end. The SDUK's successes were due almost entirely to its publisher, Charles Knight (1791–1873). Individual entrepreneurs – not just new men like Knight but gentlemanly publishers in the Regency mould – produced innovations that became permanent features of science publishing.

Those who wished to counter domination of cheap publishing by the reforming Whigs moved to meet the challenge. John Murray (1778–1848) of London's fashionable West End, publisher of the Tory *Quarterly Review*, issued fifty-three volumes of original non-fiction between 1829 and 1834 in a small octavo format. The series, called the Family Library, included the natural philosopher David Brewster's *Life of Isaac Newton* (1833) and *Letters on Natural Magic* (1832).[16] These works need to be seen not just as part of their authors' intellectual trajectories, but within the wider programme of publishing embodied in the Family Library. Even after Murray ceased commissioning for the Family Library, he continued to issue books in the same format. These included Davy's *Consolations in Travel* (1830), Mary Somerville's *On the Connexion of the Physical Sciences* (1834) and the third through sixth editions of Charles Lyell's *Principles of Geology* (1834–40).

Murray's aim in adopting this form was counter-revolutionary, to bridge widening class divides. Other publishers soon tackled the same market through the Edinburgh Cabinet Library, the Naturalist's Library, the Library of Sacred Literature and similar series. Longman, the largest firm in the trade, issued the 133-volume Cabinet Cyclopædia edited by the Irishman Dionysius Lardner (1793–1859), including 'preliminary discourses' on natural philosophy and natural history. These were prefatory to works on astronomy, botany, zoology, geology, optics, probability and so forth which made up the rest of the set. The series format meant that individual treatises could be bought for six shillings; purchasers need not be tied – as they had been with Longman's previous venture in this field – to purchasing forty-five expensive quartos. As Dibdin put it, 'A whole army of Lilliputians, headed by Dr. Lardner, was making glorious progress in the Republic of Literature.'[17]

Even publishers sceptical about the virtues of cheapness were caught up in the changes. William Pickering (1796–1854), known for exquisitely printed volumes of literature, picked up the rights to the Bridgewater Treatises on 'the Power, Wisdom, and Goodness of God, as manifested in the Creation', perhaps the most celebrated series of reflective treatises on science. Pickering expected the series to sell no better than most works of dogmatic theology. He issued only 1,000 copies of the early titles, no more than required by the Bridgewater bequest. They were hand-printed with fine illustrations, large margins and widely spaced lines of type between which 'the Earl of Bridgewater might almost have driven his cab!' Only gradually did Pickering realise that he had runaway bestsellers on his hands, 'which bid fair to traverse the whole civilized portion of the globe'.[18]

At the height of the Reform agitation, the new formats threatened to take over entirely. Murray predicted that in a few years 'scarcely any other description of books will be published' than the five- or six-shilling small octavo.[19] Expectations of success are revealed in ambitious print-runs and razor-thin profit margins. For example, Murray printed 12,500 copies of Brewster's *Newton*; of these, just over half were sold and the rest had to be remaindered, at a huge loss, to Thomas Tegg (1776–1845), remaindering jackal of the Regency book trade.[20] Sales of the later numbers of the Library of Useful Knowledge also proved disappointing, and most appear to have been bought not by working people but by the middle classes. The same also appears to have been true of other SDUK productions. In that sense, Knight achieved only limited success in using cheap factory-based printing to replace the 'foul trash' spewing forth from the pauper presses.

Progress and providence

What held the sciences together in these new circumstances of production? Answering this question presented acute problems for authors who were carving up nature into specialities. It was all very well to focus on strata-hunting or stellar mapping within the meeting rooms of the Geological (f. 1807) or Astronomical (f. 1820) societies; but this in no way met demands by those who bought five-shilling treatises. Such readers demanded general concepts and simple laws. Defining these audiences and creating appropriate forms for addressing them were central to the making of the new sciences as authoritative, specialised activities in the first place. How was this to be done?

There was no agreed answer to this question. What an introductory book on science ought to look like was open to very different views. Price, format, length, illustrations, religious orientation and demands on previous knowledge all varied dramatically. Such issues had a vital practical dimension which has often been ignored. Philosophical systems in science gained prominence because of a market-led demand for synthesis.

The new genres of reflective science were unified by a common debate about progress. Progress had been central to the Enlightenment encyclopaedic tradition, but had become associated with materialism and human perfectibility, so that it needed to be used with care in presenting the specialist sciences. Geology, the newest and most controversial of the sciences, became closely identified with the progressive history of the earth and of life (Figure 20.5). Davy's *Consolations in Travel*, published posthumously in 1830 by

Figure 20.5 The geological record as a series of books, hand-coloured plate from [James Rennie], *Conversations on Geology* (London: 1828). (Reproduced by permission of the Syndics of Cambridge University Library)

Murray, offered dialogues on the meaning of science for metaphysics and religion. In the third dialogue, the 'Unknown', a stranger describes his vision of the geological past. It is a story of progress, beginning with the Earth 'in the first state in which the imagination can venture to consider it', cooling from original fluidity to become habitable. Tropical animals and plants of simple character are succeeded by shells, fish, reptiles, mammals and finally human beings.[21]

Similar narratives provide the organising principle for a wide range of geological works. Progress became central to reflective books on geology – although not to field or museum practice – by providing a narrative to replace literal readings of Genesis. The Revd William Buckland's Bridgewater Treatise on *Geology and Mineralogy* (1836) demonstrated that divine providence could be seen at work in the deep past. Pickering sold 10,000 copies, even though the book had nearly ninety plates and cost £1 15s. The surgeon Gideon Mantell's *Wonders of Geology* (1838), based on popular lectures in fashionable Brighton, offered a cheaper alternative, as did numerous works by commercial hacks. Mary Shelley (1797–1851) proposed a book on geology to Murray, although she was unsure 'how far such a history would be amusing'.[22] Most influential of all were the geological essays in the *Penny Magazine*, at the height of its celebrity in the early 1830s.

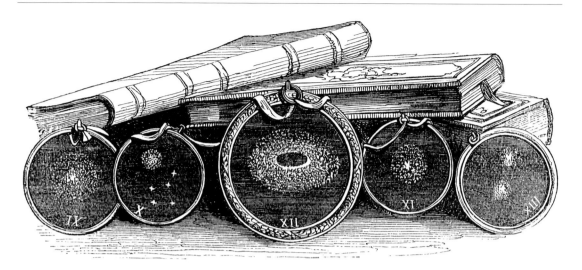

Figure 20.6 Nebulae supported by books, a wood engraving from one of the most popular early Victorian works of general science (T. Milner, *The Gallery of Nature: A Pictorial and Descriptive Tour through Creation* (London, 1846), p. 192). (Author's collection)

Progress could also be evoked in presenting other newly reconstituted disciplines, notably astronomy (Figure 20.6). In 1833 William Whewell's Bridgewater Treatise on Astronomy and General Physics coined the term 'the nebular hypothesis'. This combined the suggestion of John Herschel's father, William Herschel, that nebulae might be new sidereal systems or stars being born, with the Laplacian theory of the formation of the solar system. It was the astronomer and political economist John Pringle Nichol's *Views of the Architecture of the Heavens* (1837), however, that made the evolving nebulae into the symbol of astronomy as a progressive science. Issued by the Edinburgh publisher William Tait (1793–1864), this work described in vivid and accessible language the evolution of the universe and the formation of galaxies and stars. It sold 3,000 copies in its first two editions, with three further ones with the same publisher in the next decade. The *Architecture* presented itself as akin to a paper instrument, with numerous fine plates 'substituting for the want of powerful telescopes' and offering readers 'something of the emotion' felt by expert astronomers. The nebular hypothesis showed that progress was written in 'splendid hieroglyphics' across the sky.[23]

In other publishing projects, there were sharp differences about whether progress applied to nature. John Herschel, for example, denied that nebular hypotheses provided any basis for understanding astronomy. His Cabinet Cyclopædia volume on the science had taken a more cautious view of his father's speculations than even Whewell had done. The nebulae furnished 'an inexhaustible field of speculation and conjecture', but it was easy to become 'bewildered and lost'. Herschel privately condemned Nichol's *Architecture* as an attention-grabbing potboiler, which led him to take 'a very humble opinion' of its author's astronomical abilities.[24]

The geologist Charles Lyell (1797–1875) based his challenge to progress on an analogous concern for the foundations of knowledge. He compared the experience of reading Lamarck's evolutionary views to the pleasures of light fiction, and his *Principles of Geology* (1830–3) argued that geologists would advance their science only when they stopped assuming that the Earth itself had a progressive history. Instead, progress should centre on the human observer; drawing on Dugald Stewart and Scottish common-sense philosophy, Lyell argued for a reform of the earth sciences which depended on the witnessing of visible causes. Pursued in this way, geology would join the other sciences in exemplifying the progressive character not of nature, but of reason.[25]

Fears of association with Enlightenment cosmologies tempered enthusiasm for progress as a structuring device, for a progressive narrative backed up by material causation could be read as making humans no more than better beasts. Moreover, a strong narrative line put scientific exposition uncomfortably close to novel-writing. The emotional engagement of readers might lead to a suspension of judgement and unlicensed speculation. Introductory science books stressed the divine direction of their story lines, and that progression had nothing to do with human origins or the evolution of species. This was an important challenge for authors of the Bridgewater Treatises, who were not being paid to cast brute matter as the hero of romance. Only Buckland organised parts of his text around progress in nature, and it is not surprising that his book raised the most criticism.[26] Bridgewater authors who discussed astronomy faced the special dilemma that Laplace had used nebular condensation in the service of a godless mechanical universe. However, they successfully purified the nebulae into signs of divine progress emblazoned on the heavens.

The dangers of being tarred with the brush of speculation were most apparent in systematic accounts of living beings. Any work that discussed the laws of nature could be read as 'infidel'. The best-selling of all the books of reflective science, the Edinburgh phrenologist George Combe's *Constitution of Man Considered in Relation to External Objects* (1828), was often attacked as infidel determinism. 'Materialism' was a term of abuse rather than a coherent doctrine. Systematic treatises on philosophical anatomy downplayed or rejected evolutionary implications because these could be associated with freethought, pornography, or the unstamped press.[27] Such associations were fatal to any campaign to establish a safe science of life, or a medical profession based on technical scientific expertise. A causal account of the generation of higher species was, even in the most liberal medical circles, dangerous speculation.

The Book of Nature in the age of the machine

By the end of the 1820s, controversies about the future shape of the polity were embodied in the physical size and shape of printed materials. Many works reflected on the mechanisms of their own production. At one end of the scale, James Bateman's *Orchidaceæ of Mexico and Guatemala* (1837–43), among the largest books ever published, featured a caricature of itself being raised from the ground (Figure 20.7). A gentleman in a topcoat holds a speaking trumpet and directs a gang of working men, while printer's devils dance on the sidelines. Even books for the wealthy (the volume cost sixteen guineas, a year's wages for a servant) exhibited an awareness of the labour relations involved in their making.

Cheap books and journals described how tens and even hundreds of thousands of copies of an eight-page weekly could be produced for only two or three halfpennies. The mathematician and inventor Charles Babbage's *On the Economy of Machinery and Manufactures* (1832; 4th edn. 1835, p. vi) reflected on the role of machinery in industrial processes by referring to 'objects of easy access to the reader'. Among these processes was book-making, and Babbage (1792–1871) asked readers to contemplate the physical object in their hands. Much to the consternation of booksellers, among whom such matters were trade secrets, the *Economy of Machinery* provided a full breakdown of costs for paper, printing, binding, payments to the author, and so forth. Babbage wished to show that the price was determined not by production costs, but by taxes and restrictive trade. Booksellers engaged in a combination which 'operates upon the price of the very pages which are now communicating information respecting it' (p. 314). In citing pounds and pence, and allowing his book to 'speak' directly to the public, Babbage was uniquely explicit about these issues – so much so that he had to switch publishers at the last moment.

Babbage's interest in the political economy of print was not unusual, however, nor was his concern for the rewards and institutions of science. His book typified the new reflective works. It was priced at six shillings, about half of what a similar text would have cost a decade earlier. The format was a small octavo, printed on large sheets that could go through steam-presses. It had a case binding, which was coming into common use as a means for providing an attractive cloth cover for those who could not afford rebinding in calf. The book was issued in large numbers, with a first edition of 3,000 copies selling out within two months, and three further editions soon following. This success, Babbage claimed, was the product of public demand for information about 'the pursuits and interests of that portion of the people which has recently

Figure 20.7 Wood engraving from a drawing by George Cruikshank, in J. Bateman, *The Orchidaceæ of Mexico and Guatemala* (London: 1837–43), p. 8. (Reproduced by permission of the Syndics of Cambridge University Library)

acquired so large an accession of political influence' (p. vi). The publisher was Knight, rapidly gaining a reputation as a leading exponent of mass publishing. The physical character of the *Economy of Machinery* was integral to its message. Like Babbage's celebrated mechanical engine for calculating, the book was a demonstration device in political economy.[28]

The intimate relation between cheap production and useful knowledge was hard to miss, even in books less insistent on their mode of production. John Herschel's *Preliminary Discourse on the Study of Natural Philosophy* (1830) was commissioned, written, printed, advertised and stocked in shops as part of an encyclopaedic series, the enterprise of Lardner and Longman. Spine, half-title and first title-page give as much prominence to the *Cabinet Cyclopædia* as to the author. Purchasers were also made aware of cost: the six-shilling price could be seen from the paper label on the spine, and the size and cloth case binding identified it as a product of innovations in machine production. Opening the covers, the first thing the reader saw was a sixteen-page Longman's catalog, including other *Cyclopædia* volumes. The paper was made from cast-off rags, a byproduct of the transformation of the textile industry. The *Preliminary Discourse*, widely praised as the most high-minded scientific philosophy ever published, bore the marks of industrial progress as clearly as a railway journey or a yard of factory-produced cotton.

The *Economy of Manufactures* and *Preliminary Discourse* were early products of the industrial era of print, which now, in retrospect, can be seen to have lasted until the 1980s. Four developments dominated the next century and a half in relation to the sciences: first, the impact on printing and distribution of new science-based technologies (including photography, machine engineering, and scientific management); second, the demand of schools and universities for scientific textbooks and reference works; third, the rise of 'popular science' periodicals and books targeting the literate masses; and fourth, an explosive growth in the publication of specialist scientific journals. By the mid-nineteenth century, almost all new science appeared in periodicals, later to be diffused in textbooks and popular works. Reflective, accessible, original books such as Charles Darwin's *Origin of Species* (1859) became exceptions that proved the rule. A late, poignant expression of the optimism of the machine age was the eleventh edition of the *Encyclopædia Britannica* (1910–11), published on the eve of the First World War. This was the last time that any set of books in the English language surveyed all the sciences in their full range and depth.

No one knew better than Thomas Carlyle that scientific systems were temporary, human constructs made from rags. As Professor Teufelsdrîck wondered in Carlyle's great 'Clothes-Philosophy' of 1831, *Sartor Resartus*, why did so many believe that the 'Domestic-Cookery Book' of mechanical recipes served up by science was all there was to nature? Carlyle raged at the age in which he lived, but his words were the clearest expression of its spirit. 'We speak of the Volume of Nature: and truly a Volume it is, – whose Author and Writer is God. To read it! Dost thou, does man, so much as well know the Alphabet thereof?'[29]

Notes

Parts of this essay appear in a different form in chapters one and two of James A. Second, *Victorian Sensation: The Extraordinary Publication, Reception and Secret Authorship of* Vestiges of the Natural History of Creation (Chicago, 2000).

1 [J. G. Lockhart], *Peter's Letters to his Kinsfolk*, 3 vols., 2nd edn. (Edinburgh, 1819), vol. II, p. 174; on the 'Supplement', see R. Yeo,

'Reading encyclopaedias: science and the organisation of knowledge in British dictionaries of arts and sciences', *Isis*, 82 (1991): pp. 24–49.
2. M. Twyman, *Printing 1770–1970: An Illustrated History of its Development and Uses in England* (London, 1998), which has an updated bibliography; see also the overly schematic argument in L. Erickson, *The Economy of Literary Form: English Literature and the Industrialization of Publishing, 1800–1850* (1995).
3. R. Carlile, 'An address to men of science', in B. Simon (ed.), *The Radical Tradition in Education* (London, 1972), pp. 91–137, at p. 109.
4. W. H. Brock and A. J. Meadows, *The Lamp of Learning: Two Centuries of Publishing at Taylor & Francis*, 2nd edn. (London, 1998), p. 74.
5. Thomas Milner-Gibson, president of the Association for the Promotion of the Repeal of the Taxes on Knowledge, quoted in J. Curran and J. Seaton, *Power Without Responsibility* (London, 1991), p. 29.
6. 'Mechanism of Chambers's Journal', *Chambers's Edinburgh Journal* (6 June 1835), pp. 149–50, at p. 150.
7. [T. Carlyle], 'Characteristics', *Edinburgh Review*, 54 (1831): 351–83, at p. 365.
8. Quoted in Yeo, 'Reading encyclopaedias', p. 32.
9. See the articles by C. Lawrence and S. Schaffer in A. Cunningham and N. Jardine (eds.), *Romanticism and the Sciences* (Cambridge, 1990); and J. Golinski, *Science as Public Culture: Chemistry and Enlightenment in Britain, 1760–1820* (Cambridge, 1992).
10. See esp. M. Benjamin, 'Elbow room: women writers on science, 1790–1840', in M. Benjamin (ed.), *Science and Sensibility: Women and Scientific Enquiry, 1780–1945* (Oxford, 1991), pp. 27–59; M. Poovey, *The Proper Lady and the Woman Writer: Ideology and Style in the Works of Mary Wollstonecraft, Mary Shelley, and Jane Austen* (Chicago, 1984); and A. Shteir, *Cultivating Women, Cultivating Science: Flora's Daughters and Botany in England, 1760 to 1860* (Baltimore, 1996).
11. S. T. Coleridge, *On the Constitution of the Church and State* (1830), ed. J. Colmer (London, 1976), p. 69, and J. F. W. Herschel, 'An address to the subscribers to the Windsor and Eton Public Library and Reading Room, delivered... on Tuesday, Jan. 29, 1833', in Herschel, *Essays from the Edinburgh and London Quarterly Reviews* (London, 1857), pp. 1–20, at p. 9.
12. [T. F. Dibdin], *Bibliophobia. Remarks on the Present Languid and Depressed State of Literature and the Book Trade* (London, 1832), pp. 18–19.
13. The revisionist view was suggested in J. Sutherland, 'The British book trade and the crash of 1826', *The Library*, 6th ser., 9 (1987): 148–61, and largely confirmed in S. Eliot, *Some Patterns and Trends in British Publishing 1800–1919* (London, 1994), pp. 16–18. For Constable as Napoleon, J. G. Lockhart, *Memoirs of the Life of Sir Walter Scott, Bart.*, 7 vols. (Edinburgh, 1837–8), vol. VII, pp. 126–30.
14. S. Shapin and B. Barnes, 'Science, nature and control: interpreting Mechanics' Institutes', *Social Studies of Science*, 7 (1977): 31–74, at p. 56.

15 [Brougham], *Objects, Advantages, and Pleasures of Science* (London, 1827), p. 2; for further analysis, R. K. Webb, *The British Working Class Reader, 1790–1848: Literacy and Social Tension* (London, 1955), p. 69, and J. Topham, 'Science and popular education in the 1830s: the role of the *Bridgewater Treatises*', *British Journal for the History of Science*, 25 (1992): 397–430, at pp. 413–19.

16 S. Bennett, 'John Murray's Family Library and the cheapening of books in early nineteenth century Britain', *Studies in Bibliography*, 29 (1976): 139–66.

17 T. F. Dibdin, *Bibliophobia*, p. 18; M. Peckham, 'Dr. Lardner's Cabinet Cyclopaedia', *Papers of the Bibliographical Society of America*, 45 (1951), pp. 37–58. Other cheap series are discussed in S. Sheets-Pyenson, 'War and peace in natural history publishing: the Naturalist's Library, 1833–43', *Isis*, 72 (1981): 50–72 and in R. Altick, *The English Common Reader: A Social History of the Mass Reading Public 1800–1900* (Chicago, 1957), pp. 273–5.

18 The quotations are from T. F. Dibdin and the *Medico-Chirurgical Review*, both quoted in J. Topham, 'Beyond the "common context": the production and reading of the Bridgewater Treatises', *Isis*, 89 (1998): 233–62, at pp. 244 and 245. Topham provides a full account of the production of the Bridgewaters and their place in the publishing world of the 1830s.

19 J. Murray to C. Knight, April 1829, in S. Smiles, *A Publisher and his Friends: Memoir and Correspondence of the Late John Murray*, 2 vols. (London, 1891), vol. II, p. 296.

20 S. Bennett, 'John Murray's Family Library', pp. 161, 164.

21 H. Davy, *Consolations in Travel* (London, 1830), pp. 133–37.

22 M. Shelley to J. Murray, 8 Sept. 1830, in B. T. Bennett (ed.), *The Letters of Mary Wollstonecraft Shelley*, 3 vols. (Baltomore 1980–88), vol. II, p. 115. I am grateful to Marilyn Butler for drawing this letter to my attention.

23 J. P. Nichol, *Views of the Architecture of the Heavens* (Edinburgh, 1837), pp. vii, viii, 206–7; the sales figure is mentioned in the third edition of 1839, p. viii. On Nichol, see S. Schaffer, 'The nebular hypothesis and the science of progress', in J. R. Moore (ed.), *History, Humanity and Evolution* (Cambridge, 1989), pp. 131–64.

24 J. F. W. Herschel to W. H. Smyth, 20 Feb. 1846, *Royal Society Herschel Letters*, vol. XXII, f. 268; M. Hoskin, 'John Herschel's cosmology', *Journal for the History of Astronomy*, 18 (1987): 1–34.

25 J. A. Secord, 'Introduction', in C. Lyell, *Principles of Geology* (London, 1997), pp. ix–xliii.

26 N. Rupke, *The Great Chain of History: William Buckland and the English School of Geology (1814–1849)* (Oxford, 1983), pp. 209–18; J. Topham, 'Beyond the "common context"', pp. 249–61.

27 I. D. McCalman, *Radical Underworld: Prophets, Revolutionaries and Pornographers in London, 1795–1840* (Cambridge, 1988); A. Desmond, 'Artisan resistance and evolution in Britain, 1819–1848', *Osiris*, 2nd ser., 3 (1987): 77–110.

28 S. Schaffer, 'Babbage's dancer and the impresarios of mechanism', in

F. Spufford and J. Uglow (eds.), *Cultural Babbage: Technology, Time and Invention* (London, 1997), pp. 53–80.
29 T. Carlyle, *Sartor Resartus* (London, 1904), pp. 205–6.

Further reading

R. D. Altick, *The English Common Reader: A Social History of the Mass Reading Public 1800–1900* (Chicago, 1957)

J. Chandler, *England in 1819: The Politics of Literary Culture and the Case of Romantic Historicism* (Chicago, 1998)

A. Desmond, *The Politics of Evolution: Morphology, Medicine, and Reform in Radical London* (Chicago, 1989)

J. P. Klancher, *The Making of English Reading Audiences, 1790–1832* (Madison, 1987)

J. B. Morrell and A. Thackray, *Gentlemen of Science: Early Years of the British Association for the Advancement of Science* (Oxford, 1981)

M. Poovey, *Making a Social Body: British Cultural Formation, 1830–1864* (Chicago, 1995)

S. Schaffer, 'The nebular hypothesis and the science of progress', in J. R. Moore (ed.), *History, Humanity and Evolution* (Cambridge, 1989), pp. 131–64

J. A. Secord, *Victorian Sensation: The Extraordinary Publication, Reception and Secret Authorship of* Vestiges of the Natural History of Creation (Chicago, 2000)

J. Topham, 'Scientific publishing and the reading of science in early nineteenth-century Britain: an historiographical survey and guide to sources', *Studies in History and Philosophy of Science*, 31A (2000)

R. Yeo, 'Reading encyclopaedias: science and the organisation of knowledge in British dictionaries of arts and sciences', *Isis*, 82 (1991): 24–49

  Defining Science: William Whewell, Natural Knowledge, and Public Debate in Early Victorian Britain (Cambridge, 1993)

Afterwords

The Owl of Minerva. (Courtesy of Les Belles Lettres, Paris)

Books, texts, and the making of knowledge

> The man-hero is not the exceptional monster,
> But he that of repetition is most master.
>   (Wallace Stevens, *Collected Poems*, London, 1984, p. 406)

History of the book/theory of the text

In his Panizzi Lectures of 1985 Don McKenzie pleaded for a historical bibliography that would cover not just signs on paper, but rather 'texts as recorded forms, and the processes of their transmission, including production and reception'; and he insisted that the new discipline should draw on literary theory as well as studying 'the social, economic and political motivations of publishing, the reasons why texts were written and read as they were, why they were rewritten and redesigned, or allowed to die . . .'[1] Seven years later Roger Chartier sketched as follows the 'conditions of possibility' of a new history of reading:

Its space would be defined by three centres of attention that the academic tradition has usually kept separate: first, the analysis of writings, be they canonical or ordinary, to discern their structures, their themes, and their aims; second, the history of books and, beyond that, the history of all objects and all forms that bear writings; third, the study of practices that may take possession of these objects and these forms in a variety of ways and produce differentiated uses and meanings.[2]

That the prospects are excellent for profitable interaction between, on the one hand, literary, critical and philosophical studies of the production and reception of meanings – for want of a better term, let us call all this 'hermeneutics' – and, on the other hand, the history of the book, is suggested by the striking parallels in the development of the two fields. Thus McKenzie noted how both traditional analytical bibliography and the so-called 'New Criticism' which dominated American academia in the 1950s and 1960s treated texts as autonomous entities abstracted from the material reality of books, and without regard to the processes of their production and reception.[3] And Robert Darnton has written of the parallel growth of concern in the two fields with the roles of readers in the constitution of meanings.[4]

Alas, the image of parallels, by definition fated never to meet, may be all too apt. McKenzie himself set a superb example for the fruitful interplay of literary criticism and history of the book in his studies of the expressive functions of the material forms, typographies and layouts – *mise-en-livre* and *mise-en-page* – of Congreve's works.[5] Other historians of the book have, however, expressed reservations about literary theory. Adrian Johns is understandably suspicious of its tendency to depend on speculation rather than empirical evidence; and Roger Chartier doubts the usefulness to historians of 'a phenomenology that misses out all concrete modalities of the act of reading', going on to describe that phenomenology in terms which point to Gadamer's *Truth and Method*.[6] Likewise Darnton, whilst conceding that literary theorists' accounts of projected and implied readerships have something to offer to historians of the book, criticises them because they 'seem to assume that texts have always worked on the sensibility of readers in the same way'.[7] And, more extremely, Jonathan Rose endorses the view that recent debates over literary theory and the canon are laughable.[8]

In what follows I seek to mitigate such suspicions of literary and hermeneutic theory by sketching some consequences of a question-based account of historical meaning (partly inspired by Gadamer) for one of the central historiographical issues raised in the present volume, namely: how can study of the history of the book contribute to a historical understanding of the production of new meanings and new knowledge? After sketching the account, I look at two lively areas of the history of the book – the role of books in the conveyance and authentication of testimony; and the forms of routine authorship, that is, emendation, editing, anthologising, etc. – that are, according to that theory, crucial for the explanation of innovation in the sciences. I conclude with some optimistic reflections on the centrality of the history of the book to the history of the sciences.

Questions, meanings and styles of inquiry

Gadamer's *Truth and Method* concentrates on the educated reading of canonical works as a route to self-understanding and moral improvement.[9] It is also, contrary to the drift of Chartier's remarks, concerned with the ways in which readers' responses to the questions posed by past works are historically conditioned by the traditions in which they have been educated. Gadamer focuses on the questions arising from our literary heritage that remain real and provocative for us. But he does touch on how we understand questions no longer real for us, which we can do only to the extent that we can come to see how they were once real. The account of meaning and interpretation that I set out in *The Scenes of Inquiry* resembles Gadamer's in taking the understanding of questions as primary, and in relating the interpretation of questions no longer real for us to apprehension of what once made them real.[10] It differs, however, in assigning no special priority to elite readers' cultivated understanding of canonical works.

In its very barest outline, the theory goes like this. A question is 'real' for a group or community just in case they can, in accordance with their style of inquiry, get to grips with it. Assertions are real if the corresponding questions are. We understand a question real in a community to the extent that we grasp how they would get to grips with the question, that is, to the extent that we see what they would take to be relevant to its resolution, and why. We understand an assertion real in a community when we understand the question it answers.

The account satisfies certain basic requirements of historical interpretation. In particular, for purposes of the editing and translation of works it licenses the privileging of original meanings, that is, the interpretations that were, or would have been, made by competent readers of the period fully conversant with the genre and discipline of the work. For other kinds of approach to works, for example those involved in their dramatic performance, the theory licenses the privileging of quite different responses, for example those of the audiences to whom the dramatisation is directed. But for general historical purposes the account grants no general privileges – for the readings of works that may be historically consequential at different times, in different cultures and in different disciplines are of no one sort: they may be original or later readings, competent or incompetent, disinterested or politically and morally motivated, metropolitan or provincial, elite or 'vulgar', in the original language or in translation, in the original book or in a later product, in full or abridged and anthologised; the possibilities are endless.

For present purposes, however, two further consequences of the account are crucial. One is that the historical interpretation of claims issuing from past styles of inquiry very different from our own involves far more than simply having to get our heads around unfamiliar forms of reasoning. To understand such alien claims we need to take into account the entire ranges of beliefs, conventions and practices that affect the resolution of questions. Thus, in the case of scientific disciplines, styles of inquiry typically include local practices of 'production': all the activities that lead up to written, verbal, or dramatic presentation of findings and hypotheses. Here belong protocols for the use of instruments, together with all the various processes of control, replication, simulation and calibration involved in the separation of genuine effects from background noise and interference. Here too belong the heuristic methods that guide the cogitations of theoreticians, along with the habits and regulations that govern the division of labour and the daily routines of work in places of inquiry – whether they be alchemists' closets and naturalists' cabinets, or scientists' observatories and laboratories. Styles of inquiry include also local practices of 'presentation': techniques of argument and narration; conventions of genre; rhetorical and aesthetic strategies; technologies of replication and distribution of texts and images. Under this heading we may include also the demonstrative and dramatic devices involved in live displays of models and experimental findings. Styles of inquiry involve, finally, methods of 'adjudication': all the procedures and conventions of adjudication of testimony, refereeing, assignment of priority, and conduct and management of controversy that mediate the often long and tortuous passages of findings and claims from public appearance to general acceptance, rejection, or (most frequently of all) loss of interest.

The second implication has specifically to do with innovation in the sciences. On this account of meaning and understanding there is the most intimate link between understanding of the past meanings of works in the sciences and explaining scientific innovations. For understanding the past meanings of questions and assertions requires us to appreciate the various ways in which past practitioners resolved, or sought to resolve, questions. And that, of course, provides ingredients essential for the explanation of the emergence of new questions and new doctrines in the sciences.[11]

Bookish credit

Perhaps the most obvious way my question-based account of meaning and interpretation links history of the book with the making of knowledge has to do with credit and testimony. For that account ties interpretation to the apprehension of the routes to consensus; and the production of books and the adjudication of the testimonies to be found in them loom large indeed in the various modes of resolution of questions in the sciences.

The ways in which testimony has been and ought to be adjudicated in the sciences is currently a matter of lively debate. Steven Shapin, for example, in his *A Social History of Truth*, has sought to show how the judgement of testimony in early modern natural philosophy was governed by gentlemanly codes of honourable conduct, elevated social status of the witness providing, then as now, the ultimate warrant of truthfulness.[12] Peter Lipton, by contrast, has argued that we should, and in fact do, habitually assess testimony in terms of competence as well as honesty, balancing explanations of testimony which appeal to honest and competent witnessing against those which invoke dishonesty and/or incompetence.[13]

All parties to the recent debates on testimony are agreed on the pervasiveness of testimony, both direct and indirect, and the crucial role of its evaluation in the creation of knowledge in the sciences. Now it is very often through books that the competence and trustworthiness of witnesses have to be assessed. Accordingly, the study of the ways in which the credibility of books is established and adjudicated is a matter of major import for historians of the sciences. Indeed, in certain fields – meteorology, natural history, ethnology, for example – the very possibility of the discipline is dependent on the accreditation of published reports of variously questionable travellers, collectors and informants.[14]

In accounting for the credit of books and the testimonies they bear we face a very general predicament, which, following Anthony Grafton, may be called 'the problem of duplicity' – how to tell the real thing from the impostor.[15] There is an Italian saying 'Fatta la legge, trovato l'inganno', that is, roughly, 'the law made, the loophole found'. Thus Grafton in *Forgers and Critics* shows how the techniques of forgery and the tools of critical scholarship developed hand-in-hand, often in fact the left and right hands of the same people. And Michael McKeon's *The Origins of the English Novel* demonstrates the formation of the narrative devices of this genre through the appropriations and subversions of those very literary techniques that had been designed to discriminate true histories from mere fictions and romances.[16] In the domain of the sciences, too, it seems that the problem remains unabated, no matter how thoroughly we look into the narrative, demonstrative and rhetorical devices used to persuade readers of the reliability and trustworthiness of testimonies. For surely all such devices are imitable.

If at the level of text and narrative there is no principled way of discerning authentic works from pretenders, perhaps we may fare better by looking instead at the processes of production of books. In general this involves far more than the evaluation of the creditworthiness of authors. Today in the sciences, credibility often depends more on the reputation of the journal and its editorial board than on an author's name. And with multiple authorship increasingly the norm, it is frequently the credit of the scientific institution from which the work emanates rather than that of the authors that is crucial.[17] Further, as Foucault famously suggested, and as Mark Rose, Carla Hesse and Martha Woodmansee have demonstrated in full historical detail, the author as the locus of legal responsibility, proprietary rights, and textual authority is a relatively recent conception, originating in very different ways in the different parts of Europe around the turn of the eighteenth century.[18]

Robert Iliffe has shown how in late seventeenth- and eighteenth-century Britain it was often editors rather than authors who bore prime responsibility for the authenticity and credibility of texts.[19] And Adrian Johns has argued that in early-modern England, when the textual stability and creditworthiness of books were being consolidated, the primary 'gate-keepers of credit' were not authors or their works but the stationers and booksellers.[20] Unfortunately, here too the problem of duplicity surfaces. As Johns has shown, the struggle to stabilise and make credible the content of natural philosophical and natural historical books was protracted and laborious. Nor were mathematicians, natural philosophers and natural historians mere bystanders; rather they actively participated in producing, correcting, editing, marketing and distributing books. However, the various attempts at licensing, registering and certifying by the natural philosophers, their printers and their booksellers succeeded in establishing only insecure authenticity for a limited range of books emanating from approved printing houses and bookshops. A striking instance of the difficulties they faced has to do with the rival attempts following the death of Henry Oldenburg, first editor of the *Philosophical Transactions of the Royal Society*, to establish a credible public record of contributions to the Royal Society; attempts which were complicated by the challenge of its parodist, the popular *Athenian Mercury*, in turn emulated by the *Lacedaemonian Mercury*, a spoof of the spoof.[21]

What is genuinely alarming about Johns' account is its apparent demonstration that even at this material level of printing and distribution, of what is engagingly called by Americans 'plant', the principle of duplicity applies – credited and creditworthy places and performances could be and were successfully faked. Yet the *Philosophical Transactions* is, we all trust, an authentic production of a real Royal Society. Erasmus Darwin was surely right in noting of 'the learned labours of the immortal Press' that though 'new impositions have been perpetually produced, the arts of detecting them have improved with greater rapidity'.[22] What is less clear is by just what processes impostors have eventually been exposed and credit established for authentic publications.

In tackling this question it is, I suggest, necessary to engage in the history of the book in the broad McKenzian sense, combining study of the changing textual devices used to secure credit with investigation of the changing ways in which procedures of registration, licensing, reviewing and refereeing were used to guarantee authenticity. For Lipton is surely right to associate adjudication of testimony with inference to the best explanation of that testimony, and hence, very often, with the tracing of its causal ancestry.[23] As Silvia De Renzi's work on testimony in early modern natural history indicates, judgements concerning the reliability of transmission from original sources to passages in print involved not just questions of the competence and sincerity of witnesses, but also questions concerning Nature herself (Are the characteristics of New World animals as trustworthily indicative of their essences as are those of Old Europe?), concerning media (Is oral testimony better than written? Are descriptions better than illustrations?), and concerning the competences of authors, commentators and editors (Are classical authorities to be preferred to modern ones? How is the authenticity of the first author's original, informal and circumstantial descriptions to be balanced against the authority of the commentators' formal glosses on them?).[24] In such cases, adjudication of testimony inevitably involves, for us as for the original readers, both attention to the content of the testimony and the warrants offered for it, and investigation of the whole series of processes leading from original observations to the printed passages in question. Thus the credibility of texts, and hence, on my question-based hermeneutics, the formation of new meanings, is inextricable from the credit of books.

Authorship dispersed

Romantic conceptions of genius, discovery and authorship in the sciences are deeply suspect. From the indisputable fact that certain authors, Kepler, Newton, Darwin, Freud, have played decisive roles in the realisation of new questions and the creation of new knowledge, it is all too easy to infer that it is through minute study of their lives and thoughts that insight into the problem of innovation in the sciences is primarily to be gained. This very natural view is greatly reinforced by Cartesian notions of pure inquiry according to which such *savants* build up knowledge working on their own from scratch. And it is further aided and abetted by the Romantic notion that such geniuses possess, or are possessed by, some special poetic faculty different in quality rather than quantity from all the faculties of ordinary persons like ourselves.

Of course, following the so-called 'Death of the Author', few historians today would admit to such prepossessions. However, it seems to me that more recently fashionable styles of interpretation still often mystify the genesis of meanings. As McKenzie has observed, much literary theory treats texts as autonomous locations of meaning: this cuts the making of meanings off from the production and reception of books.[25] And the mystery is, if anything, deepened by those who in the Heideggerian manner ground novelty of meaning in the generative power of a language that speaks through authors or, *à la* Foucault, in the discursive forms to which their works belong. Further, much literary reception theory seems simply to transfer the mysteries of creating meaning from authors to readers.

Echoing Mark Twain, we may note that the report of the death of the author was an exaggeration.²⁶ The account of meaning and interpretation on which the present reflections are premised is at odds with the privileging of authors as unique creators of, or authorities on, meanings. It is, however, equally at odds with accounts which eliminate authors in favour of autonomous texts or autonomous readers. Rather, by relating meaning to the whole range of processes of consensus formation, it enjoins us to consider the production of meanings as involving transactions between authors, readers and others involved in the production and distribution of books.

As Chartier has observed of the un-dead author:

> As he returns in literary criticism or literary sociology the author is both dependent and constrained. He is dependent in that he is not the unique master of the meaning of the text, and his intentions, which provided the impulse to produce the text, are not necessarily imposed either on those who turn his text into a book (bookseller-publishers or print workers) or on those who appropriate it by reading it. He is constrained in that he undergoes the multiple determinations that organize the social space of literary production and that, in a more general sense, determine the categories and the experiences that are the very matrices of writing.²⁷

Indeed, the revived author is not merely constrained and dependent; in many cases, as McKenzie has observed, authorship is fragmented: 'an author disperses into his collaborators, those who produced his texts and their meanings'.²⁸ To the extent that meanings, the responses of communities of readers, are affected by the physical form, layout and typography of books, the makers of books are indeed part-authors. And to the extent that readers' responses are determined by other works invoked or produced by the work in question – by works it cites, echoes or emulates; by reviews and works which cite, echo or emulate it – authorship is yet more promiscuously dispersed.

In grappling with the making of meanings in the sciences, it is, I suggest, a good strategy to concentrate on cases in which authorship is manifestly constrained, dependent and dispersed. Thus, instead of heading at once for the production of meanings by the Newtons, Darwins and Freuds, we should in the first instance look at the workaday shufflings and shiftings of meaning brought about by typesetting, translating, proof-correcting, editing, annotating, commentating, reviewing, anthologising, popularising, compiling encyclopaedias, composing textbooks, etc. – precisely those activities apt to be dismissed as mechanical, derivative, second-rate or inauthentic according to Cartesian and Romantic conceptions of authorship.

The present volume bears ample witness to a recent explosion of interest in the forms of routine authorship. McKitterick, Blair and Clark reveal the varied roles of indexing and cataloguing in the maintenance of disciplines. Grafton, Kassell, Johns, Frasca-Spada and Terrall testify to the creative powers of readers 'with a pen in hand', showing how translators, annotators and commentators joined forces with the original authors in the production of meanings. Broman points to critical reviewing as a crucial factor in the extraordinary rise in the status of journals as authoritative vehicles of learning in the course of the eighteenth century. Yeo and Secord spell out the strategies through which systematic surveys of the sciences in encyclopaedias and general textbooks were used to counter the fragmentation and panic induced by 'the horrible mass of books'. Topham shows how the attempt to found a whole new school in English mathematics was centred on a single textbook; and Roldán Vera considers the dispersion of authorship, investigating the parts played by publishers, translators, commercial agents and reformer-diplomats in the creative 'customisation' of English textbooks for the newly independent Latin-American republics.

Turning from such strictly historical studies, we may note that literary theorists have recently treated issues highly relevant to understanding routine authorship under the heading of 'intertextuality', that is, the diverse modes in which texts may cite, evoke, imitate, emulate, parody, etc., each other.[29] Most of this theorising concentrates on canonical works; much of it is premised on a view of meaning as the product of the interplay of autonomous texts; and much of it proceeds with little regard to the history of the conventions and genres of writing and reading. However, there is one notable work on intertextuality, Gérard Genette's wonderfully scholarly, creative and witty *Palimpsests*, on which historians of routine authorship in the sciences could profitably draw.[30] Yet Fernand Hallyn's *The Poetic Structure of the World*, which appeals to Genette in its treatment of narrative forms, figurative language, and patterns of emulation in early modern cosmology, is the only work in the history of the sciences known to me to use notions of intertextuality effectively.[31]

Routine writing and new knowledge

On the account of historical interpretation sketched above, study of routine authorship is on several counts fundamentally important for understanding the formation of new questions and new doctrines in the sciences.

1. Understanding of innovation in the sciences depends on a grasp of the specificities of the past practices of the sciences; and it is routinely authored works – instrument handbooks, instruction manuals, observatory and laboratory protocols, etc. – that are the crucial primary sources for such practices.

2. The forms of routine authorship are themselves specific local

practices, often playing central roles in the formation and stabilisation of questions and discoveries in the sciences. One such role, celebrated by Hermann von Helmholtz, is the rendering accessible of knowledge:

> One of the reasons why we can so far surpass our predecessors in each individual study is that they have shown us how to organise our knowledge. This organisation consists, in the first place, of a mechanical arrangement of materials, such as is to be found in our catalogues, lexicons, registers, indexes, digests, scientific and literary annuals, systems of natural history, and the like. By these appliances thus much at least is gained, that such knowledge as cannot be carried about in the memory is immediately accessible to anyone who wants it.[32]

Further, the communication, standardisation, calibration and replication of the practices of the sciences are largely dependent on instrument manuals. Other indispensable roles are in the coordination of the activities and methods of the sciences with each other, and with more general codes and patterns of social behaviour. Consider, for example, the range of routinely authored materials associated with museums – guidebooks, catalogues, registers, visitors' books, instructions for collectors – which coordinate the patterns of conduct not only of travellers, collectors and curators, but also of visitors, sponsors and entrepreneurs.[33] Most importantly of all, perhaps, it is largely through routinely authored works – textbooks, commentaries, reading lists, exam papers, etc. – that the sciences are passed on from generation to generation. Close study of such works makes possible reconstructions of the appropriations and elaborations of meanings by teachers and students. They allow the historian to get behind the official programmes of instruction as embodied in syllabuses, so as to see these meaning-making processes from the diverse perspectives of students, professors, administrators, even parents and townspeople.[34] That places of instruction, as the primary sites of reproduction of disciplines, are crucial for study of meaning change in the sciences is pretty obvious. But that they may be sites of radical innovation is perhaps less widely appreciated. Yet a glance at the history of education reveals pedagogy as the most contested of fields, with its full quota of revolutions, battles over methods, fads and crazes. Further, certain pedagogic movements – Ramism in the sixteenth and seventeenth centuries, Comenianism in the seventeenth and eighteenth centuries, the proliferation of research seminars in the German Lands in the eighteenth and nineteenth centuries – have brought about major transformations in the sciences.[35]

3. Understanding of the ways in which testimony about objects and findings was assessed is crucial in accounting for the adoption of new questions and doctrines in the sciences. As I have already

indicated, the adjudication of such testimony has often depended on the credit assigned to forms of routine authorship – editing, commenting, etc.

4. A little more speculatively, I suggest that study of routine authorship may provide leads to the understanding of innovation at the hands of the Big Boys, the 'Heroes of Knowledge'. For example, Alain Segonds and I (partly inspired by Hallyn's *The Poetic Structure of the World*) have examined the stages by which Kepler, becoming dissatisfied with the translations of Aristotle's *De caelo* that he had to hand, so emended, reworked and glossed them as to excavate from the work a corrupted and enigmatically expressed primordial Pythagorean cosmology prefiguring his own.[36] We argue that it would be a mistake to suppose that Kepler is here merely providing literary embellishment or retrospective legitimation of his cosmological constructions. Rather, these philological exercises are of a piece with his mathematical and physical ventures in astronomy. Both his philology and his astronomy involve creative re-enactments – of the way in which the most ancient astronomy engendered those texts of Aristotle in which it is concealed/revealed; of the way in which God produced the world, at once concealing and revealing His nature. These are, for Kepler, playful origins – the Pythagorean encoding of esoteric doctrines was playful, even capricious, and God played as He created. Accordingly, for Kepler, it is through playful conjecture that the original mysteries are to be sought out. By unveiling the divine archetypes, whether indirectly through philological scholarship, or directly through astronomical calculation, we participate in the Divine Mind. Kepler's case, in which his routine authorial activities are of a piece with his other, less visible meaning-producing activities – collation of celestial phenomena, their mathematical analysis, contrivance of physical explanations, metaphysical speculations and constructions – is, I suggest, an entirely typical one. Specifically, I would argue that the handling of textual materials in routine authorship is often representative of patterns for the handling of materials generally – the ordering of specimens, the cobbling together of experimental systems from bits and pieces, the marshalling of arguments, the fitting together of hypotheses into theories.[37]

5. Still on the subject of the 'Heroes of Knowledge', it should be noted that it is above all through reviewing, editing and commenting that works come to be regarded as canonical, their authors vaunted as geniuses.[38] As indicated in the chapters of Frasca-Spada and of Stewart and L. A. Jardine, these routine canon-forming activities are frequently associated with the toning-down and systematisation of once distressingly novel and dissonant doctrines, such 'normalisation' being often indispensible for the effective

achievement of consensus, and hence for the consolidation of new questions and new doctrines in the sciences.[39]

Conclusion

With the fading away of the chronicles of accumulation of positive knowledge that once dominated the Anglo-American history of science, it may seem that the discipline has lurched from fad to fad, being idealised and philosophised in the 1950s and 1960s, sociologised in the late 1960s and 1970s, anthropologised, materialised, gendered, you name it, in the 1980s, and in the 1990s swamped by culture. There are some who would argue that the new cultural history of the sciences is not merely another new bottle for the old wine, but a step in the long-delayed integration of the history of the sciences into history proper. However, this new historiography – as epitomised by such works as Mario Biagioli's *Galileo, Courtier* (1993), and Lorraine Daston and Katherine Park's *Wonders and the Order of Nature* (1998) – has some worrying features. Faced with its decentring, its emphasis on the local and the specific, on the multiplicities of the cultural significances of deeds and works, one may well wonder how it can come to terms with larger explanatory questions concerning the coming-into-being of scientific disciplines and the consolidation of new questions and new doctrines within those disciplines. It is, or so I have tried to argue, precisely the history of the book, generously conceived in the manner of McKenzie as inclusive of both empirical and theoretical studies of the production and reception of works, as at once a history of books and a history of texts, that can enable the history of the sciences to engage, or rather to re-engage, with the making of scientific knowledge.

Notes

For helpful advice, I thank Silvia De Renzi, Marina Frasca-Spada, Nick Hopwood, Lauren Kassell, Eugenia Roldán Vera, Jim Secord, and the other members of the Cambridge Historiography Group and of Jim Secord's Book History Reading Group.

1 D. F. McKenzie, *Bibliography and the Sociology of Texts* (London, 1986), pp. 4–5.
2 R. Chartier, *The Order of Books* [1992], trans. L. G. Cochrane (Oxford, 1994), pp. 2–3 (translation modified).
3 McKenzie, *Bibliography*, p. 7.
4 R. Darnton, *The Kiss of Lamourette: Reflections in Cultural History* (New York, 1990), pp. 179ff., 302.
5 D. F. McKenzie, 'Typography and meaning: the case of William Congreve', *Wolfenbütteler Schriften zur Geschichte des Buchwesens*, 4 (1981): 81–125.

6 A. Johns, 'Science and the book in early modern cultural historiography', *Studies in History and Philosophy of Science*, 29A (1998): 167–94; Chartier, *Order of Books*, pp. 3–4.
7 Darnton, *Kiss of Lamourette*, pp. 131, 181.
8 J. Rose, 'The history of books: revised and enlarged', in H. T. Mason (ed.), *The Darnton Debate: Books and Revolution in the Eighteenth Century* (Oxford, 1998), pp. 83–104.
9 H.-G. Gadamer, *Truth and Method* (1960), trans. of 2nd edn anon. (London, 1975); my reading of Gadamer is much indebted to G. Warnke, *Gadamer: Hermeneutics, Tradition and Reason* (Oxford, 1987).
10 N. Jardine, *The Scenes of Inquiry: On the Reality of Questions in the Sciences* (Oxford, 1991), ch. 3.
11 I explore these implications more fully in 'Original meanings and historical interpretation', and 'Original significances and historical explanation', both appended to *The Scenes of Inquiry*, 2nd edn (Oxford, 2000).
12 S. Shapin, *A Social History of Truth: Civility and Science in Seventeenth-Century England* (Chicago, 1994).
13 P. Lipton, 'The epistemology of testimony', *Studies in History and Philosophy of Science*, 29A (1998): 1–31.
14 See S. De Renzi, q.v., ch. 8; also P. G. Adams, *Travellers and Travel Liars, 1660–1800* (Berkeley, 1963); N. Rennie, *Far-Fetched Facts* (Oxford, 1995), chs. 2 and 3; D. Carey, 'Compiling nature's history: travellers and travel narratives in the early Royal Society', *Annals of Science*, 54 (1997): 269–92; L. Daston, 'The language of strange facts in early modern science', in T. Lenoir (ed.), *Inscribing Science* (Stanford, CA, 1998), pp. 20–38.
15 A. Grafton, *Forgers and Critics: Creativity and Duplicity in Western Scholarship* (Princeton, NJ, 1990).
16 M. McKeon, *The Origins of the English Novel, 1600–1740* (Baltimore, 1987).
17 See, for example, M. Biagioli, 'Aporias of scientific authorship: credit and responsibility in contemporary biomedicine', in M. Biagioli (ed.), *The Science Studies Reader* (New York, 1999), pp. 12–30.
18 M. Foucault, 'What is an author?' (1969), trans. D. F. Bouchard and S. Simon, in *Language, Counter-Memory, Practice: Selected Essays and Interviews* (Oxford, 1977), pp. 113–38; M. Rose, *Authors and Owners: The Invention of Copyright* (Cambridge, Mass., 1993); C. Hesse, *Publishing and Cultural Politics in Revolutionary Paris, 1789–1810* (Berkeley, 1991); M. Woodmansee, *The Author, Art, and the Market: Rereading the History of Aesthetics* (New York, 1994). Where Rose and Woodmansee relate the birth of the modern author in Britain and the German lands to the commercialisation of the book trade, Hesse establishes an entirely different trajectory in France, where exclusive authorial privileges originated under the Ancien Régime and were weakened in the revolutionary and Napoleonic periods.
19 R. Iliffe, 'Author-mongering: the "editor" between producer and consumer', in A. Bermingham and J. Brewer (eds.), *The Consumption of Culture, 1600–1800: Image, Object, Text* (London, 1995), pp. 166–92.

20 A. Johns, *The Nature of the Book* (Chicago, 1998), ch. 2.
21 Ibid., ch. 7.
22 E. Darwin, *The Temple of Nature; or, the Origin of Society* (London, 1803), p. 270.
23 Lipton, 'Epistemology of testimony'.
24 See De Renzi's chapter; also her 'Il drago di Aldrovandi e le vipere di Redi', *Intersezioni*, 17 (1997): 123–30.
25 McKenzie, *Bibliography*, pp. 7ff.
26 Mark Twain, in *Oxford Dictionary of Quotations*, 3rd edn (Oxford, 1979), p. 554.
27 Chartier, *Order of Books*, p. 28.
28 McKenzie, *Bibliography*, p. 18.
29 In this sense the term is said to derive from Julia Kristeva's 'Bakhtine, le mot, le dialogue et le roman', *Critique*, 33 (1967): 438–65. A useful collection of essays on intertextuality is H. G. Plett (ed.), *Intertextuality* (Berlin, 1991).
30 *Palimpsests: Literature in the Second Degree* (1982), trans. C. Newman and C. Doubinsky (Lincoln, NB, 1997). *En passant*, I cannot resist noting the delightful new word 'palimpsestuous' generated by Genette's book: see Gilbert Adair, *Surfing the Zeitgeist* (London, 1997), p. 75. Thus, for example, we may say that Chartier's *The Order of Books*, for all its striking originality, palimpsestuously relates to McKenzie's *Bibliography and the Sociology of Texts* with regard to form and themes.
31 F. Hallyn, *The Poetic Structure of the World: Copernicus and Kepler* (1987), trans. anon. (New York, 1990).
32 *Popular Lectures on Scientific Subjects*, trans. E. Atkinson (London, 1873), p. 12.
33 On documents as coordinators of the activities associated with collections and museums, see, for example, P. Findlen, *Possessing Nature: Museums, Collecting and Scientific Culture in Early Modern Italy* (Berkeley, 1994); S. Müller-Wille, 'Collectors, system builders, and world-wide commerce', paper presented to the workshop 'The Natural History Collection and its Function in the Eighteenth Century', Potsdam, February 1998; E. Spary, 'Codes of passion: natural history specimens as a polite language in eighteenth-century France', in H. E. Bödeker, P. H. Reill and J. Schlumbohm (eds.), *Wissenschaft als kultureller Praxis, 1750–1900* (Göttingen, 1999), pp. 105–35.
34 See Bill Clark's magisterial *The Hero of Knowledge (Homo Academicus Germanicus)* (Berkeley, CA, forthcoming).
35 On Ramism in the sciences see R. Hooykaas, *Humanisme, science et réforme. Pierre de la Ramée (1515–1572)* (Leiden, 1958); K. Meerhoff and J.-C. Margolin (eds.), *Autour de Ramus* (Québec, 1996). On the rise of the research seminar see especially R. S. Turner, 'The growth of professorial research in Prussia, 1818 to 1848 – causes and context', *Historical Studies in the Physical Sciences*, 3 (1971): 137–82. On Comenian pedagogy see K. Schaller, *Die Pädagogik des Johann Amos Comenius und die Anfänge des pädagogischen Realismus im 17. Jahrhundert*, 2nd edn (Heidelberg, 1967).

36 N. Jardine and A. Segonds, 'Kepler as reader and translator of Aristotle', in C. Blackwell and S. Kusukawa (eds.), *Conversations with Aristotle* (Aldershot, 2000), pp. 205–32.
37 Cf. Anne Moss' discussion of the early-modern method of commonplaces as a general pattern for the assembly of materials: *Printed Commonplace-Books and the Structuring of Renaissance Thought* (Oxford, 1996), especially chs. 7 and 8; see also Anthony Grafton's review of the book, *Times Literary Supplement*, 31 January 1997, pp. 10–11.
38 Often, indeed, the incipient geniuses are themselves involved in the routine authorial activities which establish their heroic status: see, for example, R. Iliffe, '"*Per* this and *per* that": understanding and the authorship of the *Principia*', in M. Biagioli and P. Galison (eds.), *What Is a Scientific Author?* (Chicago, forthcoming).
39 On 'normalising' strategies to minimise the shock of novelty, see T. S. Kuhn, *The Structure of Scientific Revolutions* (Chicago, 1962).

Further reading

M. Biagioli and P. Galison (eds.), *What Is a Scientific Author?* (Chicago, forthcoming)
R. Chartier, *The Order of Books* (1992), trans. L. G. Cochrane (Oxford, 1994)
R. Darnton, *The Kiss of Lamourette: Reflections in Cultural History* (New York, 1990)
H.-G. Gadamer, *Truth and Method* (1960), trans. of 2nd edn anon. (London, 1975)
J. Genette, *Palimpsests: Literature in the Second Degree* (1982), trans. C. Newman and C. Doubinsky (Lincoln, NB, 1997)
N. Jardine, *The Scenes of Inquiry: On the Reality of Questions in the Sciences* (Oxford, 1991; 2nd edn with supplementary essays Oxford, 2000)
A. Johns, 'Science and the book in early modern cultural historiography', *Studies in History and Philosophy of Science*, 29A (1998): 167–94
D. F. McKenzie, *Bibliography and the Sociology of Texts* (London, 1986)

The past, present, and future of the scientific book

Scientific research is not now published in books. It has not been for a long time. Scientists who publish books are generally aiming at a non-specialist readership, and whilst the royalties reaped from such ventures can be substantial, the academic credit is not. The journal has ruled since certainly the late nineteenth century, and quite possibly earlier. No Darwin would today publish his major theories first as a single extended volume. The claims made in the pages of reputable scientific journals now represent the knowledge of the scientific community; peer review processes are commonly taken to guarantee its value. Distributed to universities and made freely available to working researchers, the journals constitute the body of understanding to which apprentice scientists aspire and for which they are trained. As their careers develop, they gain reputation and tenure on the basis of their participation in this kind of publishing. In important ways, then, the system of journals has come to characterise the scientific enterprise itself.[1]

Today, however, the sovereignty of the scientific journal is itself being questioned. In part, the challenge has arisen out of technological changes. Electronic publishing provides opportunities to extend the reach and transform the character of scientific communication. Yet opportunities are not all that new media generate: the advent of the World Wide Web also creates new doubts about existing practices, and to some the direction of current change seems to pose threats to a proven system of appraisal and reward. Even more pertinently, there is serious concern about the current and future state of scientific authorship irrespective of the imperatives of new media themselves. Problems of cost, ownership and accessibility are suddenly urgent even in the traditional realm of print, and carry a significance rivalling that of those posed by electronic technologies. In effect, a congeries of practice and representation that has long been reckoned fundamental to the very viability of the scientific enterprise is in the throes of profound change. No wonder the effects are disquieting.

Yet the system of communication, credit and reward established around journals has not always existed. As the unrivalled arbiter of learned excellence, it is not even particularly old. The earliest 'scientific' periodicals – the denomination is in fact anachronistic – date from the mid-1660s, and for decades they survived only with great difficulty. For a very long period after that, natural knowledge continued to be properly distributed in any number of forms: by letter, poster, pamphlet, tract, treatise or encyclopaedia as well as by journal itself. It is possible to explore historically the practices by which such diverse entities came into being, circulated, and were put to use. And just as periodicals have come to stand for a moral economy of science in the modern period, so these other objects, too, could be seen as underpinning distinct conventions for the pursuit of natural knowledge. That is why their investigation is worthwhile. The history of the scientific book can lead us to important insights into the history of scientific knowledge – and hence to a new understanding of the predicament of such knowledge today.

Modern publishing and reading practices pose particular problems for science. How we perceive those problems may be affected by considering similar issues that science has successfully addressed in the past. This chapter suggests how. It argues that the efforts of modern scientists to come to terms with fundamental questions of authorship, credit and accessibility are perhaps not entirely unprecedented. In previous centuries, natural philosophers, mathematicians and scientists found themselves exposed to the ravages of a craft of communication that could be remarkably risky for a learned writer. They had to work long and hard to shape the realm of print into one conducive to the maintenance of a learned community. They fashioned conventions of propriety and civility to secure that community, and modern scientists' regard for peer review is but one of their legacies.[2] Their successors today are having to work just as hard all over again, but with a new medium. This certainly does not mean that they are simply reiterating past efforts. But it does mean that their labours have a history. That history deserves to be recovered, for it represents a major element in the making of modern science.

Present and future

As a result of burgeoning agglomeration in the publishing industry, the few companies producing research journals today find themselves in possession of near-monopolies. They have not been hesitant to exploit their position. Subscription prices for major scientific perodicals have risen, if not quite exponentially, then at rates substantially higher than inflation. Even the best-endowed academic institutions find these prices steep. Moreover, their effects are not restricted to the sciences. In order to continue a respected participant in a given research activity, an institution simply must subscribe to its leading journals; this means that, with purchasing budgets at best remaining static, room for the increased costs must be found elsewhere. At leading research universities in the United States, journals now account for some 70–85 per cent of libraries' annual budgets. It is not humanities periodicals that are consuming these sums. Indeed, this, more than absolute cuts in funding, is why the purchasing of humanities journals and monographs is declining – in the case of monographs, by 23 per cent in the last decade at the major American institutions. That in turn helps to explain the perception of a 'crisis' in monograph publishing that pervades the ranks of historians and literary scholars. The problem is real, even if its status as a full-blown emergency remains as yet controversial. And it threatens the careers of many younger academics – scientists and humanists alike – whose prospects for tenure depend on becoming authors at a stage when they are yet unproven.[3] This is truly a case in which questions of publishing affect the conduct, shape and character of academic life in a broad range of disciplines.

These quantitative concerns about cost, production and distribution give rise also to a more qualitative and principled fear about authorship. Questions about the policies of publishers soon lead on to doubts about hitherto uncontroversial elements of science itself. What is scientific authorship? How can it be recognised, and what socio-economic rewards should it merit?[4] At present, scientists typically publish papers for which they and their institution receive no payment; copyright then becomes the property of the publisher, which can license reprints. The publisher reaps rewards from its journals without paying either for the intellectual property or for the research that created that property. What they do provide is physical distribution – not a very valuable commodity in an increasingly electronic age – and, above all, *credibility*. This, in the end, is what scientists and institutions pay for. Scientists receive accreditation by publishing. The credit they gain is underwritten by the peer review process, and that process is almost always managed by the publisher of a given journal. So valuable is this credit that institution and researcher alike have until now been prepared to fund (or at least facilitate) research, and then pay again to receive its fruits in the form of journal issues, without much of a murmur. Their very silence is a powerful indication of the centrality of journal-borne credit to scientific life.

But the silence is now being broken. Institutions and scientists are beginning to voice complaints at what suddenly seems a gross inequity built into the system. Initiatives in the realm of copyright law only accentuate their concerns. These initiatives are of global scope, and affect the conduct of science as surely as they do that of musical performance or software-authorship. The possibility of creating *exact* copies of digital information makes piracy a real concern for what are now called 'content providers' – that is, publishers, music companies, film and video corporations, and 'the media'. Faced with the possibility that existing copyright laws may be obsolescent, the response of these industries has been to lobby for their reinforcement. The World Intellectual Property Organisation and similar groups mount powerful campaigns to strengthen copyright laws within and, moreover, among nations. Their campaigns sometimes proceed even at the expense of loopholes that have long been conventionally accepted, and for which plausible public-interest cases can be made. From the point of view of academic institutions, a major example is the convention of 'fair use', according to which it is legitimate for scientists and other scholars to make use of portions of published work in constructing and displaying their own arguments in informal settings such as colloquia and classroom presentations. WIPO's proposals for international intellectual property regulation would explicitly do away with this convention. Moreover, placing such legal power in publishers' hands takes on extra significance when the changing character of library holdings is considered. Unlike printed journals, electronic publications create no paper archive at a customer institution. What is sold is not now a material object but an access licence. An electronic publisher could therefore simply refuse access to issues which an institution had already 'received', if the institution cancelled its subscription. This has potentially serious consequences for the direction and character of research, since it will place the textual raw materials of such research – the archived journals on which scientists and others depend on a daily basis – in interested and non-academic hands. We habitually forget the extent to which even the most advanced scientific investigation depends on 'historical' labour, namely the checking of archived sources like these. But we may soon remember it all too well. If universities have no real alternative but to subscribe to the leading journals in a field, their vulnerability will only increase in such a system.

In consequence, scientists are becoming acutely aware that if the future of science itself is to be secure, they may have to seize the initiative in science publishing. Lively debates about the future of publishing are suddenly taking place in many scientific fields, from physics to biomedicine. One example is the process of 'deliberative discourse' carried on in mid-1998 by the California Institute of Technology. The debate there has been spurred by Provost Steven Koonin, whose concern for proprieties was shown when he played a major role in the demolition of cold fusion in 1989, repairing what he perceived to be a gross breach of scientific peer review and publication procedures.[5] Caltech – a wealthy and prestigious institute, but not a very large one – is exploring the possibility of alliances with other major institutions in an effort to intervene and reshape the conventions of scientific publishing to preserve what it can of those procedures. There may indeed be partners to hand, for others are proposing similar strategies. Stevan Harnad, a cognitive scientist in Southampton, has likewise made a proposal for academics to seize the day and begin to distribute texts electronically themselves, thus forcing the publishing industry into line. Harnad's plans have received substantial support, including immediate offers to install the required technology.[6] In any case, it is emerging that if any response is to be effective it must be collective; no single scientist, institution, and perhaps even discipline is powerful enough to effect change. The consequences of these initiatives will also be distributed. They are very likely to extend to the humanities and social sciences as well as science itself.

Yet responses are indeed possible. One is to abandon conventional printed publication altogether. The capabilities of electronic distribution may then be appropriated by the scientific community itself and put to constructive use, without pre-existing commercial institutions standing as impediments. Already, in physics much initial 'publication' now does not occur through journals at all. Since Paul Ginsparg at Los Alamos pioneered his *E-print archive*, online announcements have become routine, and conventions of archiving and reference are by now advanced. Ginsparg himself commented in 1996 that 'the essential question at this point is no longer *whether* the scientific research literature will migrate to fully electronic dissemination, but rather *how quickly* this transition will take place'.[7] A fellow physicist told the *Los Angeles Times* in early 1999 that in his field, at least, journal publication was simply 'dead'.[8] A decade ago, workers in other sciences, and for that matter in physics itself, would have been shocked at the insouciance of such a remark, made about what was then the essential accreditation and circulation mechanism of scientific knowledge.

This kind of 'publication' need not, and in practice often does not, rely on commercial companies of the traditional kind. It promises to change some very ingrained habits and assumptions. For example, the requirement that papers in a journal be grouped together into a discrete issue need no longer obtain. It may be that electronic journals will no longer have issues at all, and that papers will simply appear on a continuous basis. Ceasing to exist as a physical entity, a journal would then become something like a 'brand', valuable for its endorsement alone (which, as already observed, is in fact the principal source of its value in any case). Papers may also continue to change even after they make the transition into the journal's authoritative space, since authors may respond to subsequent criticisms and developments by making successive improvements. Ginsparg's system already allows for this, by retaining copies of all successive versions of a given paper, and allowing access to them by the scientific community. One of the most useful aspects of Ginsparg's service, indeed, is that it provides access to *old* materials. It therefore also chips away at the problem of archiving raised by today's publishers. It is not difficult to see that the potential to preserve libraries' role as purchasers and harbourers of knowledge is substantial, albeit unintended.[9]

But such initiatives are unlikely to replace peer-reviewed journal publication in the near future. The reason lies not with technological limitations or commercial interests, but with scientific credit. The importance of this is made clear by the tendency of debates to focus, not on new technologies *per se*, but on the fortunes of peer review. Ginsparg's own view is that peer review is simply 'an awkward compromise imposed by the inadequacies of the paper medium', and that it will wither as the primacy of print declines. He regards his system as a prototype for the social mechanisms that will replace it. Authors are encouraged to revise their own papers in the light of others' comments, which are made openly and in what might be described as an electronic conversation.[10] But such is the perceived centrality of peer review that many scientists are distinctly uneasy about processes like these. 'If I drive my car across a high bridge', Fytton Rowland remarks, 'I need to be confident that the engineers who designed the bridge had access to sources of thoroughly reliable and tested information about the properties of materials, the geology of the foundations, and so on.' The same is even more true of the patient who visits a doctor for treatment. In this light, the importance of credit outweighs the demands of new technology.[11]

It is perhaps for this reason that the biomedical sciences, especially, have seen the most animated discussion of this subject.[12] In biomedicine the issues of creativity, property and responsibility out of which modern authorship itself developed are writ large.[13] Peer review is widely respected as a necessary guarantor of quality. In effect, scientists argue that it permits them to read a biomedical journal in a certain way distinctive of the enterprise. Without it, they allege, readers would have to become peers themselves, prepared to exercise pointed scepticism, and to interrogate intensively every claim in a given paper. But at the same time peer review is acknowledged to be problematic. It is potentially liable to reject radically innovative claims, and the delays it causes make the community vulnerable to priority disputes. For the same reason, peer review makes it difficult to situate the research front at any given moment, since it lies somewhere in the limbo of papers currently being reviewed. In that sense, one consequence is to remove discussion from the printed realm altogether, and place it (ironically, perhaps) in that of conversation. So defenders concede the desire to change the system.

One major possibility is to make peer review an interactive process between authors and panels of reviewers. This would be pursued online but in a secure website, and moderated by an editor. Eventually the text and comments alike would be 'published' in a Web- and/or print-based journal. Discussion would then continue even after this moment of announcement. The benefits would accrue in the currency of credibility. Far from undermining the credit vested in papers – and hence, by extension, entire disciplines – the increased openness offered by this kind of process, a representative of the World Association of Medical Editors has argued, might even 'improve the trust in the system'. Such principles, if they ever became routine, would have a number of corollaries. They would transform the role of editors from gatekeepers into moderators – recreating, perhaps, a once-central component of that role.[14] By the same token, it would become possible for non-participants not only to access research papers, but to see some trace of the negotiating processes through which such papers come into being. The process might therefore change profoundly the relation between scientist and public.

Some scientists have expressed concern at that possibility, believing that a public reared on images of science as the domain of objectivity and certainty may find its trust wavering when faced with the inevitable compromises to such an image implicit in such processes. It is an area in which the anxieties voiced during the recent 'science wars' come home to roost. Again, however, the prospect is in one sense not unprecedented. It may be seen in terms of a partial return to a less categorical distinction between scientist and laity that obtained for much of the nineteenth century, and in some fields persists today.[15] In that light, one may reasonably ask whether the public's respect for science really needs to rest on a simplistic image of the enterprise – and indeed, whether it in fact does rest on such an image.[16] Moreover, in the near future the Web is likely to remain relatively *in*accessible compared to sources that the public encounters on an everyday basis, and that are at present far more problematic in this respect. An important example, in the United States at any rate, is the superficial news broadcasting offered by many local television companies affiliated to the networks. It would simply be implausible to nominate such broadcasting as a model of informative objectivity.

Debates over the future of copyright and peer review do not directly address the changing character of scientific texts themselves. But that character is indeed in flux. Electronic distribution permits the capturing and conveying of kinds of information not suited to traditional printed communication. For example, it is possible to integrate video and sound segments into a paper. It is also possible to link the paper to the (sometimes enormous) databases on which its argument may depend; and 'readers' can then interrogate the databases themselves, interactively. In these ways, not only the social mechanisms of science communication but its very nature is changing. In short, everything about the practice of scientific publishing is being called into question. Issues of scientific authorship, of copyright, of reading, and of the character of knowledge-transfer are no longer matters for philosophers, sociologists and historians of science, but for scientists themselves.

Past and present

The transitional moment we inhabit today is characterised above all by uncertainty. It is that uncertainty that can lead to a company's share-price gyrating by 400 per cent in a single day, to the hoarding of provisions in preparation for Y2K anarchy, and to haughty optimism about an incipient age of worldwide democracy. In this context, it is tempting to assert the disjunctive character of our situation. The 'communications revolution', we are everywhere told, is just that: a profound rupture, producing social, cultural and intellectual effects that have no prior precedent. Hence their incomprehensibility. Yet this air of insecurity itself has its precedents. In particular, its counterpart certainly existed in the early years of printing. In the early modern period, during the first generations after Gutenberg's invention of the printing press, a new communications technology was available and open to appropriation, and proclamations of change were widely issued. It was clear that great alterations in scholarly life would come about, but the character of those changes remained unclear. The impact of a major new technology was very much 'up for grabs'. There was no common regulation, no copyright, no secure norms of literary or creative propriety. How those characteristics of a print culture – the characteristics that we now take for granted about print – came into being carries lessons for scientists today.

The natural philosophers and mathematicians of the early modern period exerted great efforts to *make* print into a reliable medium for the creation and distribution of natural knowledge. That we now view print as the source of certainty against which new media must measure up is a tribute to their success. In fact, our experience of the printed text – whether it be a work of literature, a volume of poetry, a scientific journal, or an academic monograph – would be very different but for the hard work of early modern naturalists. Things might have been different. Printed books originally had a strangeness about them that was quite the equal of the internet today. That strangeness gave rise to its own legends – of Faust, most strikingly, who was said to have invented the press and revealed the secret only after being condemned for necromancy in

Paris.[17] Practitioners of natural philosophy and (more particularly) the mathematical sciences played a major role in achieving the familiarity of modern print culture. The history of their work is little known, and deserves to be recovered.

One major historical strand is formed by practitioners who attempted to establish their own, privileged systems of production and distribution. The pioneer of this strategy was the Renaissance mathematician and astronomer Johann Müller (1436–76), better known as Regiomontanus. Within a generation of Gutenberg's invention, Regiomontanus was planning to use his own printing press to recreate virtually the entire corpus of ancient mathematics.[18] His ambitious project preceded and perhaps inspired the humanist endeavours, today much better known, of Aldus Manutius and other scholar-printers. Regiomontanus himself died young, his plans unrealised, but similar proposals continued to be aired into the eighteenth century. They were prominent in bishop John Fell's plans for a university press at Oxford in the late seventeenth century, for example.[19] Another major instance, and at first glance a rather more successful one, was the Danish nobleman and astronomer Tycho Brahe. Brahe is generally taken to personify more than any other figure the power of 'print culture' to transform scientific work.[20] Like Regiomontanus, he sought to make printing into a secure foundation of textual stability on which new knowledge could be built. The paper mill and printing house he maintained on the island of Hven were inspired by his desire to create records to high standards under his own personal oversight. Similarly, the Polish brewing magnate Johannes Hevelius built his own press – and went one step further by himself learning how to engrave and print.[21] In many of these cases, innovative appropriations of the very mechanisms of print helped make for important achievements in the sciences themselves. Regiomontanus helped to shape the reconstruction of ancient mathematics; Brahe furnished a new standard for instruments, accuracy and observational consistency; and Hevelius created a new 'visual language' for interpreting and recording the features of celestial objects. But as Adam Mosley argues in chapter 6, q.v., even Tycho found printing a problematic enterprise, and frequently preferred manuscript transmission.[22] Such frustrations pervaded all the projects mentioned here. Indeed, that these proposals consistently came to so little indicates the difficulty of making the book trade into a viable medium for the mathematical sciences. But at the same time it also shows the determination of successive generations of practitioners to make it into one. The trajectory that results leads, by way of many diversions, from Regiomontanus to the modern university press.

But if a natural philosopher or mathematician were not as imaginative as Regiomontanus, as privileged as Brahe, or as rich as

Hevelius, then he had to deal with the commercial book trade. That called for very different tactics. In particular, it meant negotiating with a trade community possessed of its own prized craft identity. A central part of that identity was a complex set of customs governing propriety and impropriety, that often did not correspond in any evident way to the norms of gentility or scholarly life. Some of today's most renowned scientific works depended on particular characteristics of the trade. *The Lancet*, for example, began as a distinctly scurrilous organ devoted to pirating physicians' lectures; upstairs in the bookseller's home, a conventionally privy location, was the only safe haven in which it could safely be compiled.[23] But the trade created as many problems as opportunities. Complaints were rife that booksellers refused to publish important work, and that they mangled the work that they did condescend to print. Piracy, too, came to be seen as a common hazard. The facts of bibliographical life in this realm were not tractable according to the wishes of scholars. That, after all, was why Regiomontanus, Brahe, Hevelius and Fell felt compelled to pursue their independent initiatives in the first place.

Different trade communities in Europe created different regimes to maintain what was called 'propriety'. This concept, combining as it did property and civility, would eventually transmute into copyright.[24] But such measures weren't the sole preserve of the booksellers and printers. Struggles to overcome the same problem of credit saw printed books made into vehicles for learned work by natural philosophers, physicians and mathematicians. This kind of effort took place in numerous contexts, but that of the early academies – the Royal Society in London, the Académie Royale des Sciences in Paris, and others – has peculiar significance. The Royal Society, for example, instituted conventional *ways* of reading that structured the conduct of inquiry itself. It moved to manage disagreements arising from the realm of the book in which authorship, creativity and intellectual achievement must be manifested. These would be underwritten here, not by the self-policing propriety of a trade, but by the word of a gentleman.[25] Experimental philosophy itself depended on the participants entrusting their reputations to this system. Making local practical craft (of the kind needed to perform an experiment in an academy itself) into universal knowledge (regarded as independent of place) depended on the creation, circulation and use of printed reports. Without faith in the experimentalists' conventions for reading and publishing, their very enterprise would atrophy.

Along with its counterpart academies in France and the German countries, the Royal Society manifested this reliability by establishing conventions for the proper reading and distribution of contributions. In particular, it adopted procedures of 'perusal' and 'registration'. Both signalled regard for the contributor and his claims, and also served to facilitate philosophical conversation and experimental work. Typically, two selected Fellows would be delegated to 'peruse' a given paper, and report back. When the Society was announced to have 'read' something, its 'reading' was always in fact this reported perusal. Registration then meant the subsequent recording of the contribution in a manuscript volume, to 'secure [its author] against usurpations.' The register stood as an archive of discoveries and authorship, in theory sacrosanct and immune from all tampering. That supposedly secured authorship and credit within the fellowship. Then some entries would be further directed to what was called the Society's 'public register' – a printed journal invented by the secretary, Henry Oldenburg, and entitled the *Philosophical Transactions*. By this means they could be circulated to philosophers across Europe, in a printed vehicle licensed by the Society and created by privileged printers. Register and journal together became twin certificates of a new civility, inside and outside the institution itself. True, that civility, like the customary propriety of the book trade, did have the incidental effect of creating its own criterion of transgression, namely plagiarism. In a system centred on registration of authorship, priority disputes became the archetype for scientific controversy. But these disputes, unlike their earlier equivalents, were managed so as not to threaten the very conduct of conversation itself (by contrast, sixteenth-century confrontations had sometimes culminated in actual violence). And the perceived advantages of the regime more than made up for this liability. That largely explains the virtuosi's determination not just to launch a periodical like the *Philosophical Transactions*, but to continue with it in the face of repeated adversities. The result was that the *Transactions* survived to become the oldest scientific periodical of them all.[26]

When scientists facing today's uncertain publishing environment stress the centrality of the journal to their enterprise, and express concern for its future, they are in fact reflecting the efforts of their forebears to create a learned civility in the context of equally profound changes in their own practices of communication. The journal was invented as a way of sustaining international scholarly conversation in the context of a book trade seen as indifferent at best. The security that the scientific journal achieved – or, more properly, that the journal in consort with practices of perusal and registration achieved – was what allowed early modern science to reshape itself. It partly constituted the formation of an international community dedicated to advancing knowledge irrespective of regional, national or confessional allegiances.[27] It also underpinned the oft-repeated claim to reject the testimony of books in favour of that of the senses. Books were not abandoned, to be sure; but the fact of their *being* books ceased to be material to the inquiries of a working natural philosopher, physician or mathematician. They came to be regarded as inert – as vessels for containing and transporting knowledge. That achievement was far more hard-won than in retrospect it often appears. It involved answering questions not just of production and distribution, but of the most fundamental activities of scientific scholarship, including creativity and reading. Experiment, in short, could seem to replace reading because reading itself was made safe for science.

The implications of that achievement extended beyond the sciences, to affect the history of the book in general. The modern copyright regimes that are now so contested were first instituted in the eighteenth century, in response to the perceived collapse of craft regimes of propriety. At that juncture, lawyers, booksellers and writers alike cast around for successful models of authorship and literary propriety. Among other places, they found them in the world that the experimental philosophers had stabilised around the civilities of the *Philosophical Transactions* and, by now, many other journals. To these participants, Isaac Newton and Robert Boyle were not just great philosophers and national heroes; they also served as the archetypes on which a new dispensation for the book trade itself could be built. So did Aristotle, Gassendi and Descartes. Technology and natural philosophy alike became grist to the mill of protagonists on all sides. An author might be defined as akin to the inventor of a mechanism, like Harrison with his chronometer. Or else true knowledge might be regarded as something founded in Creation, in which case 'no private right can possibly exist'. 'It is said of Pascal, that without ever seeing the book, he went as far as the 43d proposition of Euclid', remarked one judge; so if anybody had 'gone on in Newton's' book the same way, proving thereby the truth of its contents, could its proprietor really claim violation of property? 'If the first publishers of any such works are to have a perpetual monopoly, how absurd would such a position be, and how unjust to the rest of mankind.'[28] In France, Condorcet and Diderot debated on these grounds whether property in ideas was conceivable at all. Legally, bases for a resolution were reached separately in Britain, France and the German countries in the last decades of the eighteenth century and the first of the nineteenth. A series of systems providing for variously constrained property protections came into being. We are today their inheritors.[29]

These historical considerations impinge on the current predicament of scientific publishing in a number of ways. They certainly highlight the conventional character of what are regarded as central constituents of scientific authorship itself. They show that wrestling with issues of individual and collective creativity in assigning the responsibility and reward for science publication is not a new requirement. Nor is the prime importance attached to the manner of pre-publication reading novel – in the Royal Society's seventeenth-century word, perusal; in ours, peer review. Peer review itself is a product of the nineteenth-century redefinition of disciplines out of which the modern sciences emerged, and its history is contiguous rather than continuous with that told here. Yet the historical trajectory given above is nonetheless one that historians of reading should attend to, since it offers their work a rare relevance. Above all, historical reflections show that it is not new for the scientific community to appreciate the need to labour in order to *shape* the world of communication. Newton and Boyle knew that that labour was legitimate and appropriate for the sake of their enterprise. It is therefore fitting that the physicists and biotechnologists who participate in present discussions of future publishing and accreditation practices find themselves in the unfamiliar position of historians. Asking today's questions about the origins, character and justification of copyright and intellectual property leads us straight back to the history of science itself.[30]

Past and future

It is plausible to suggest that scientists' efforts in this realm will affect the future of the book as they have its past and present. One vivid example of the possibilities of technology tied to social practice is suggested by current research into electronic paper being carried out independently at MIT and Xerox's Palo Alto Research Center. Electronic paper looks and feels like conventional paper, but can display changing images in a way similar to a computer screen. Only prototypes are currently operational, but if and when the technology comes to market it could affect the culture of the book more, even, than today's electronics, because it will be intimately compatible with reading practices to an extent that even the most portable of PCs cannot match.

Neal Stephenson has strikingly imagined some of the consequences of electronic paper in his novel, *The Diamond Age*. Stephenson's narrative charts the interactions between a book of this kind and Nell, a young girl growing up in a neo-Victorian future in which nanotechnology is an everyday reality. If, as Roger Chartier is given to remarking, a conventional book is remarkable because it 'changes by the fact that it does not change when the world changes', then here is a book remarkable because it really *does* change to suit its surroundings. *The Diamond Age* is a fascinating indication of the mutual debts between the book and new media, between technology and culture, and between literary and scientific creativity. And the influence it displays does not flow only in one direction, from science to literature. Such is the appetite for inspiration of today's software and media companies that Stephenson has found himself forced seriously to consider taking out patents on the devices portrayed in what is, after all, a fiction.[31]

The Diamond Age is also percipient in setting its narrative in a revived Victorian culture. The inference is that problems and opportunities of new media derive not just from technology, but from its interaction with changing perceptions of history. That much is certainly true of today's arguments about publishing and science. The problems faced by early modern naturalists persevere today in two significant respects – significant not least because of our conviction that both are radically new. First, piracy is again a prominent practice, and one with considerable geopolitical importance. Claimed losses are running at the level of tens of billions of dollars a year. The tendentious character of such claims must, of course, be noted – they are frequently calculated on the dubious assumption that every user of a pirated program would otherwise have paid the full retail price for an authorised version.[32] Yet it cannot be denied that the phenomenon is a major cultural and commercial reality. In Hong Kong, which is apparently now the 'world's piracy capital', pirates can sell you more up-to-date copies of the latest software than legitimate dealers, and at a cheaper price; they will even have fixed the bugs.[33] This is no mark of immature technology. On the contrary, new technology, by making fidelity digital, actually facilitates the practice. But it may be a mark of an immature *understanding* of technology – or rather, to avoid the anthropomorphism, of an understanding that owes too much to unreflective adoption of the cultural resources traditional to the realm of print.

On a similar note, it is interesting that credit is once again at a premium as we labour to establish trust in communication over the World Wide Web. The *New York Times* recently published a set of rules for assessing the degree of credit to vest in claims made in websites. The newspaper itself could have used them just weeks earlier, when it reported as facts a series of claims about the film industry that had originated as web-based jokes.[34] Moreover, it is not too fanciful to compare these problems, and the efforts to solve them, with the endeavours to secure printed communication in the era of Regiomontanus, Brahe and Newton. And the efforts of the sciences in this respect may, as in that era, form archetypes for more widely distributed cultural practices. Scientists' experiments with authorship may hint at the successor to copyright, whatever that will be. In addition, citizens are indeed now beginning to trust their credit-card numbers to the Web in increasing numbers, and in taking this decision they are not reacting to any dramatic change in technology. Rather, they themselves are changing the assumptions on which they are prepared to trust to the veracity and security of the medium. The grounds for giving assent are shifting *among users*, and the Web is becoming safe by consensus, as the printed book did centuries ago. The technologies of encryption and electronic security are certainly essential components in this process, but they are not alone sufficient to explain it. There is certainly a long-term context to such decisions. To understand them, as scientists themselves are coming to realise, one must engage with history, philosophy and sociology.

So we may indeed learn something useful from the history of the book, and from the history of the scientific book in particular. Looking at the past allows us more accurately to characterise what it is that is really revolutionary in today's situation; many of the most frequently alleged transformations – even the very conviction that we are facing transformations transcending any past experience – turn out to have long historial pedigrees. It shows us that the work that needs to be done now is the latest stage in a long-term genealogy of labours of the same kind, extending back at least to Regiomontanus and probably beyond. And it confirms that we have the potential ability to shape central elements of the communications circuit. We can continue to change even quite self-evident characteristics of Web culture as they come into being and develop, just as past actors did the cultures of print that they themselves encountered. The temptation to hail our situation as revolutionary should be resisted, then, pending a more historically informed understanding of what exactly it is that is so new, and why its appearance should constitute a revolution.

Notes

1 See W. F. Bynum, S. Lock and R. Porter (eds.), *Medical Journals and Medical Knowledge: Historical Essays* (London, 1992), and D. A. Kronick, *A History of Scientific and Technical Periodicals: The Origins and Development of the Scientific and Technical Press 1665–1790* (New York, 1962). The role of peer review and journal publishing was recognised as central to science by Robert Merton, in *The Sociology of Science* (Chicago, 1973), pp. 460–96.
2 A. Goldgar, *Impolite Learning: Conduct and Community in the Republic of Letters, 1680–1750* (New Haven, CT, 1995).
3 R. Darnton, 'The new age of the book', *New York Review of Books*, 46/5 (18 March 1999), pp. 5–7. The British universities no longer have a formal tenure system, but the government's regular Research Assessment Exercises create their own peculiar pressures.
4 M. Biagioli, 'The instability of authorship: credit and responsibility in contemporary biomedicine', *FASEB Journal*, 12 (1998): 3–16, reprinted as 'Aporias of scientific authorship: credit and responsibility

in contemporary biomedicine', in Biagioli (ed.), *The Science Studies Reader* (New York, 1999), pp. 12–30.

5 http://mars2.caltech.edu/libtest/index.html; http://library.caltech.edu/publications/ScholarsForum/; T. F. Gieryn, 'The (cold) fusion of science, mass media, and politics', in Gieryn, *Cultural Boundaries of Science* (Chicago, 1999), pp. 183–232, esp. 217–24.

6 See the 1998 debate on Harnad's proposal sponsored by the Association of Research Libraries: www.arl.org/scomm/subversive/toc.html. For Harnad on the link between credit, peer review, and paper publication, see http://www.ariadne.ac.uk/issue8/harnad/

7 http://xxx.lanl.gov/. For an overview of Ginsparg's facility, see http://xxx.lanl.gov/blurb/sep96news.html

8 K. C. Cole and R. L. Hotz, 'Science, hype and profit', *Los Angeles Times*, 24 January 1999.

9 R. Lewis, 'Journals feel pressure to speed the publishing process', *The Scientist*, 8/18 (19 September 1994), pp. 21, 23. See also A. Odlyzko, 'The economics of electronic journals', in R. Ekman and R. E. Quandt (eds.), *Technology and Scholarly Communication* (Berkeley, CA, 1999), pp. 380–93 (a reference I owe to Andrew Scull).

10 K. Hafner, 'Physics on the web is putting science journals on the line', *New York Times*, 21 April 1998, B12.

11 F. Rowland, 'The four functions of the scholarly journal': http://www.ariadne.ac.uk/issue7/fytton/

12 P. Wouters (moderator), C. Bingham, T. Delamothe, J. P. Kassirer, G. Lundberg and C. Thompson, *The Future of Medical Publishing: A Debate*, at www.biomednet.com/hmsbeagle/46/cutedge/overview.htm. Note that this debate was underwritten as 'another free service from Elsevier Science'.

13 R. Chartier, *The Order of Books: Readers, Authors, and Libraries in Europe between the Fourteenth and Eighteenth Centuries* (Cambridge, 1994), pp. 29–30; Biagioli, 'Instability of authorship'.

14 R.C. Iliffe, 'Author-mongering: the "editor" between producer and consumer', in A. Bermingham and J. Brewer (eds.), *The Consumption of Culture, 1600–1800: Image, Object, Text* (London, 1995), pp. 166–92.

15 J. Secord, 'Extraordinary experiment: electricity and the creation of life in Victorian England', in D. Gooding, T. Pinch and S. Schaffer (eds.), *The Uses of Experiment* (Cambridge, 1989), pp. 337–83; S. Shapin, 'Science and the public', in R. C. Olby, G. N. Cantor, J. R. R. Christie and M. J. S. Hodge (eds.), *Companion to the History of Science* (London, 1990), pp. 990–1007.

16 S. Shapin, 'How to be anti-scientific', in H. Collins and J. Labinger (eds.), *Science Peace* (forthcoming); Shapin, 'Rarely pure and never simple: talking about truth', *Configurations*, 7 (1999): 1–14, esp. pp. 11–13.

17 A. Johns, *The Nature of the Book: Print and Knowledge in the Making* (Chicago, 1998), pp. 350–2.

18 N. M. Swerdlow, 'Science and humanism in the Renaissance: Regiomontanus's *Oration on the Dignity and Utility of the Mathematical Sciences*', in P. Horwich (ed.), *World Changes: Thomas Kuhn and the Nature of Science* (Cambridge, MA, 1993), pp. 131–68; M. Folkerts,

'Regiomontanus's role in the transmission and transformation of Greek Mathematics', in F. J. Ragep et al. (eds.), *Tradition, Transmission, Transformation* (Leiden, 1996), pp. 89–113.

19 M. Feingold, 'The mathematical sciences and new philosophies', in N. Tyacke (ed.), *The History of the University of Oxford. Vol. 4, Seventeenth-Century Oxford* (Oxford, 1997), pp. 359–448, esp. 435–6.

20 E. L. Eisenstein, *The Printing Press as an Agent of Change: Communications and Cultural Transformations in Early-Modern Europe*, 2 vols. (Cambridge, 1979), e.g., vol. II, pp. 623–6; B. Latour, *Science in Action: How to Follow Scientists and Engineers through Society* (Milton Keynes, 1987), pp. 226–7.

21 M. G. Winkler and A. Van Helden, 'Johannes Hevelius and the visual language of astronomy', in J. V. Field and F. A. J. L. James (eds.), *Renaissance and Revolution: Humanists, Scholars, Craftsmen, and Natural Philosophers in Early Modern Europe* (Cambridge, 1993), pp. 97–116.

22 A. Mosley, 'Astronomical books and courtly communication', q.v., ch. 6.

23 J. F. Clarke, *Autobiographical Recollections of the Medical Profession* (London, 1874), pp. 12–20.

24 M. Rose, *Authors and Owners: The Invention of Copyright* (Cambridge, MA, 1993); J. Feather, *Publishing, Piracy and Politics: An Historical Study of Copyright in Britain* (London, 1994).

25 S. Shapin, *A Social History of Truth: Civility and Science in Seventeenth-Century England* (Chicago, 1994), pp. 302–9. There are interesting reflections on the place in the Royal Society of another kind of reading practice associated with alchemy in L. M. Principe, *The Aspiring Adept: Robert Boyle and his Alchemical Quest* (Princeton, 1998), pp. 139–49. Boyle, on many occasions the advocate of openness in experimental philosophy, employed a series of codes, ciphers and nomenclatures to mask his alchemical writings, even when taking notes on his reading of other authors. Some of their keys remained unknown even to his amanuenses. In published works, too, he sometimes employed sophisticated dissimulations.

26 Johns, *Nature of the Book*, ch. 7.

27 There is a very extensive literature on the republic of letters and the public sphere. See, for example, A. Goldgar, *Impolite Learning: Conduct and Community in the Republic of Letters, 1680–1750* (New Haven, 1995), and L. Daston, 'The ideal and reality of the republic of letters in the Enlightenment', *Science in Context*, 4 (1991): 367–86.

28 *Speeches or Arguments of the Judges of the Court of King's Bench* (Leith, 1771), pp. 50, 93, 124; Rose, *Authors and Owners*, p. 87; Hesse, *Publishing and Cultural Politics in Revolutionary Paris, 1789–1810* (Berkeley, 1991), p. 104. These are only a few of the many sources that could be cited.

29 Rose, *Authors and Owners*; Hesse, *Publishing and Cultural Politics*; M. Woodmansee, *The Author, Art, and the Market: Rereading the History of Aesthetics* (New York, 1994), pp. 35–86.

30 For discussions of the modern predicament of the learned book in relation to its history, see Biagioli, 'Instability of authorship', Darnton,

'New age of the book', and C. Hesse, 'Books in time', in G. Nunberg (ed.), *The Future of the Book* (Berkeley, 1996), pp. 21–36.
31 N. Stephenson, *The Diamond Age: Or, A Young Lady's Illustrated Primer* (London, 1995); interview with Stephenson: www.hotwired.com/talk/club/special/transcripts/95-01-19.stephenson.html; Chartier, *Order of Books*, p. 16. It is now possible to obtain a few classic works in the history of science, for example Robert Hooke's *Micrographia* (1665), in digitised, hypertextualised form: www.octavo.com
32 J. Boyle, *Shamans, Software, and Spleens: Law and the Construction of the Information Society* (Cambridge, MA, 1996), pp. 2–3.
33 'Software piracy: better than the real thing', *The Economist*, 24 January 1998.
34 T. Kelley, 'Whales in the Minnesota River', *New York Times*, 4 March 1999; J. Sterngold, 'Lost, and gained, in the translation', *New York Times*, 15 November 1998; 'Corrections', *New York Times*, 4 December 1998.

Further reading

M. Biagioli, 'Aporias of scientific authorship: credit and responsibility in contemporary biomedicine', in Biagioli (ed.), *The Science Studies Reader* (New York, 1999), pp. 12–30

J. Boyle, *Shamans, Software, and Spleens: Law and the Construction of the Information Society* (Cambridge, MA, 1996)

W. F. Bynum, S. Lock, and R. Porter (eds.), *Medical Journals and Medical Knowledge: Historical Essays* (London, 1992)

R. Chartier, *Forms and Meanings: Texts, Performances, and Audiences from Codex to Computer* (Philadelphia, 1995)

R. Chartier and H.-J. Martin (eds.), *Histoire de l'edition française*. 2nd edn, 4 vols. (Paris, 1989–91)

R. Darnton, *The Business of Enlightenment: A Publishing History of the Encyclopédie, 1775–1800* (Cambridge, MA, 1979)

  'The new age of the book', *New York Review of Books*, 46/5 (18 March 1999), pp. 5–7

L. Daston, 'The ideal and reality of the republic of letters in the Enlightenment', *Science in Context*, 4 (1991): 367–86

E. L. Eisenstein, *The Printing Press as an Agent of Change: Communications and Cultural Transformations in early Modern Europe*, 2 vols. (Cambridge, 1979)

A. Goldgar, *Impolite Learning: Conduct and Community in the Republic of Letters, 1680–1750* (New Haven, 1995)

C. Hesse, *Publishing and Cultural Politics in Revolutionary Paris, 1789–1810* (Berkeley, 1991)

A. Johns, *The Nature of the Book: Print and Knowledge in the Making* (Chicago, 1998)

D. A. Kronick, *A History of Scientific and Technical Periodicals: The Origins and Development of the Scientific and Technical Press 1665–1790* (New York, 1962)

C. C. Mann, 'Who will own your next good idea?', *Atlantic Monthly* (September 1998), pp. 57–82

D. F. McKenzie, *Bibliography and the Sociology of Texts* (London, 1985)
  The Cambridge University Press 1696–1712: A Bibliographical Study, 2 vols. (Cambridge, 1966)
D. J. McKitterick, *A History of Cambridge University Press*, 2 vols. to date (Cambridge, 1992–8)
M. McLuhan, *The Gutenberg Galaxy* (London, 1962)
  Understanding Media: The Extensions of Man (London, 1964)
G. Nunberg (ed.), *The Future of the Book* (Berkeley, 1996)
D. Schiller, *Digital Capitalism: Networking the Global Market System* (Cambridge, Mass., 1999)
S. Sheets-Pyenson, 'Popular science periodicals in Paris and London: the emergence of a low scientific culture, 1820–1875', *Annals of Science*, 42 (1985): 549–72
N. Stephenson, *The Diamond Age: or, A Young Lady's Illustrated Primer* (London, 1995)
J. Topham, 'Scientific publishing and the reading of science in early nineteenth-century Britain', *Studies in History and Philosophy of Science*, 31A (2000)
M. G. Winkler and A. Van Helden, 'Johannes Hevelius and the visual language of astronomy', in J. V. Field and F. A. J. L. James (eds.), *Renaissance and Revolution: Humanists, Scholars, Craftsmen, and Natural Philosophers in Early Modern Europe* (Cambridge, 1993), pp. 97–116
P. Wouters (moderator), C. Bingham, T. Delamothe, J. P. Kassirer, G. Lundberg and C. Thompson, *The Future of Medical Publishing: A Debate*. www.biomednet.com/hmsbeagle/46/cutedge/overview.htm

Notes on contributors

ANN BLAIR is John L. Loeb Associate Professor of History and of History and Literature at Harvard University. She is the author of *The Theater of Nature: Jean Bodin and Renaissance Science* (Princeton University Press, 1997) and is working on a new project on the development of consultation reading and reference tools in early modern Europe.

THOMAS BROMAN is an Associate Professor in the Department of History of Science at the University of Wisconsin, Madison. His first book, *The Transformation of German Academic Medicine*, was published by Cambridge University Press in 1996. He is currently working on the history of literary criticism and the publishing business in eighteenth-century Germany.

JERRY BROTTON is a Lecturer in the Department of English at Royal Holloway, University of London, where he teaches Renaissance literature and culture and Shakespeare. He is the author of several articles on early modern drama, geography and travel, and of *Trading Territories: Mapping the Early Modern World* (1997) and, with Lisa Jardine, of *Global Interests* (2000), both books being published by Reaktion, London, and Cornell University Press, Ithaca.

WILLIAM CLARK taught at the Institut für Wissenschaftsgeschichte, Göttingen, from 1991–7; was a Research Fellow at the Max-Planck-Institut für Wissenschaftsgeschichte, Berlin in 1997–8; and has recently been teaching in the Department of History and Philosophy of Science, Cambridge. He is coeditor with Jan Golinski and Simon Schaffer of *The Sciences in Enlightened Europe* (University of Chicago Press, 1999), and with Peter Becker of *Little Tools of Knowledge* (Ann Arbor: University of Michigan Press, 2000). His *The Hero of Knowledge (Homo Academicus Germanicus)* will apper with University of California Press.

SILVIA DE RENZI is a Lecturer in History of Medicine at the Open University. She has her doctorate from the University of Bologna, has held a Wellcome Postdoctoral Fellowship, and has worked as a rare books researcher in the Whipple Library, Cambridge. She has published articles on collecting, natural history and medicine in Baroque Rome, and is currently writing a book on that subject.

MARINA FRASCA-SPADA is a historian of philosophy and the author of a book on David Hume, *Space and the Self in Hume's 'Treatise'* (Cambridge University Press, 1998). She is a Fellow of St Catharine's College, Cambridge, an Affiliated Lecturer of the Department of History and Philosophy of Science and the Associate Editor of *Studies in History and Philosophy of Science* and *Studies in History and Philosophy of Biological and Biomedical Sciences*. Her current research interests include epistemological issues in Hume's writings, eighteenth-century theories of human nature, and the teaching and reception of logic and metaphysics in eighteenth-century England.

AILEEN FYFE is a Lecturer in the Department of History, National University of Ireland, Galway. She is interested in the publication of books on the sciences for non-elite audiences in the nineteenth century, and the effects of religious debates on such publishing. She has written on children's science books, and her most recent work has focussed on the cheap scientific publications of the Religious Tract Society in the mid-nineteenth century.

ANTHONY GRAFTON teaches history and history of science at Princeton University. His books include *Joseph Scaliger* (Oxford, 1983–93), *Defenders of the Text* (Cambridge, MA, 1991), *The Footnote* (Cambridge, MA, and London, 1997), and *Cardano's Cosmos* (Cambridge, MA, 2000).

LISA JARDINE is Professor of Renaissance Studies at Queen Mary, University of London and an Honorary Fellow of King's College, Cambridge. Her most recent book is *Ingenious Pursuits: Building the Scientific Revolution*, published by Little, Brown.

NICK JARDINE is a Fellow of Darwin College and Professor of History and Philosophy of the Sciences at the University of Cambridge. His most recent books are *Cultures of Natural History*, ed. with J. Secord and E. C. Spary (Cambridge, 1996), and *The Scenes of Inquiry: On the Reality of Questions in the Sciences* (2nd edn. with supplementary essays, Oxford, 2000). He is Editor of *Studies in History and Philosophy of Science* and *Studies in History and Philosophy of Biological and Biomedical Sciences*. His current research projects are, in collaboration with Alain Segonds, on priority disputes in early modern cosmology, and on historiography of the sciences.

ADRIAN JOHNS is an Associate Professor of History at the California Institute of Technology. His *The Nature of the Book: Print and Knowledge in the Making* (University of Chicago Press, 1998) was the winner of the 1999 Book Prize awarded by the Society for the History of Authorship, Reading and Publishing (SHARP), and of the 1998/99 Louis Gottschalk prize for the best book in eighteenth-century history awarded by the American Society for Eighteenth-Century Studies. His researches centre on science in the early-modern period and on the history of the book. He is currently working on the history of intellectual piracy, and experimenting with the use of multimedia in science studies.

LAUREN KASSELL is an Assistant Lecturer in the Department of History and Philosophy of Science, and a Fellow of Pembroke College, Cambridge. She is completing a book on Simon Forman, medicine, astrology and alchemy (a revision of her D. Phil) for Oxford University Press, and is beginning a study of alchemy and medicine in early-modern England.

SACHIKO KUSUKAWA is Fellow and College Lecturer in History and Philosophy of Science, Trinity College, Cambridge. Her publications include *The Transformation of Natural Philosophy: The Case of Philip Melanchthon* (Cambridge University Press, 1995), 'Incunables and sixteenth-century books', in A. Turner (ed.), *Scientific Books, Libraries and Collectors* (Aldershot: Ashgate, 1999) and 'Leonhart Fuchs on the Importance of Pictures', *Journal of the History of Ideas*, 58/3 (1997). She is currently preparing a monograph on early modern scientific illustrations.

ROSAMOND MCKITTERICK is Professor of Medieval History in the University of Cambridge and Fellow of Newnham College. She has published books and articles on the early medieval culture and book production, including *The Frankish Church and the Carolingian Reforms, 789–895* (London, 1977), *The Frankish Kingdoms under the Carolingians 751–987* (London, 1983), *The Carolingians and the Written Word* (Cambridge, 1989), *Books, Scribes and Learning in the Frankish Kingdoms, 6th–9th Centuries* (Aldershot, 1994), *The Frankish Kings and Culture in the Early Middle Ages* (Aldershot, 1995), and the edited volumes *The Uses of Literacy in Early Medieval Europe* (Cambridge, 1990), *Carolingian Culture: Emulation and Innovation* (Cambridge, 1994) and *The New Cambridge Medieval History, vol. II c. 700–c. 900* (Cambridge, 1995).

ADAM MOSLEY is a Junior Research Fellow of Trinity College, Cambridge. His current research, into the instrument literature of the sixteenth century, extends one of the principal themes of his doctoral thesis 'Bearing the Heavens: Astronomers, Instruments and the Communication of Astronomy in Early-Modern Europe'.

EUGENIA ROLDÁN VERA is about to take up a Lectureship in the Department of History, National Autonomous University of Mexico. Her University of Cambridge Ph.D. dissertation focuses on the publishing enterprise of Rudolph Ackermann for Latin America in the 1820s. She has published a handful of articles on various topics in the history of education and school textbooks in nineteenth-century Latin America.

JAMES SECORD is a Reader in the Department of History and Philosophy of Science at the University of Cambridge, where he teaches the history of science and science communication. He edited *Cultures of Natural History* (with Nick Jardine and Emma Spary, Cambridge University Press, 1996) and is the author of *Controversy in Victorian Geology: The Cambrian-Silurian Dispute* (Princeton, 1986) and *Victorian Sensation: The Extraordinary Publication, Reception and Secret Authorship of* Vestiges of the Natural History of Creation (Chicago, 2000).

EMMA SPARY is a Research Fellow of the Max-Planck-Institut für Wissenschaftsgeschichte, Berlin. Together with Nicholas Jardine and James Secord, she has edited *Cultures of Natural History* (Cambridge University Press, 1996) and has published several articles on the history of natural history. She is the author of *Utopia's Garden: French Natural History from Old Regime to Revolution* (Chicago University Press, 2000).

ALAN STEWART is a Lecturer in Late Medieval and Renaissance English at Birkbeck, University of London. He is the author of *Close Readers: Humanism and Sodomy in Early Modern England* (Princeton, 1997); *Hostage to Fortune: The Troubled Life of Francis Bacon 1561–1626*, co-authored with Lisa Jardine (London: Gollancz, 1998); and *Philip Sidney: A Double Life* (London: Chatto and Windus, 2000).

MARY TERRALL is an Assistant Professor in the Department of History at the University of California, Los Angeles, where she teaches the history of early-modern science. She is the author of articles on Emilie de Châtelet, on Maupertuis' theory of generation, and on issues of gender in Enlightenmnent science. She is working on a biographical study of Maupertuis.

JONATHAN TOPHAM is an AHRB Research Fellow on the collaborative project 'Science in the Nineteenth-Century Periodical', based at the Universities of Sheffield and Leeds. From 1993 to 1997 he served as an editor of the *Correspondence of Charles Darwin* (Cambridge, 1985–99). He is currently working on a book about scientific publishing and the readership for science in early nineteenth-century Britain.

RICHARD YEO is Reader in the History and Philosophy of Science at the School of Humanities, Griffith University, Brisbane, Australia. He has written on the cultural aspects of science in the eighteenth and nineteenth centuries. Recent books are *Defining Science: William Whewell, Natural Knowledge and Public Debate in Early Victorian Britain* (Cambridge, 1993), *Telling Lives in Science: Essays on Scientific Biography*, co-edited with Michael Shortland (Cambridge, 1996) and *Encyclopaedic Visions: Scientific Dictionaries and Enlightenment Culture* (Cambridge, 2001).

Index

Abu'l-Fida, 35
Ackermann, Rudolph, 339–46, 348, 351
Adam, 109, 137, 141
Adanson, Michel, 7, 266–8
Addison, Joseph, 232
Aethicus Ister, 24
Agrippa von Nettesheim, Heinrich Cornelius, 50–1
Aikin, John, 279–80, 282–8
Albohali, 59
Albrecht of Hohenzollern, Duke of Prussia, 57, 117
Alcabitius, 59
Alcuin, 24
Aldrovandi, Ulisse, 69
Algarotti, Francesco, 248, 250
Alsted, Johann Heinrich, 82, 85, 208
Amalric, Abbot, 26
Amirutzes, 45
Anderson, Benedict, 46
Antonius de Montulmo, 57
Apian, Peter, 114
Apicius, 20
Appleyard, E. S., 318
Aratus, 20, 21
Archimedes, 1, 69
Argenterio, Giovanni, 49, 51
Aristotle, 19, 28, 69, 311, 403, 419
Ashmole, Elias, 132, 135–7, 139–44, 147
Aslaksson, Cort (Aslachus), 119
Aubrey, John, 64
Augustine, Saint, Bishop of Hippo, 64
Austin, Alfred, 309
Averroes, 19
Avicenna, 75
Avienus, Rufus Festus, 21

Babbage, Charles, 317, 325, 326, 328–9, 331–3, 384–5
Babington, Thomas (Lord Macaulay), 357–61, 365
Backhouse, William, 132
Bacon, Francis, 73–4, 213, 354–66
Bacon, Roger, 69
Baer, Nicolai Reymers (Ursus), 121, 124, 125
Baillet, Adrian, 69–70
Bakhtin, Mikhail M., 171
Barbauld, Anna, 278–80, 282–8
Barrow, Isaac, 358
Bateman, James, 384
Bauer, Barbara, 62
Bayezid II, Ottoman Sultan, 43, 44
Bayle, Pierre, 171, 176, 208, 212, 221, 231–2, 233
Beck, Strasburg printer, 109
Bede, the Venerable, 16, 22, 25
Bell, George, 319
Bellini, Gentile, 44
Belon, Pierre, 92, 97, 108
Benedetti, Alessandro, 108
Benedicht, Laurentius, 116
Behn, Aphra, 241, 243
Bentley, Richard, 183, 358
Berengario da Carpi, 101, 108
Berkeley, George, 85, 175
Berlinghieri, Francesco, 43–4, 46
Bernard, Jacques, 176
Bernard of Chartres, 28
Biagioli, Mario, 404
Birch, Thomas, 363
Blackbourne, Richard, 363
Blair, Ann, 5, 6, 7, 400
Blakesley, Joseph Williams, 356
Blanchard *see* Blankaart, Stephan
Blanco White, Joseph, 342, 343–4, 347
Blankaart, Stephan (Blanchard), 210, 213
Blount, Thomas, 208
Blumenbach, Johann Friedrich, 373
Bodin, Jean, 74
Boethius, Anicius Manlius Severinus, 20, 26, 28, 225

Bolivar, Gregorio, 153, 160–2, 164
Bonnier de la Mosson, Joseph, 272
Borges, Jorge Luis, 190
Borja Migoni, Manuel, 342
Bossuet, Jacques Bénigne, 357
Bostocke, Richard, 142
Boucher, François, 261, 262
Bouillet, 363
Boyle, Robert, 132, 133, 228, 299, 302, 419, 420
Brahe, Tycho, 115–28, 416–17, 421
Brewer, John, 286
Brewster, David, 358, 379, 380
Broman, Thomas, 4, 6, 401
Bromhead, Edward, 317, 325, 329
Brotton, Jerry, 3
Brougham, Henry, 358–9, 374, 377–8
Browne, Peter, 175
Browne, Sir Thomas, 132, 145–6, 147
Brunfels, Otto, 92, 97, 99, 109
Buckland, Revd. William, 381, 383
Buffon, Georges Louis Leclerc, comte de, 162
Bülow, Heinrich von, 196
Butterfield, Herbert, 1, 3
Byrom, John, 304

Cagi Acmet, *see* Hajji Ahmed
Calcidius, 24
Cardano, Fabio, 55, 59
Cardano, Girolamo, 49–52, 54–65, 81
Carey, Hilary, 54
Carlile, Richard, 371, 373
Carlyle, Thomas, 356, 365, 374, 386
Carolus, Johann, 227
Casaubon, Meric, 300–1
Cassiodorus, 17
Castellus, Bartholomaeus, 210
Castiglione, Bonaventura, 54
Cem, son of Mehmed the Conqueror, 44
Censorinus, 20
Cesalpino, Andrea, 107
Cesi, Federico, 154–5
Chambers, Ephraim, 177, 185, 207–8, 210, 212–21
Charlemagne, 20, 23, 24
Charles I, King of England, 300
Charles V, Holy Roman Emperor, 54, 55, 60
Charles the Bald, King of France, 20, 23
Charleton, Walter, 300
Charlotte, Queen of England, 278

Chartier, Roger, 5, 393–4, 400, 420
Chaucer, Geoffrey, 140
Chauvin, Stephen, 210
Cicero, 21, 24, 26, 28, 64, 119
Clark, William, 4, 400
Clarke, John, 177, 178
Clarke, Dr Samuel, 171, 174, 176–8, 181, 183, 185, 186
Cocles, Bartolomeo, 60
Coleridge, Samuel Taylor, 375
Colliber, Samuel, 183
Collins, Wilkie, 293
Colonna, Fabio, 155, 156, 159–60
Columbus, Christopher, 42
Combe, George, 383
Condorcet, Jean-Antoine-Nicolas de Caritat, marquis de, 419
Congreve, William, 394
Constable, Archibald, 369, 377
Constantine of Fleury, 26, 27, 28
Constantinus, 25
Contreni, John J., 25
Cooper, William, 133
Copernicus, Nicolaus, 1, 50, 57, 58, 65, 117
Corbetta, Gualtiero, 55–6
Cornarius, Janus, 58, 60, 105, 107
Costanzo de Ferrara, 45
Courtonne, Jean, 272
Cowley, Abraham, 292–3
Craig, John, 125
Crell, Lorenz, 234–5
Crick, Francis, 226
Cromwell, Oliver, 228
Cudworth, Ralph, 177, 183
Cuvier, Georges, 375

d'Alembert, Jean le Ronde, 208, 216
da Colle, Paolo, 44
Dacier, Anna, 303
Dal Pozzo, Cassiano, 155, 158
Darnton, Robert, 5, 6, 393, 394
Darwin, Charles, 226, 386, 399, 400, 408
Darwin, Erasmus, 1, 3, 398
Daston, Lorraine, 404
Davy, Sir Humphry, 375, 381
d'Espargnet, Jean, 295
de Laet, Johan, 156
de Mendebil, Pablo, 344
de Mora, José Joaquin, 342, 345
De Renzi, Silvia, 4, 5, 398
de Sallo, Denys, 229

de Soto, Domingo, 82
de Urcullu, José, 342, 345
de Villanueva, Joaquin Lorenzo, 342
Dee, Arthur, 132, 139, 146, 147
Dee, John, 132–3, 135, 136, 140, 142, 144–7
Deighton, John, 320–5, 326–7, 332–5
Delesserts, Benjamin, 7
della Porta, Giambattista, 81, 101
Desaguliers, John Theophilus, 249
Descartes, René, 177, 243, 298, 299, 303, 419
Dezallier d'Argenville, Antoine-Joseph, 262–6, 272
Dibdin, Thomas Frognall, 377
Dickens, Charles, 377
Diderot, Denis, 208, 216, 419
Diogenes Laertius, 62
Dionysius Exiguus, 21, 22
Dioscorides, 25, 69, 101, 107
Donne, William Bodham, 356
Dryden, John, 358
Duhem, Pierre, 292
Duns Scotus, 82
Dunstan, St, Archbishop of Canterbury, 133, 135–7, 139–47
Dürer, Albrecht, 92, 109

Edgeworth, Charles Sneyd, 276, 280
Edgeworth, Maria, 280–1, 282, 284
Edgeworth, Richard, 280–1, 282, 284, 287
Edwardus Generosus Anglicus Innominatus, 136–7, 139–41, 144
Egenolff, Christian, 109
Einhard, 17
Eisenstein, Elizabeth, 3
Elliotson, John, 373
Ellis, Robert Leslie, 362, 365
Erasmus, Desiderius, 58, 77, 80
Essex, Robert Devereux, Earl of, 355
Euclid, 20, 91, 292, 419
Eunapius, 62
Eustachio, Bartholomaeo, 104, 108
Eve, 109, 141

Faber, Johannes, 153, 155–62, 164
Faraday, Michael, 282
Faventinus, 20
Federigo da Montefeltro, Duke of Urbino, 44
Fell, John, 416–17

Ferdinand, King of Bohemia and Hungary, 54
Ficino, Marsilio, 82
Filarete, Antonio, 45
Filelfo, Francesco, 55–6
Fine, Oronce, 37
Firmicus Maternus, 53, 56
FitzGerald, Edward, 356, 362, 364
Flamsteed, John, 115
Fludd, Robert, 296, 297
Fontenelle, Bernard de, 240–5, 248, 250
Forman, Simon, 141
Francis I, King of France, 55, 60
Frasca-Spada, Marina, 4, 5, 6, 401, 403
Frederick II, King of Denmark, 115
Freud, Sigmund, 399, 400
Fuchs, Leonhard, 92, 100–1, 105, 109
Furetière, Antoine, 208
Fyfe, Aileen, 4, 6

Gadamer, Hans-Georg, 394–5
Galen, 25, 51, 101, 107, 108, 311
Galilei, Galileo, 119, 122–3, 124, 128, 151, 162, 243, 346
Gallois, Abbé Jean, 229
Garcaeus, Johannes, 59
Gassendi, Pierre, 64, 181, 419
Gaurico, Luca, 54–5, 59, 61, 62, 64
Geber, 295
Genette, Gérard, 401
Georgius Trapezuntius (George of Trebizond), 45–6, 55–6
Gerbert (Pope Sylvester II), 17, 26–8, 30
Germanicus, 21
Gersaint, Edmé-François, 261, 262, 272
Gesner, Conrad, 70, 77, 92, 97, 99, 101, 108
Ghisi, Giorgio, 101
Ginsparg, Paul, 412–13
Giovio, Paolo, 62
Giustinian, Marc'Antonio, 37, 39, 46
Gladstone, William Ewart, Prime Minister, 365
Goclenius, Rudolph, 59
Goethe, Johann Wolfgang von, 201–2
Gower, John, 140
Grafton, Anthony, 4, 6, 397, 401
Green, Robert, 177
Greig, John, 344
Grosseteste, Robert, 19, 30
Guazzo, Stefano, 162
Gutenberg, Johann, 114, 415, 416

Habermas, Jürgen, 8
Hajji Ahmed (Cagi Acmet), 35–9, 41, 46
Hale, John, 43
Hale, Sir Matthew, 183
Haller, Albrecht von, 196
Hallyn, Fernand, 401, 403
Haly, 51
Harnad, Stevan, 412
Harris, John, 207, 208, 213, 217, 244–5, 249
Hartley, David, 304–5, 307, 311
Harvey, Gabriel, 61–2
Hayek, Thaddaeus, 121
Hayes, Charles, 185
Hazlitt, William, 374
Heath, Douglas Denon, 362, 365
Helmholtz, Hermann von, 402
Henry VIII, King of England, 60
Herbert of Cherbury, Baron Edward, 358
Hermes, 137
Hernandez, Francisco, 153–6, 160, 161, 164
Herschel, John, 325, 326, 329, 331–4, 375, 382–3, 385
Herschel, William, 382
Hesse, Carla, 5, 397
Hevelius, Johannes, 417
Heyne, Christian Gottlob, 197–201
Hickman, Bartholomew, 145
Hippocrates, 25, 64, 311
Hobbes, Thomas, 292
Home, Henry (Lord Kames), 311
Homer, 53
Horace, 28
Horne, George, 179
Hoskin, Michael, 176
Hraban Maur, 16, 24
Huet, Pierre-Daniel, 225
Humboldt, Alexander von, 349, 350
Hume, David, 176, 179
Husey, John, 132
Hyginus (Gromaticus), 20, 21

Iliffe, Robert, 397
Illyricus, Matthias Flaccius, 52
Ingoli, Franciscus, 128
Irving, Christopher, 344
Isidore of Seville, 25
Ivins, William H., 38

Jackson, John, 178
James IV, King of Scotland, 121, 124
James of Venice, 19
Jardine, Lisa, 4, 403
Jardine, Nick, 4
Jefferson, Thomas, 85
Jerome, St, 119
John of Salisbury, 28
Johns, Adrian, 4, 7, 258, 398, 400
Johnson, Dr Samuel, 185, 357
Johnson, Thomas, 177
Johnson, William, 210
Jones, William, 179
Jonston, Johann, 69
Joyce, Jeremiah, 282
Juvenal, 28

Kames, Lord, *see* Home, Henry
Kassell, Lauren, 3, 4, 6, 7, 296, 400
Kayser, Albrecht, 203
Keill, John, 177, 185
Kelley, Edward, 132, 135, 136, 140, 142, 144–5, 147
Kepler, Johannes, 2, 114, 119, 121, 123, 124, 125, 128, 292–3, 399, 403
Kerrigan, Anthony, 190
Khunrath, Heinrich, 296
Kierkegaard, Søren, 176
Kimball, Fiske, 270
King, William, Archbishop of Dublin, 171, 173–7, 179–81, 185
Kippis, Andrew, 210
Klein, Lawrence, 239
Knight, Charles, 379, 380, 385
Koonin, Provost Steven, 411
Koyré, Alexandre, 1
Kunoff, Hugo, 194
Kusukawa, Sachiko, 5

Lacroix, Silvestre Francois, 326, 332, 333, 334
Lafayette, Mme de, 241
Lajoüe, Jacques de, 261
Lamarck, Jean-Baptiste, 374, 383
Laplace, Pierre Simon, 374
Lardner, Dionysius, 379, 385
Larpent, Anna, 286
Las Torres, Alfonso, 155
Latour, Bruno, 3
Lavoisier, Antoine, 308
Law, Edmund, 172–81, 183, 185–6
Leibniz, Gottfried Wilhelm von, 176, 177, 181, 183, 195, 212, 332
Leonardo da Vinci, 292

Leopold I, Holy Roman Emperor, 132
Leowitz, Cyprian, 64
Lessing, Gotthold Ephraim, 233
Lichtenberg, Johannes, 54–5
Linnaeus, Carl, 7
Lipton, Peter, 396, 398
Lister, Anna, 258
Lister, Martin, 258
Lister, Susanna, 258
Locke, John, 85, 171, 172, 173, 177, 180–1, 183, 186,, 220, 303
Lokman, 35
Lombard, Peter, 18
Longman, Thomas, 319, 327, 334, 379, 385, 386
Lucan, 28
Lucretius, 69
Lupus of Ferrières, 17, 26, 28
Luther, Martin, 54, 58, 60, 61
Lydus, John, 53
Lyell, Charles, 379, 383

Macaulay, Lord, *see* Babington, Thomas
Machiavelli, Niccolò, 62
Mack, Pamela, 298
Maclean, Ian, 55
Macmillan, Daniel, 319
Magnus, Olaus, 92
Maier, Michael, 296
Manilius, 53
Mantell, Gideon, 381
Manutius, Aldus, 416
Marcellus, 25
Marcet, Jane, 282
Markgraf, Georg, 151, 156
Marmontel, Jean-François, 306
Martin, Benjamin, 245, 249
Martin, Henri-Jean, 5
Mästlin, Michael, 124, 125
Mattioli, Pier Andrea, 107
Maupertuis, Pierre-Louis Moreau de, 249–50
Mauritz von Nassau, Governor of Brazil, 151, 156
Mavor, William, 355
Mawman, Joseph, 334
Maximilian I, Holy Roman Emperor, 59
McKenzie, Don, 5, 393–4, 398, 399, 400, 404
McKeon, Michael, 397
McKitterick, Rosamond, 3, 4, 7, 400
Medici, Lorenzo de', 44

Mehmed the Conqueror, Ottoman Sultan, 43–6
Melanchthon, Philipp, 54, 58, 60, 82, 227
Membre, Michel, 37, 38
Mendelssohn, Moses, 233
Mendes da Costa, Emanuel, 267
Mercator, Gerard, 38
Michelangelo, 101
Michiel, Pietro Antonio, 107
Mill, John Stuart, 293, 305–10, 374
Milton, John, 357, 358
Mondino de' Luzzi, 101
Montague, Basil, 356, 358, 363
Montaigne, Michel de, 72, 302, 303
Montuus, Sebastianus, 105
More, Henry, 298–9, 300
More, Thomas, 158
Moreri, Louis, 208, 212
Morhof, Daniel Georg, 70, 132
Moritz, Karl Philipp, 225
Moritz, Landgrave of Hesse-Kassel, 120, 122, 125
Moses, 137
Mosley, Adam, 4, 7, 416
Müller, Johannes *see* Regiomontanus
Münchhausen, Gerlach Adolph von, 196–8
Murray, John, 379, 380, 381

Napier, Macvey, 358
Naudé, Gabriel, 70
Newbery, John, 276
Newton, Isaac, 135, 142–4, 147, 171, 172, 173, 177, 181, 185, 186, 226, 248, 250, 251, 252, 308, 369, 379, 380, 399, 400, 419, 420, 421
Nichol, John Pringle, 382–3
Nicolai, Friedrich, 200, 233
Nieremberg, Juan Eusebio, 154, 158
Nollet, Abbé Jean-Antoine, 249, 252
Norton, Thomas, 136, 140, 142
Nostradamus, 64
Nuñez de Arenas, José, 342, 345

Ockham, William of, 19
Oldenburg, Henry, 228–30, 398, 418
Oldys, William, 210
Oresme, Nicole, 64
Oribasius, 25
Orsmar of Tours, Archbishop, 26
Ortelius, Abraham, 38, 42

Osborne, Francis, 302, 303
Ozanam, Jacques, 210

Paracelsus, Theophrastus, 133, 142
Park, Katherine, 404
Parker, John William, 319
Parkes, Samuel, 344, 346
Pascal, Blaise, 302, 419
Paul III, Pope, 55, 58, 60, 61
Peacock, George, 331–4
Pearson, John, 358
Peckham, John, Archbishop of Canterbury, 56
Percival, Thomas, 282, 287
Persius, 28
Peter Lombard, 18
Petrarch (Francesco Petrarca), 55–6
Petreius, Johannes, 57–8
Petrus Severinus, 142
Peucer, Caspar, 121
Peurbach, Georg, 65
Philip IV, King of Spain, 151, 153–4
Phillips, Edward, 210
Phillips, Richard, 319
Philostratus, 62
Pickering, William, 379–80, 381
Pico della Mirandola, Giovanni, 64
Pinnock, William, 319, 344–6, 348–9
Pinon, Laurent, 80
Piso, Francisco, 156
Plato, 24, 35, 53
Pliny the Elder, 24, 25, 69, 71, 75, 82, 101, 119
Plutarch, 53
Poliziano, Angelo, 51–2, 60
Pomponius Mela, 24, 25
Pope, Alexander, 176, 355, 357
Porphyrius, 28
Postel, Guillaume, 38
Priestley, Joseph, 279, 286
Ptolemy, 39–45, 50, 51, 53, 59, 61, 64, 118, 298
Puschner, Johann, 190

Racine, Jean, 357
Ramus, Petrus (Pierre de la Ramée), 128, 217
Ramusio, Giovan Battista, 38
Rantzov, Heinrich, 117, 121, 123
Raphson, Joseph, 181
Ratdolt, Erhard, 91, 99
Reboul, Marie-Thérèse, 268

Recchi, Antonio Leonardo, 154–6
Rees, Abraham, 220–1
Regiomontanus (Johannes Müller), 59, 65, 117, 125, 416–17, 421, 422
Reinhold, Erasmus, 55, 57, 58, 59, 117, 125
Rhasis, 295
Rheticus, Georg Joachim, 57–8, 119
Rhodiginus, Caelius, 77
Richardson, Samuel, 311
Richer of Saint-Rémy, 26–8
Ripley, George, 133, 136, 140
Rivadavia, Bernardino, 342
Rocafuerte, Vicente, 342–3, 347
Rohault, Jacques, 171, 176–7
Roldán Vera, Eugenia, 4, 6, 287, 401
Romulus, 53
Rondelet, Guillaume, 92, 108
Rose, Hugh James, 334
Rose, Jonathan, 394
Rose, Mark, 5, 397
Rosenburg, Lord, 145
Rösslin, Eucharius, 92, 109
Rothmann, Christoph, 116, 119, 122, 124, 125, 128
Rouse, Mary, 76
Rouse, Richard, 76
Rousseau, Jean-Jacques, 292–3
Rowland, Fytton, 413
Rozier, Jean Baptiste, 235
Rudolph II, Holy Roman Emperor, 115, 122, 124, 132, 145
Rumpf, Georg Eberhard, 261
Ryff, Walter, 109

'sGravesande, Willem Jacob van, 249
Sacrobosco, Johannes de, 99
Sarmiento, Domingo F., 338, 340
Sarton, George, 1
Scaliger, Joseph Justus, 80, 81
Schöner, Johannes, 59, 65, 125
Schott, Johannes, 97, 109
Schreck, Johannes, 155, 156
Scott, Katie, 272
Scott, Patrick, 300
Scott, Sir Walter, 377
Scribonius Largus, 25
Secord, James, 6, 7
Segonds, Alain, 403
Sellich, Count, 132
Serenus, Quintus, 25
Seyler, Wenzel, 132

Sforza, Ludovico, 55, 56
Shapin, Steven, 159, 396
Shelley, Mary, 381
Sixtus ab Hemminga, 64
Skelton, Raleigh Ashlin, 44
Smith, Adam, 179, 369
Smith, John, 327–9, 334
Smith, Nigel, 298
Smith, Sir Thomas, 58
Smith, W. H., 309
Socrates, 35
Solinus, 25
Solomon, 70
Somerville, Mary, 379
Soranus, 25
Solinus, 25
Somerville, Mary, 379
Southey, Robert, 374
Spary, Emma, 5, 7
Spedding, James, 354–65
Spenser, Edmund, 293, 357
Staphilus, Fridericus, 58
Starkey, George (Eirenaeus Philalethes), 133, 293–7, 300, 301
Statius, 28
Steele, Richard, 232
Stephen, Sir James, 356
Stephen, Leslie, 357
Stephenson, Neal, 420–1
Stevens, Wallace, 393
Stewart, Alan, 4, 403
Stewart, Dugald, 374, 383
Stillingfleet, Edward, 183
Stöberlein, Johann, 192–3
Strabo, Walafrid, of Reichenau, 23
Suleiman I, Ottoman Sultan, 55
Sutton, Geoffrey, 241
Swicher, monastic scribe, 15
Sylvius, Jacobus, 104, 108

Tait, William, 382
Tannery, Paul, 1
Tarrutius, 53
Taylor, Sir Henry, 356, 364
Tegg, Thomas, 380
Telescope, Tom, 276
Tennyson, Alfred, 355
Terence, 28
Terrall, Mary, 6, 401
Theophilus, 30
Thomas Aquinas, 18
Thompson, W. H., 356, 364

Thorndike, Lynn, 1
Timpler, Clemens, 82
Topham, Jonathan, 4, 6, 401
Tragus, Hieronymus, 97
Trimmer, Sarah, 278, 282–8
Tuvill, Daniel, 303
Twain, Mark, 399

Ursus *see* Baer, Nicolai Reymers

van Calcar, Jan, 92
van den Brock, Henry, 123
van Helmont, Jan Baptista, 294
Varro, 52
Vesalius, Andreas, 1, 49, 92, 101, 104, 108
Vettius Valens, 53
Victorinus, 28
Vince, Samuel, 320
Vincent of Beauvois, 75
Virgil, 28
Vitellius Caesar, Roman Emperor, 73
Vitruvius, 20
Voltaire (Arouet, François Marie), 250–1

Wagenseil, Johann Christoph, 191–3
Wakefield, Priscilla, 278, 282, 286
Waldseemüller, Martin, 42
Warwick, Andrew, 317
Waterland, Daniel, 172, 177
Watson, James, 226
Watt, Isaac, 180
Watteau, Antoine, 261
Weber, Carl Maria, 307
Weiditz, Hans, 92, 97
Whewell, William, 328, 334, 358, 382
Whittaker, George Byrom, 319, 334
Wilhelm IV, Landgrave of Hesse-Kassel, 116–18, 120, 121, 123
Wilkins, John, 228
Willughby, Francis, 302
Wittich, Paul, 117, 125
Wood, James, 320, 329
Woodhouse, Robert, 331
Woodmansee, Martha, 5, 397
Wordsworth, William, 307

Yates, Frances, 142
Yeo, Richard, 4, 6

Zetzner, Lazarus, 294
Zwinger, Theodor, 73